Advances in Sedimentology and Coastal and Marine Geology—2nd Edition

Advances in Sedimentology and Coastal and Marine Geology—2nd Edition

Advances in Sedimentology and Coastal and Marine Geology—2nd Edition

Guest Editor

Gemma Aiello

MDPI

Basel • Beijing • Wuhan • Barcelona • Belgrade • Novi Sad • Cluj • Manchester

Guest Editor
Gemma Aiello
CNR ISMAR Sezione
Secondaria di Napoli
Napoli
Italy

Editorial Office
MDPI AG
Grosspeteranlage 5
4052 Basel, Switzerland

This is a reprint of the Special Issue, published open access by the journal *Journal of Marine Science and Engineering* (ISSN 2077-1312), freely accessible at: https://www.mdpi.com/journal/jmse/special_issues/PPIXWWRZ6Q.

For citation purposes, cite each article independently as indicated on the article page online and as indicated below:

Lastname, A.A.; Lastname, B.B. Article Title. *Journal Name* **Year**, *Volume Number*, Page Range.

ISBN 978-3-7258-7480-4 (Hbk)
ISBN 978-3-7258-7481-1 (PDF)
https://doi.org/10.3390/books978-3-7258-7481-1

Contents

About the Editor

Gemma Aiello

Gemma Aiello studied Geological Sciences and graduated from the University Federico II of Naples, Italy. At the same university, she completed a PhD in Sedimentary Geology in 1993 and a Post-Doctoral Fellowship in Earth Sciences from 1993 to 1995. Since then, she has developed her activities as a researcher at the National Research Council of Italy where her work is evidenced by more than 150 publications in the field of geology, as well as her participation in many research projects. Her main fields of interest are sedimentary geology and stratigraphy, marine geology and marine geophysics.

Preface

The unique perspective of the Reprint "Advances in Sedimentology and Coastal and Marine Geology – 2nd Edition" is a multi-scale geodynamic interpretation, linking tectonic processes (structural evolution, deformation, and rifting) with sedimentary and paleobiological responses through Earth history. This perspective describes integrated Earth system science or biogeodynamics, which connects the deep-seated mechanisms of the solid Earth (tectonic evolution, rifting, and deformation) to surface processes (sediment transport, sediment provenance, and landscape evolution) and their subsequent impact on the biosphere (extinction, adaptation, and evolution).

This Reprint brings together selected contributions that reflect the latest developments and practical applications in sedimentology and coastal and marine geology. It presents continental margin research, sedimentological models, statistic innovations in geology, and case histories that address solid earth and basin development, surface processes (sediment transport and provenance), and earth system responses and biospheric interactions.

By joining these works into a single volume, this Reprint offers readers a comprehensive overview of the current progress and emerging trends in sedimentology and coastal and marine geology. It serves as a valuable reference for researchers and academics in the field of Earth Sciences, policymakers, and industry practitioners for the advanced monitoring and planning of the continental shelf zones.

Gemma Aiello
Guest Editor

Journal of
*Marine Science
and Engineering*

MDPI

Editorial

Advances in Sedimentology and Coastal and Marine Geology—2nd Edition

Gemma Aiello

Istituto di Scienze Marine (ISMAR), Consiglio Nazionale delle Ricerche (CNR), Sezione Secondaria di Napoli, 80133 Napoli, Italy; gemma.aiello@cnr.it

The eleven papers in this Special Issue "Advances in Sedimentology and Coastal and Marine Geology—2nd Edition", as listed subsequently, contribute to our geological knowledge on areas both inside and outside of the Mediterranean significantly. They cover Italy and beyond. In Italy, areas off shore of Somma–Vesuvius and Campi Flegrei (Naples Bay; Contribution 1); the Eastern Adventure Plateau (Sicily Channel, Southern Tyrrhenian Sea, Italy; Contribution 4); and the Volturno basin and the Massico structure (the Campania–Latium continental margin; Southern Tyrrhenian Sea, Italy; Contribution 11) are studied.

Outside of Italy, this Special Issue covers the Zhu III Depression in the Pearl River Mouth Basin (southeastern China; Contribution 2); Cangnan and the southeastern area of sea near Xiangshan in Zhejiang Province (eastern China; Contributions 3 and 5); Shandong Peninsula (China; Contribution 7); the lower reaches of the Changhua River and its estuary on Hainan Island (China; Contribution 8); the Central Indian Ocean Basin (CIOB; Contribution 9); and a shallow section of water (Wuke section) located in the western part of Sichuan Province in southern China (Contribution 10). Research outside of the Mediterranean predominates in this volume.

To outline the geological settings of the areas studied in this Special Issue, firstly, Naples Bay is one of the most hazardous active volcanic areas in the world, with three volcanic complexes: the Somma–Vesuvius, Campi Flegrei, and Ischia volcanoes (atop Mt. Epomeo). These volcanic complexes have been studied in depth, but only the most recent references on their geology, volcanology, and morpho-bathymetry are considered herein [1–8]. In this region, the geological hazard is very high [9–16]. Significant bradyseismic crises occurred in Campi Flegrei from 1969 to 1972 and from 1982 to 1984, and a new bradyseismic crisis is still in course. The bradyseism in Pozzuoli is part of ground uplift and subsidence affecting the Campi Flegrei area. This phenomenon is currently characterized by ground uplift that has accelerated in recent months and an ongoing seismic swarm. On 18 July 2025, a magnitude 4.0 seismic event was recorded that shook the entire area, with its epicenter between Pozzuoli and Bagnoli, but numerous earthquakes have previously occurred in Naples (https://terremoti.ingv.it/events, accessed on 18 July 2025). The situation is constantly being monitored by the Vesuvius Observatory, which has declared a yellow alert level and an "attention" operational phase.

The Volturno basin and the Massico structure in the Northern Campania Volcanic Zone (NCVZ) represent the seaward prolongation of Mt. Massico on the shore [17]. In this area, the stratigraphic architecture of the ignimbrites is constrained by wells and geological sections [17–24]. The Volturno basin consists of an emerged portion, including the Volturno delta, and a submerged portion, whose stratigraphic architecture consists of several seismo-stratigraphic units [20–23,25,26]. The Massico structure is the seaward prolongation of Mt.

Massico, bounding the Volturno Plain towards the northwest. Off shore of Volturno, the Massico horst consists of Meso-Cenozoic carbonate sequences formed of flower structures.

The Sicily Channel is located in an intraplate rift where Sicily is displaced away from the African continent in the NE direction [27,28]. The structural lows in the Sicily Channel have also been interpreted as large and discrete pull-apart basins involving deep crustal levels within a wrench zone that has developed in front of the Africa–Europe collisional belt [29]. This rift in the Sicily Channel was created by mantle convection as the African lithosphere slab rolled back beneath the Tyrrhenian basin [30,31]. In particular, the seismo-stratigraphic setting [32], the paleo-shoreline configuration at the last glacial maximum [33], and the fluid uprising [34] in the Adventure Plateau have been analyzed in depth.

The Pearl River Mouth Basin is an extensional basin whose geodynamic relationships with the southeastern Eurasian margin have been studied extensively, together with its formation mechanisms and petroleum geology [35,36]. Its tectonic setting is controlled by two episodes of rifting of the attenuated continental crust along the South China margin, ranging in age between the late Cretaceous and the early Oligocene. This rifting has controlled the individuation of various segments of the basin, Zhu 1, Zhu 2, and Zhu 3, which are separated by local uplifts and are characterized by horst and graben structures, respectively, controlled by regional normal faults. In particular, the Zhu 3 Depression of the Pearl River Mouth Basin is studied in this Special Issue with respect to the formation mechanisms of normal faults and their meaning to the geological evolution of the basin [37].

Zhejiang Province is characterized by different geotectonic units, including the Yangtze Plate to the northwest, in tectonic contact with the South China fold system to the southeast along the Jiangshan–Shaoxing fault [38]. The South China fold system is divided further into the Chencai–Suichang uplift belt and the Wenzhou–Linhai geotectogene belt, whose boundary is the Yuyao–Lishui fault. This area has been extensively studied, both regarding the volcanic dykes in the offshore sector [39] and the occurrence of hazardous geological features [40]. In this volume, the geomorphological problems in its embayed sandy beaches have been studied [41].

Shandong Peninsula, located in eastern China, has a complex geological history influenced by its position within the North China Craton and its interactions with the Pacific Plate. The peninsula is characterized by a Precambrian basement, overlain by Mesozoic-Cenozoic sedimentary and magmatic rocks and affected by faulting and orogenic events. It has been studied from various angles, including its iron formations and the related geochemical data [42]; the stratigraphic and sedimentary evolution of its Holocene deposits, characterized by a clinoform pattern [43,44]; and its gold mineralization and deposits [45]. In this volume, the problems that have arisen during analyses of the provenance of the sediments deposited in the Northern Offshore Mud Area during the last 2000 years are considered [46].

The contributions to this Special Issue discuss mass extinctions, in particular that which occurred during the Late Ordovician [47]. This mass extinction has been interpreted as genetically related to a climatic and ecological crisis [48]. In the geological record, this represents one of the five greatest mass extinctions, genetically related to the growth and decay of the ice cap of the Gondwana continent. Benthic and planktonic organisms were involved in strong decay affecting the continental shelves of the Ordovician period [48]. These ecological communities returned to their pre-extinction levels during the Silurian, reaching levels similar to those in the Ordovician [48]. During the Phanerozoic, five main periods of mass extinction occurred [49]: the Ordovician–Silurian, the Late Devonian, the Permian–Triassic, the Triassic–Jurassic, and the Cretaceous–Paleogene. These events were characterized by significant and relatively rapid decreases in biodiversity. In particular, the Ordovician–Silurian extinction occurred around 440 million years ago and primarily impacted marine life. The Late Devonian extinction occurred 365 My ago, affecting marine organisms. The Permo–Triassic extinction was the most drastic and occurred at 252 My ago, with massive losses of both marine and terrestrial species. About 201 million years ago, the Triassic–Jurassic extinction paved the way for the rise of dinosaurs. The Cretaceous–Paleogene extinction, the most recent of the five, occurred 66 million years ago, famously wiping out non-avian dinosaurs [49].

Contribution 1 in this Special Issue analyzes the seismo-stratigraphic setting of the buried Somma–Vesuvius volcano (Contribution 1 in the list of contributions) [50]. Seven seismic units have been recognized and partly calibrated with well data for the Trecase 1 and Palazzo Reale wells. Its volcanic structures, previously interpreted as cryptodomes [51], have been interpreted as tuff rings based on their height/weight ratio on seismic interpretation and correlated with the tuff rings on shore at Campi Flegrei. A new and important volcanic seismic unit—a 5 m branch in the isopach of the Pomici di Avellino pyroclastic deposits—is recognized and interpreted, representing the important PDCs at Somma–Vesuvius [52].

Contribution 2 in the Special Issue takes the formation mechanisms of the NW-SE-trending normal faults bounding the Zhu 3 sub-basin of the Pearl River Mouth Basin as its subject [53]. Based on a structural analysis of their seismic profiles, these faults have been thoroughly structurally characterized, filling a previous gap in the research. In this way, extensional faults, dextral transtensional faults, and sinistral strike–slip faults are distinguished. The corresponding structural implications for the basin's evolution have been analyzed, identifying tectonic inversion triggered by the reactivation of faults involving the basement.

Contribution 3 in this Special Issue is methodological in nature, establishing a novel method for estimating the undrained shear strength of marine soil based on CPTU tests (piezocone penetration tests) [54]. Key parameters and tests in the field of marine geotechnics have been considered, including the piezocone penetration test, the soil behavior type index, the undrained shear strength, and the consolidated quick direct shear test. Using data obtained from direct shear tests, a regression model has been constructed to explain the functional relationship between the soil behavior type index and the cone factor. The performance of this model was assessed through a comprehensive evaluation of metrics including the coefficient of determination, the root mean square error (RMSE), and the mean absolute error (MAE). The proposed methodology was tested in nearby marine project sites to confirm its applicability.

Contribution 4 in this Special Issue discusses the submarine volcanism in the Sicily Channel [55]. On the eastern side of the Adventure Plateau, a collection of five volcanoes known as the "Tetide volcanic cluster" creates a NW-SE alignment and is interpreted as the morphological expression of a failed rift. This volcanic alignment might be the remains of an unsuccessful attempt to create a rift that failed to produce another tectonic trough. Three volcano–tectonic phases have been distinguished, including a magmatic intrusion in the early Pliocene associated with a NW-SE normal fault that formed during the opening of the Pantelleria Graben; a Plio-Quaternary phase of quiescence; and a renewed magmatic intrusion through fissures or cracks that resulted in the late Quaternary volcanoes.

Contribution 5 in this Special Issue covers the coastal geomorphology and sedimentology of the sandy beaches in Zhejiang Province (eastern China) [56]. Field measurements on 75 accessible sandy beaches in Zhejiang with lengths greater than 100 m, including 30 mainland beaches and 45 island beaches, were carried out. Beach morphology was characterized according to the beach length (l), defined as the coastal length of a sandy beach, and averaged dry beach width (w), calculated as the arithmetic average of the width between the base (the foot of landward rocky cliff or the seawall) and the waterline at low tide, measured from each beach's profile. The distribution patterns and geomorphological characteristics indicate that waves and tides have a prevalent influence to the north and south, respectively. The distribution patterns and geomorphological settings on these beaches are mainly affected by the embayment characteristics.

Contribution 6 in this Special Issue is more methodological and deals with marine geotechnics [57], including a geotechnical analysis of the physical and mechanical parameters of sediments and typical clay during a natural dewatering process. Four sediments were sampled from French river dams, and kaolin clay was used. The test procedures included monitoring the natural dewatering and undrained shear strength. Dewatering was characterized according to changes in the water content and calculation of the void ratio. The geotechnical characterization included calculating the variation in dry unit weight and the shear strength relationships. While the percentage of water drained followed a sigmoidal trend, a unique and linear correlation was found relating dry unit weight to solid particle weight as a function of the water content, as the sediments tested had a low clay content.

Contribution 7 in this Special Issue analyzed the area of mud off shore of the northern part of Shandong Peninsula, representing a source-to sink system in the continental shelf of China [58]. A source-to-sink analysis was used to study the sedimentary systems from areas of erosion (sources) to areas of deposition (sinks) and the processes of sediment transport along the way. A holistic approach that considers the relationships between the different segments of the system, such as catchments, continental shelves, slopes, and submarine fans, is used [59,60]. The mud wedge in Shandong Peninsula represents a subaqueous delta where sea level oscillations, shelf circulation patterns, monsoon evolution, sediment transport, and depositional processes have been recorded [61–63]. The comparative analysis of the clay mineral composition of samples from the surrounding rivers, along with ternary diagrams and cross plot analyses, suggests that these clay minerals mainly originate from the Yellow River, the rivers of the Shandong Peninsula, and the Yangtze River, highlighting the sediment's multi-source characteristics.

Contribution 8 in this Special Issue studied the sedimentary dynamics in the lower reaches of the Changhua River and its estuary on Hainan Island [64]. Its sedimentological analysis facilitated the construction of a sediment distribution map in which gravelly sands and gravelly mud sands were predominant. Based on grain size parameters, topography, sediment sources, hydrodynamics, and sediment transport, the study area is divided into four sedimentary regions: the Northern Offshore Depositional Zone, the Central Offshore Depositional Zone, the Southern Offshore Depositional Zone, and the Downstream River Channel Depositional Zone. The trends in the sediment transport in the area were also distinguished. The sediments in the lower reaches of the Changhua River are mainly transported towards the south and the southwest according to Eulerian residual flow patterns, particularly in offshore areas.

Contribution 9 in this Special Issue deals with the morpho-bathymetric and structural setting of the Central Indian Ocean Basin, in particular the intraplate deformation zone [65], characterized by a block structure. Detailed morphotectonic studies were previously carried out in this area, dealing with its structural and geodynamic setting [66]. In this article [65], a linear block at 0.2–0.6 s with a branch-like shape in plain view is morphologically distinct from the other blocks. It represents a system of structural elements of different scales (folds, flexures, ruptures) forming a dextral transpressive tectonic regime. Gravimetric and magnetic data have been integrated with structural data, highlighting the close relationships that exist between the tectonic setting and gravimetric and magnetic anomalies. Multichannel seismic profiles portray the seismo-stratigraphic setting of the area, where key regional unconformities emerged in the Late Miocene and the Early Pliocene.

Contribution 10 in this Special Issue explains the mass extinction in the Late Ordovician as a result of anoxic events [67,68]. In particular, geochemical proxies were used to reconstruct the event stratigraphy in the Late Ordovician period. An enhanced chemical weathering rate (increased $^{87}Sr/^{86}Sr$ ratios) and a decreased intensity of chemical weathering (decreased CIA values) characterized the late Katian. The Hirnantian Glaciation was characterized by high $^{87}Sr/^{86}Sr$ ratios in carbonates and extremely low CIA values in shales, genetically related to the emersion of continents during lowstand periods and the glacial grinding of unweathered rocks. $\delta^{13}C$ records and climate events for the Late Ordovician have been analyzed, comparing new data with previous data for the area. Climatic shifts and mass extinctions in the Late Ordovician were significantly impacted by changes in weathering. Increased weathering contributed to Katian extinction by inducing high primary production and the spread of anoxic seawater. The greatest rate of denudation and the least amount of chemical weathering during the Hirnantian Glaciation probably led to increased CO_2 release and helped to end glaciation.

Finally, Contribution 11 in this Special Issue provides seismo-stratigraphic data and geological interpretations on the filling of the Volturno basin in relation to the Massico structure. It describes the stratigraphic setting of the Campania continental margin, focusing on the ignimbrites in the Campania Volcanic Zone (CVZ) [17]. A geological section of the Volturno basin is integrated with the seismic profiles of Zone E, showing flower structures involving the Meso-Cenozoic carbonate basement. The stratigraphic correlations in the lithostratigraphic data for the wells are indicative of pyroclastic and alluvial deposits, which constitute the sedimentary material within the Volturno Plain, alongside volcanic lava (andesites and basalts) originating from the Villa Literno volcanic complex.

Funding: This research received no external funding.

Conflicts of Interest: The author declares no conflicts of interest.

List of Contributions:

1. Aiello, G. Submarine Stratigraphy of the Eastern Bay of Naples; New Seismo-Stratigraphic Data and Implications for the Somma-Vesuvius and Campi Flegrei Volcanic Activity. *J. Mar. Sci. Eng.* **2022**, *10*, 1520. https://doi.org/10.3390/jmse10101520.
2. Zhu, P.; Zhao, L.; Zhang, J.; Mu, D.; Chen, Y.; Rong, P. Formation Mechanism of NW-Trending Faults and Their Significance on Basin Evolution in Zhu III Depression of the Pearl River Mouth Basin, SE China. *J. Mar. Sci. Eng.* **2024**, *12*, 858. https://doi.org/10.3390/jmse12060858.
3. Fu, S.; Shen, Y.; Jia, X.; Zhang, Z.; Li, X. A Novel Method for Estimating the Undrained Shear Strength of Marine Soil Based on CPTU Tests. *J. Mar. Sci. Eng.* **2024**, *12*, 1019. https://doi.org/10.3390/jmse12061019.
4. Civile, D.; Mangano, G.; Micallef, A.; Lodolo, E.; Baradello, L. A Failed Rift in the Eastern Adventure Plateau (Sicilian Channel, Central Mediterranean). *J. Mar. Sci. Eng.* **2024**, *12*, 1142. https://doi.org/10.3390/jmse12071142.
5. Guo, J.; Shi, L.; Zhang, M.; Gong, Z.; Chen, W.; Xia, X. Regional Difference in Distribution Pattern and Morphological Characteristics of Embayed Sandy Beaches in Zhejiang Province, Eastern China. *J. Mar. Sci. Eng.* **2024**, *12*, 1223. https://doi.org/10.3390/jmse12071223.
6. Azaiez, D.; Boullosa Allariz, B.; Levacher, D. Study of Physical and Mechanical Relationships during the Natural Dewatering of River Sediments and a Kaolin. *J. Mar. Sci. Eng.* **2024**, *12*, 1354. https://doi.org/10.3390/jmse12081354.
7. Wu, S.; Liu, J.; Feng, Y. Provenance Analysis of the Northern Offshore Mud Area of the Shandong Peninsula, China, Spanning the Last 2000 Years. *J. Mar. Sci. Eng.* **2024**, *12*, 1501. https://doi.org/10.3390/jmse12091501.
8. Wu, Y.; Li, X.; Zhao, E.; Wang, Y.; Zhang, S.; Xu, Z.; Wang, Q.; Jiang, D.; Xing, Z. Analysis of Downstream Sediment Transport Trends Based on In Situ Data and Numerical Simulation. *J. Mar. Sci. Eng.* **2024**, *12*, 1982. https://doi.org/10.3390/jmse12111982.

9. Yutsis, V.V.; Levchenko, O.V.; Tevelev, A.V.; Marinova, Y.G.; Veklich, I.A.; Del Razo Gonzalez, A. A typical Linear Tectonic Block of the Intraplate Deformation Zone in the Central Indian Ocean Basin. *J. Mar. Sci. Eng.* **2024**, *12*, 2231. https://doi.org/10.3390/jmse12122231.
10. Tang, P.; Yang, X.; Yan, D. Enhanced Continental Weathering Triggered the Anoxia of Seawater and Mass Extinctions During the Late Ordovician. *J. Mar. Sci. Eng.* **2024**, *12*, 2237. https://doi.org/10.3390/jmse12122237.
11. Aiello, G. Regional Geological Data on the Volturno Basin Filling and Its Relationship to the Massico Structure (Southern Tyrrhenian Sea, Italy). *J. Mar. Sci. Eng.* **2025**, *13*, 241. https://doi.org/10.3390/jmse13020241.

References

1. Aiello, G.; Iorio, M.; Molisso, F.; Sacchi, M. Integrated Morpho-Bathymetric, Seismic-Stratigraphic, and Sedimentological Data on the Dohrn Canyon (Naples Bay, Southern Tyrrhenian Sea): Relationships with Volcanism and Tectonics. *Geosciences* **2020**, *10*, 319. [CrossRef]
2. Milia, A.; Aiello, G.; Iannace, P.; Torrente, M.M. Complex stratigraphic relationships between volcanic features and sedimentary deposits in a submarine environment: The northern offshore Holocene Ischia volcanic field (Italy). *J. Volcanol. Geoth. Res.* **2021**, *419*, 107379. [CrossRef]
3. Aiello, G.; Caccavale, M. Quaternary Evolution of Ischia: A Review of Volcanology and Geology. *Appl. Sci.* **2023**, *13*, 3554. [CrossRef]
4. Sbrana, A.; Marianelli, P.; Pasquini, G. The Phlegrean Fields volcanological evolution. *J. Maps* **2021**, *17*, 557–570. [CrossRef]
5. Foglini, F.; Rovere, M.; Tonielli, R.; Castellan, G.; Prampolini, C.; Budillon, F.; Cuffaro, M.; Di Martino, G.; Grande, V.; Innangi, S.; et al. A new multi-grid bathymetric dataset of the Gulf of Naples (Italy) from complementary multi-beam echo sounders. *Earth Syst. Sci. Data* **2025**, *17*, 181–203. [CrossRef]
6. Aiello, G.; Caccavale, M. Marine Geohazards of the Bay of Naples (Southern Tyrrhenian Sea, Italy): A Review Integrating Morpho-Bathymetric and Seismo-Stratigraphic Analysis. *GeoHazards* **2024**, *5*, 393–414. [CrossRef]
7. Sbrana, A.; Cioni, R.; Marianelli, P.; Sulpizio, R.; Andronico, D.; Pasquini, G. Volcanic evolution of the Somma-Vesuvius Complex (Italy). *J. Maps* **2020**, *16*, 137–147. [CrossRef]
8. Carlino, S.; Sbrana, A.; Pino, N.A.; Marianelli, P.; Pasquini, G.; De Martino, P.; De Novellis, V. The Volcano-Tectonics of the Northern Sector of Ischia Island Caldera (Southern Italy): Resurgence, Subsidence and Earthquakes. *Front. Earth Sci.* **2022**, *10*, 730023. [CrossRef]
9. Parascandola, A. *I Fenomeni Bradisismici del Serapeo di Pozzuoli*; Genovese: Napoli, Italy, 1947; pp. 1–117.
10. Berrino, G.; Corrado, G.; Luongo, G.; Toro, B. Ground deformation and gravity changes accompanying the 1982 Pozzuoli uplift. *Bull. Volcanol.* **1984**, *44*, 187–200. [CrossRef]
11. Cinque, A.; Russo, M.; Pagano, M. La successione di terreni di età post-romana delle terme di Miseno (Napoli): Nuovi dati per la storia e la stratigrafia del bradisisma puteolano. *Boll. Soc. Geol. Ital.* **1991**, *110*, 231–244.
12. Dvorak, J.J.; Mastrolorenzo, G. The mechanisms of recent vertical crustal movements in Campi Flegrei caldera, southern Italy. *Geol. Soc. Am. Spec. Pap.* **1991**, *263*, 1–47.
13. Morhange, C.; Marriner, R.; Laborel, J.; Todesco, M.; Oberlin, C. Rapid sea-level movements and noneruptive crustal deformations in the Phlegrean Fields caldera, Italy. *Geology* **2005**, *34*, 93–96. [CrossRef]
14. Aiello, G. Seismo-stratigraphic data of the Gulf of Pozzuoli (Southern Tyrrhenian Sea, Italy): A review and their relationships with the new bradyseismic crisis. *GeoHazards* 2025, submitted.
15. Guidoboni, E.; Ciuccarelli, C. The Campi Flegrei caldera: Historical revision and new data on seismic crises, bradyseisms, the Monte Nuovo eruption and ensuing earthquakes (twelfth century 1582 AD). *Bull Volcanol.* **2011**, *73*, 655–677. [CrossRef]
16. Somma, R.; Musto, P.; Benini, A.; Fedele, A.; Trocciola, A. Evoluzione del paesaggio costiero della Penisola di Miseno (Campi Flegrei, Campania). Evolution of the coastal landscape of the Miseno Peninsula (Campi Flegrei, Campania). *Mem. Descr. Della Carta Geol. D'italia* **2023**, *110*, 487–506.
17. Rolandi, G.; De Natale, G.; Kilburn, C.R.J.; Troise, C.; Somma, R.; Di Lascio, M.; Fedele, A.; Rolandi, R. 8—The 39 ka Campanian Ignimbrite eruption: New data on source area in the Campanian Plain. In *Vesuvius, Campi Flegrei, and Campanian Volcanism*; De Vivo, B., Belkin, H.E., Rolandi, G., Eds.; Elsevier: Amsterdam, The Netherlands, 2020; pp. 175–205. ISBN 9780128164549. [CrossRef]
18. Milia, A.; Torrente, M.M. The influence of paleogeographic setting and crustal subsidence on the architecture of ignimbrites in the Bay of Naples (Italy). *Earth Planet. Sci. Lett.* **2007**, *263*, 192–206. [CrossRef]

19. Torrente, M.M.; Milia, A.; Bellucci, F.; Rolandi, G. Extensional tectonics in the Campania Volcanic Zone (eastern Tyrrhenian Sea, Italy): New insights into the relationship between faulting and ignimbrite eruptions. *Ital. J. Geosci.* **2010**, *129*, 297–315. [CrossRef]
20. Amorosi, A.; Pacifico, A.; Rossi, V.; Ruberti, D. Late Quaternary incision and deposition in an active volcanic setting: The Volturno valley fill, southern Italy. *Sediment. Geol.* **2012**, *282*, 307–320. [CrossRef]
21. Aiello, G.; Cicchella, A.G.; Di Fiore, V.; Marsella, E. New seismo-stratigraphic data of the Volturno basin (northern Campania, Tyrrhenian margin, southern Italy): Implications for tectono-stratigraphy of the Campania and Latium sedimentary basins. *Ann. Geophys.* **2011**, *54*, 265–283.
22. Aiello, G.; Marsella, E.; Cicchella, A.G.; Di Fiore, V. New insights on morpho-structures and seismic stratigraphy along the Campania continental margin (Southern Italy) based on deep multichannel seismic profiles. *Rend. Lincei Sci. Fis. E Nat.* **2011**, *22*, 349–373. [CrossRef]
23. Ruberti, D.; Buffardi, C.; Sacchi, M.; Vigliotti, M. The late Pleistocene-Holocene changing morphology of the Volturno delta and coast (northern Campania, Italy): Geological architecture and human influence. *Quat. Int.* **2022**, *625*, 14–28. [CrossRef]
24. Aiello, G. Submarine Stratigraphy of the Eastern Bay of Naples: New Seismo-Stratigraphic Data and Implications for the Somma-Vesuvius and Campi Flegrei Volcanic Activity. *J. Mar. Sci. Eng.* **2022**, *10*, 1520. [CrossRef]
25. Bruno, P.P.G.; Putignano, M.L.; Cella, F.; Florio, G. Investigating tectonic links between the deep Volturno plain and Campanian volcanism using vintage seismic and potential field data. *Sci. Rep.* **2025**, *15*, 21602. [CrossRef]
26. Corrado, G.; Amodio, S.; Aucelli, P.P.C.; Pappone, G.; Schiattarella, M. The Subsurface Geology and Landscape Evolution of the Volturno Coastal Plain, Italy: Interplay between Tectonics and Sea-Level Changes during the Quaternary. *Water* **2020**, *12*, 3386. [CrossRef]
27. Beccaluva, L.; Gabbianelli, G.; Lucchini, F.; Rossi, P.L.; Savelli, C. Petrology and K/Ar ages of volcanics dredged from the Eolian seamounts: Implications for geodynamic evolution of the southern Tyrrhenian basin. *Earth Planet. Sci. Lett.* **1985**, *74*, 187–208. [CrossRef]
28. Finetti, I. Geophysical study of the Sicily Channel rift zone. *Boll. Geofis. Teor. Appl.* **1984**, *26*, 3–28.
29. Boccaletti, M.; Cello, G.; Tortorici, L. Transtensional tectonics in the Sicily Channel. *J. Struct. Geol.* **1987**, *9*, 869–876. [CrossRef]
30. Argnani, A. The strait of sicily rift zone: Foreland deformation related to the evolution of a back-arc basin. *J. Geodyn.* **1990**, *12*, 311–331. [CrossRef]
31. Catalano, R.; Di Stefano, P.; Vitale, F.P. Structural trends and palaeogeography of the central and western Sicily belt: New insights. *Terra Nova* **1995**, *7*, 189–199. [CrossRef]
32. Civile, D.; Lodolo, E.; Alp, H.; Ben-Abraham, Z.; Cova, A.; Baradello, L.; Accettella, D.; Burca, M.; Centonze, J. Seismic stratigraphy and structural setting of the Adventure Plateau (Sicily Channel). *Mar. Geophys. Res.* **2014**, *35*, 37–53. [CrossRef]
33. Lodolo, E.; Loreto, M.F.; Melini, D.; Spada, G.; Civile, D. Palaeo-Shoreline Configuration of the Adventure Plateau (Sicilian Channel) at the Last Glacial Maximum. *Geosciences* **2022**, *12*, 125. [CrossRef]
34. Maiorana, M.; Spatola, D.; Todaro, S.; Caldareri, F.; Parente, F.; Severini, A.; Sulli, A. Seismo-stratigraphic and morpho-bathymetric analysis revealing recent fluid-rising phenomena on the Adventure Plateau (northwestern Sicily Channel). *Mar. Geophys. Res.* **2024**, *45*, 15. [CrossRef]
35. Yu, H. The Pearl River Mouth Basin: A rift basin and its geodynamic relationship with the southeastern Eurasian margin. *Tectonophysics* **1990**, *183*, 177–186. [CrossRef]
36. Mi, L.; Zhang, X.; Pang, X.; Zheng, J.; Zhang, L. Formation mechanism and petroleum geology of Pearl River Mouth Basin. *Acta Pet. Sin.* **2019**, *40*, 1–10.
37. Zhu, P.; Zhao, L.; Zhang, J.; Mu, D.; Chen, Y.; Rong, P. Formation Mechanism of NW-Trending Faults and Their Significance on Basin Evolution in Zhu III Depression of the Pearl River Mouth Basin, SE China. *J. Mar. Sci. Eng.* **2024**, *12*, 858. [CrossRef]
38. Zeng, Q.; Wang, Y.; Zhang, S.; Liu, J.; Qin, K. Types, features, and prospecting potential for Mesozoic metal ore deposits in Zhejiang Province, southeast China. *Int. Geol. Rev.* **2011**, *54*, 1031–1051. [CrossRef]
39. Chen, N.; Dong, J.; Chen, J.; Dong, C.; Shen, Z. Geometry and emplacement of the Late Cretaceous mafic dyke swarms on the islands in Zhejiang Province, Southeast China: Insights from high-resolution satellite images. *J. Asian Earth Sci.* **2014**, *79*, 302–311. [CrossRef]
40. Qiu, J.; Liu, J.; Yue, N.; Wang, S.; Mai, D. Distribution and Characteristics of Hazardous Geological Features in the Marine Coastal and Offshore Areas of Zhejiang Province, East China Sea. *J. Ocean Univ. China* **2018**, *17*, 1318–1324. [CrossRef]
41. Guo, J.; Shi, L.; Zhang, M.; Gong, Z.; Chen, W.; Xia, X. Regional Difference in Distribution Pattern and Morphological Characteristics of Embayed Sandy Beaches in Zhejiang Province, Eastern China. *J. Mar. Sci. Eng.* **2024**, *12*, 1223. [CrossRef]
42. Moon, I.; Lee, I.; Yang, X. Geochemical constraints on the genesis of the Algoma-type banded iron formation (BIF) in Yishui County, western Shandong Province, North China Craton. *Ore Geol. Rev.* **2017**, *89*, 931–945. [CrossRef]

43. Liu, J.; Saito, Y.; Wang, H.; Yang, Z.; Nakashima, R. Sedimentary evolution of the Holocene subaqueous clinoform off the Shandong Peninsula in the Yellow Sea. *Mar. Geol.* **2017**, *236*, 165–187. [CrossRef]
44. Xue, C.; Qin, Y.; Ye, S.; Laws, E.A.; Wang, Z. Evolution of Holocene ebb-tidal clinoform off the Shandong Peninsula on East China Sea shelf. *Earth Sci. Rev.* **2018**, *177*, 478–496. [CrossRef]
45. Guo, P.; Santosh, M.; Li, S. Geodynamics of gold metallogeny in the Shandong Province, NE China: An integrated geological, geophysical and geochemical perspective. *Gondwana Res.* **2013**, *24*, 1172–1202. [CrossRef]
46. Wu, S.; Liu, J.; Feng, Y. Provenance Analysis of the Northern Offshore Mud Area of the Shandong Peninsula, China, Spanning the Last 2000 Years. *J. Mar. Sci. Eng.* **2024**, *12*, 1501. [CrossRef]
47. Tang, P.; Yang, X.; Yan, D. Enhanced Continental Weathering Triggered the Anoxia of Seawater and Mass Extinctions During the Late Ordovician. *J. Mar. Sci. Eng.* **2024**, *12*, 2237. [CrossRef]
48. Brenchley, P.J.; Marshall, J.D.; Underwood, C.J. Do all mass extinctions represent an ecological crisis? Evidence from the Late Ordovician. *Geol. J.* **2001**, *36*, 329–340. [CrossRef]
49. Sepkoski, J.J. Patterns of Phanerozoic Extinction: A Perspective from Global Data Bases. In *Global Events and Event Stratigraphy in the Phanerozoic*; Walliser, O.H., Ed.; Springer: Berlin/Heidelberg, Germany, 1996. [CrossRef]
50. Santacroce, R.; Cioni, R.; Marianelli, P.; Sbrana, A.; Sulpizio, R.; Zanchetta, G.; Domahue, D. Age and whole rock-glass compositions of proximal pyroclastics from major explosive eruptions of Somma-Vesuvius: A review as a tool for distal tephrostratigraphy. *J. Volcanol. Geotherm. Res.* **2008**, *177*, 1–18. [CrossRef]
51. Milia, A.; Mirabile, L.; Torrente, M.M.; Dvorak, J.J. Volcanism offshore of Vesuvius volcano in Naples Bay. *Bull. Volcanol.* **1998**, *59*, 404–413. [CrossRef]
52. Gurioli, L.; Sulpizio, R.; Cioni, R.; Sbrana, A.; Santacroce, R.; Luperini, W.; Andronico, D. Pyroclastic flow hazard assessment at Somma-Vesuvius based on the geological record. *Bull. Volcanol.* **2010**, *72*, 1021–1038. [CrossRef]
53. Jin, Z.; Yuan, G.; Zhang, X.; Cao, Y.; Ding, L.; Li, X.; Fu, X. Differences of tuffaceous components dissolution and their impact on physical properties in sandstone reservoirs: A case study on Paleogene Wenchang Formation in Huizhou-Lufeng area, Zhu I Depression, Pearl River Mouth Basin, China. *Pet. Explor. Dev.* **2023**, *50*, 111–124. [CrossRef]
54. Fu, S.; Shen, Y.; Jia, X.; Zhang, Z.; Li, X. A Novel Method for Estimating the Undrained Shear Strength of Marine Soil Based on CPTU Tests. *J. Mar. Sci. Eng.* **2024**, *12*, 1019. [CrossRef]
55. Civile, D.; Mangano, G.; Micallef, A.; Lodolo, E.; Baradello, L. A Failed Rift in the Eastern Adventure Plateau (Sicilian Channel, Central Mediterranean). *J. Mar. Sci. Eng.* **2024**, *12*, 1142. [CrossRef]
56. Shi, L.; Guo, J.; Chen, S.; Chang, Y.; Zhang, D.; Gong, Z. Morphodynamic response of an embayed beach to different typhoon events with varying intensities. *Acta Oceanol. Sin.* **2023**, *42*, 51–63. [CrossRef]
57. Azaiez, D.; Boullosa Allariz, B.; Levacher, D. Study of Physical and Mechanical Relationships during the Natural Dewatering of River Sediments and a Kaolin. *J. Mar. Sci. Eng.* **2024**, *12*, 1354. [CrossRef]
58. Zhu, Y.; Feng, X.; Zhu, L.; Zhong, W. Origin and geochemistry of surface sediments in the mud deposit area offshore the Shandong Peninsula, China. *J. Ocean. Limnol.* **2021**, *39*, 483–499. [CrossRef]
59. Amorosi, A.; Sammartino, I.; Dinelli, E.; Campo, B.; Guercia, T.; Trincardi, F.; Pellegrini, C. Provenance and sediment dispersal in the Po-Adriatic source-to-sink system unraveled by bulk-sediment geochemistry and its linkage to catchment geology. *Earth-Sci. Rev.* **2022**, *234*, 104202. [CrossRef]
60. Xu, C.; Gong, C. Predictive stratigraphy: From sequence stratigraphy to source-to-sink system. *Oil Gas Geol.* **2023**, *44*, 521–538. [CrossRef]
61. Milliman, J.D.; Qin, Y.-S.; Ren, M.; Saito, Y. Man's influence on the erosion and transport of sediment by Asian rivers; the Yellow River (Huanghe) example. *J. Geol.* **1987**, *95*, 751–762. Available online: https://www.jstor.org/stable/30063822 (accessed on 22 July 2025). [CrossRef]
62. Liu, J.P.; Milliman, J.D.; Gao, S. The Shandong mud wedge and post-glacial sediment accumulation in the Yellow Sea. *Geo-Mar. Lett.* **2001**, *21*, 253–254. [CrossRef]
63. Liu, J.P.; Milliman, J.D.; Gao, S.; Cheng, P. Holocene development of the Yellow River's subaqueous delta, North Yellow Sea. *Mar. Geol.* **2004**, *209*, 45–67. [CrossRef]
64. Wu, Y.; Li, X.; Zhao, E.; Wang, Y.; Zhang, S.; Xu, Z.; Wang, Q.; Jiang, D.; Xing, Z. Analysis of Downstream Sediment Transport Trends Based on In Situ Data and Numerical Simulation. *J. Mar. Sci. Eng.* **2024**, *12*, 1982. [CrossRef]
65. Yutsis, V.V.; Levchenko, O.V.; Tevelev, A.V.; Marinova, Y.G.; Veklich, I.A.; Del Razo Gonzalez, A. A typical Linear Tectonic Block of the Intraplate Deformation Zone in the Central Indian Ocean Basin. *J. Mar. Sci. Eng.* **2024**, *12*, 2231. [CrossRef]
66. Yatheesh, V.; Dyment, J.; Bhattacharya, G.C.; Royer, J.Y.; Kamesh Raju, K.A.; Ramprasad, T.; Chaubey, A.K.; Patriat, P.; Srinivas, K.; Choi, Y. Detailed structure and plate reconstructions of the Central Indian Ocean between 83.0 and 42.5 Ma (Chrons 34 and 20). *J. Geophys. Res. Solid Earth* **2019**, *124*, 4305–4322. [CrossRef]

67. Latyshev, A.V.; Fetisova, A.M.; Veselovskiy, R.V. Linking Siberian Traps LIP Emplacement and End-Permian Mass Extinction: Evidence from Magnetic Stratigraphy of the Maymecha-Kotuy Volcanic Section. *Geosciences* **2020**, *10*, 295. [CrossRef]
68. Aiello, G. Regional Geological Data on the Volturno Basin Filling and Its Relationship to the Massico Structure (Southern Tyrrhenian Sea, Italy). *J. Mar. Sci. Eng.* **2025**, *13*, 241. [CrossRef]

Journal of
Marine Science and Engineering

MDPI

Article

Submarine Stratigraphy of the Eastern Bay of Naples: New Seismo-Stratigraphic Data and Implications for the Somma-Vesuvius and Campi Flegrei Volcanic Activity

Gemma Aiello

Institute of Marine Sciences (ISMAR), National Research Council of Italy (CNR), 80133 Naples, Italy; gemma.aiello@cnr.it

Abstract: The submarine stratigraphy of the eastern Bay of Naples is studied through seismo-stratigraphic data correlated with borehole data. Multichannel seismic profiles are interpreted in order to reconstruct the stratigraphic relationships between the Quaternary marine seismic units and the volcanic acoustic substratum. Seven seismic units are recognized based on the geological interpretation of seismic profiles and using seismo-stratigraphic criteria. The top of the lowest seismic unit was correlated with the Campanian Ignimbrite (Southern Campania Volcanic Zone). The stratigraphic setting of the eastern Bay of Naples is characterized by NE-SW trending seismic structures, probably corresponding with tuff rings. These tuff rings can be compared with the Porto Miseno, the Archiaverno and Averno, and the Astroni tuff rings (Campi Flegrei). Offshore, the Somma-Vesuvius a seismic unit was interpreted as the fallout deposits representing the base of the AD 79 eruption. However, since a branch of the isopach of 5 m of the "Pomici di Avellino" pyroclastic deposits is very close to the Tyrrhenian coastline and near our GRNA01 and GRNA03 seismic profiles, we cannot exclude that the seismic unit could be also correlated with the deposits of this eruption.

Keywords: eastern Bay of Naples; high-resolution seismic reflection profiles; seismic units; tuff rings; pyroclastic flow deposits; Somma-Vesuvius; Campi Flegrei

1. Introduction

The submarine structure of the Somma-Vesuvius volcano and the buried volcanic structures of the Naples Bay were extensively studied [1–7]. One important previously identified feature is a seismic reflector interpreted as evidence of a southwestwards lateral collapse of the volcano, probably between 35 and 11 ky B.P. ago [2]. In contrast, the eastern side of the volcanic complex is characterized by the occurrence of buried parasitic craters and pyroclastic deposits [1].

This paper interprets the submarine stratigraphy of the Gulf of Naples, focusing on the eastern Bay of Naples, based on multichannel seismic data. These data highlight the stratigraphic relationships between the Quaternary marine and volcanic deposits, with thick seismic units identified as the Campanian Ignimbrite (CI seismic unit) [8–15] and the Neapolitan Yellow Tuff (NYT seismic unit) [16–21].

This paper will show the seismic units corresponding with the Campanian Ignimbrite and the Neapolitan Yellow Tuff (NYT) deposits, seismic units interpreted as tuff rings, and an important new seismic unit, recognized offshore as the Somma-Vesuvius, which correlates with the fallout deposits representing the base of the AD 79 eruption sequence [22–28].

Gurioli et al. [25] delineated the area impacted by PDCs (pyroclastic density currents) at Somma-Vesuvius, focusing on the dispersal, thickness, and extent of the PDC deposits, generated during seven plinian and sub-plinian eruptions (Pomici di Base, Greenish Pumice, Pomici di Mercato, Pomici di Avellino, Pompeii Pumice, AD 472 Pollena, and AD 1631 eruptions. The maps of the total thickness of the PDC deposits for each eruption were constructed. Among the seven eruptions, five of them have dispersed PDCs radially, showing a preferential direction controlled by the location of the vents and by the palaeotopography. The PDCs from the Pomici di Avellino and Pompei Pumice eruptions have shown the most dispersed, broadest distributed deposits. Both the isopach maps and the features of the deposits have revealed that the PDC dispersal was controlled by the intensity of the eruption, by the location of the vent with respect to the Somma caldera wall and by the pre-existing topography.

2. Datasets and Methods

A multibeam bathymetric dataset was previously acquired as part of the CARG project of the Naples and Salerno Bays (http://www.isprambiente.gov.it/Media/carg/campania.html, accessed on 16 October 2022; Figure 1) [29]. The DEM (Digital Elevation Model) merges onshore and offshore data and shows the location of the seismic lines analyzed in this paper (Figure 1). Onshore data display the physiographic domains of the Somma-Vesuvius volcanic complex, the Campi Flegrei volcanic complex, the Ischia and Capri islands, and the Sorrento Peninsula (Figure 1). The seismic lines analyzed in this paper are located in the eastern sector of the Bay of Naples, from the Somma-Vesuvius offshore to the Naples town (Posillipo), and from Sorrento Peninsula offshore, covering an inner shelf to outer shelf physiographic domain.

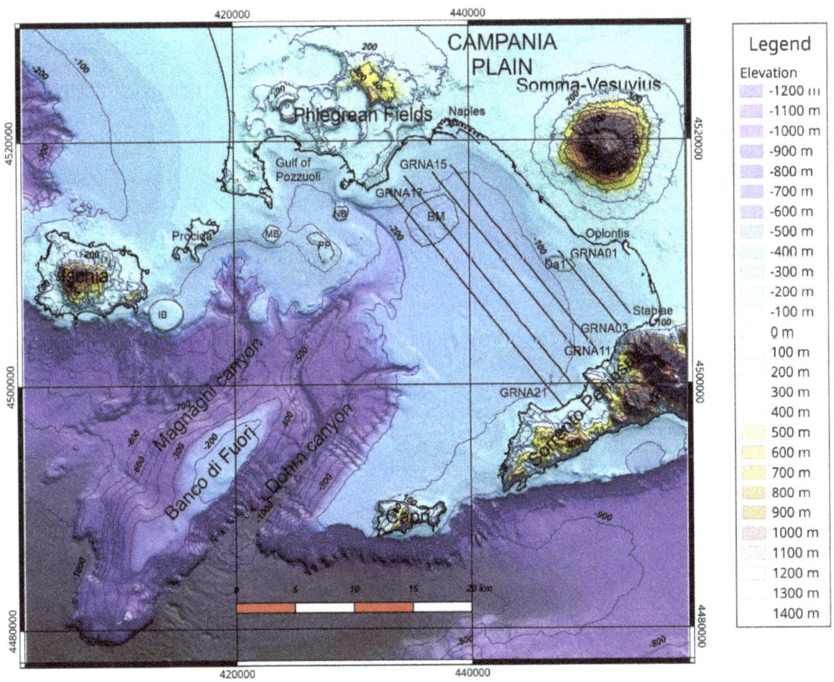

Figure 1. Sketch map showing the location of the seismic profiles superimposed on the Multibeam bathymetric map of Naples Bay. The main morpho-structural features of the Bay of Naples are also reported (Dohrn canyon, Magnaghi canyon, Banco di Fuori). Da1: debris avalanche; BM: Banco della Montagna; NB: Nisida Bank; PP: Pentapalummo Bank; MB: Miseno Bank; IB: Ischia Bank.

The seismic profiles were processed and interpreted with the aim of improving the geological knowledge of the eastern sector of Naples Bay, focusing in particular on the submarine flank of the Vesuvius volcano. The seismic data were collected previous to this study in the oceanographic cruise GMS00_05 [5], which took place during October–November 2000 in the Southern Tyrrhenian sea onboard of the R/V Urania (National

Research Council of Italy). The seismic profiles GRNA01, GRNA03, GRNA11, GRNA17, and GRNA21, from the same cruise, are unpublished, and are herein processed and interpreted for the first time. The seismic profile GRNA15 was previously processed and interpreted by Aiello et al. [5] but is re-processed and re-interpreted here. Aiello et al. [5] provide additional details on the seismic acquisition parameters of the survey and methods for processing and interpretation of the seismic data.

The processing of the seismic lines was carried out using Seismic Unix software [30], including the extraction of the first channel from the shot gathers, application of Automatic Gain Control (AGC), and a spectral analysis of the seismic traces with a Fourier Transform and the application of a Butterworth bandpass filter (f = 40 Hz, 45 Hz, 240 Hz, 250 Hz).

The processing of the seismic data consisted of five steps. The first step was quality control and the assignment of the field geometry. The second step was editing the seismic traces to identify and remove spikes from the seismic record. The third step was top muting, to eliminate signals before the first arrivals of the seismic traces. The fourth step was the application of Automatic Gain Control (AGC), improving the normalization of the seismic traces. The fifth step was a velocity analysis aimed at removing move-out on the CDP (Common Depth Point) families and defining the velocity of the different seismic horizons. This yielded stacked seismic sections ready for the geological interpretation.

Geological interpretation of the seismic profiles was carried out using criteria of seismic stratigraphy [31–33]. A key concept is that the geometry of the seismic reflectors provides information on depositional geometry. Seismic reflectors correspond with significant contrasts in acoustic impedance, allowing for the calculation of transmitted vs. reflected acoustic energy. Seismo-stratigraphic analysis is carried out by identifying stratigraphic discontinuities and depositional sequences, reconstructing of the original geometry of the sedimentary bodies, and chronostratigraphic correlation. Planke et al. [34] developed the concept of the volcanic seismo-stratigraphy, which is the analysis of the seismic facies by mapping and geological interpretation of the units having the same seismic facies. However, individual volcanic deposits are sometimes difficult to image, requiring a knowledge of the seismic response of the different volcanic deposits.

In this paper, the identified seismic units are designated with letters, avoiding any lithostratigraphic or chronostratigraphic interpretation. Time-to-depth conversions of significant seismic reflectors were made, using an average velocity of the overlying Quaternary marine deposits of 1650 m/s. An average seismic wave velocity of 1650 m/s 1 within the first seismic units beneath the seafloor was assumed, as a result of test-calibration carried out in previous papers [35]. This conversion is necessary to allow the seismic reflector corresponding to the top of the seismic unit of the Campanian Ignimbrite, for example, to be correlated with the depth of these deposits at the "Trecase 1" exploration well [36] (Figure 2).

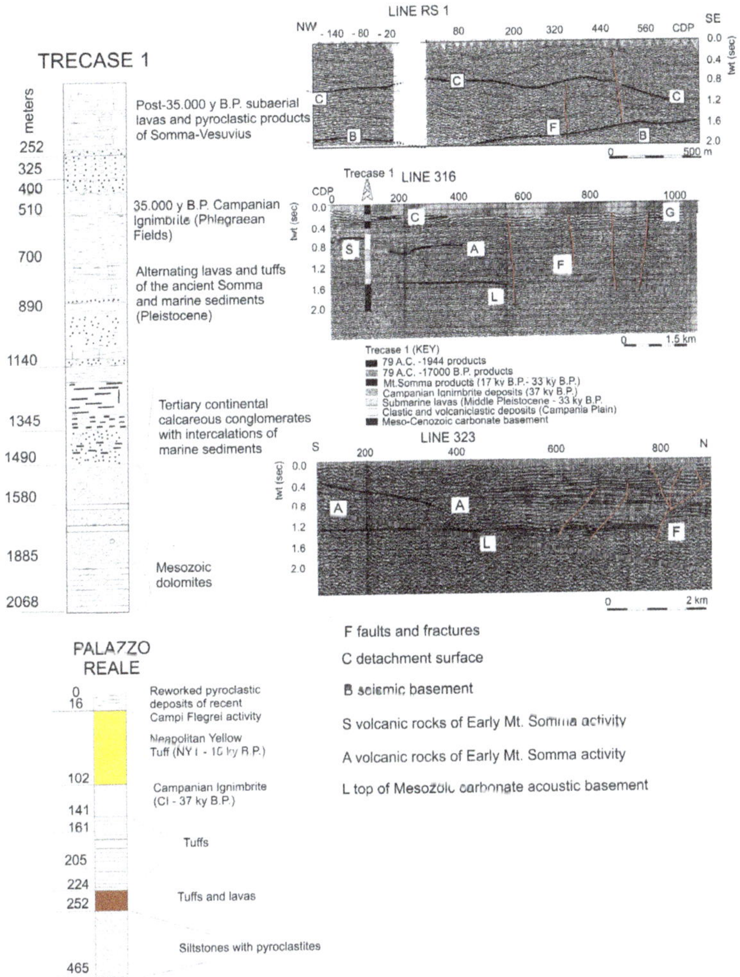

Figure 2. Sketch stratigraphy of the exploration well "Trecase 1" (modified after [36]), integrated with the seismic profiles recorded onshore in the Vesuvius area and calibrated by the Trecase 1 well (modified after [2]). The stratigraphy of the "Palazzo Reale" borehole (modified after [37]) has also been reported.

3. Geological Setting

The Bay of Naples lies in the southern part of a tectonic depression, namely the Campania Plain, produced by the regional tectonic subsidence that affected the Southern Tyrrhenian margin during the Pleistocene (Figure 1). It corresponds to a half-graben structure controlled by a NE-SW master fault running parallel to the "Banco di Fuori" structural high. The pre-Pleistocene basement crops out in the Sorrento Peninsula structural high. The age of basin formation is roughly Pleistocene based on the "Trecase 1" exploration well on land (Figure 2) [36]. A seismic profile located onshore in the Vesuvius area shows the main seismic reflectors (Figure 2) [2], corresponding with the acoustic basement, the faults mapped onshore in the volcanic complex [38], the seismic reflectors with a complex morphology, and the diffractions near the volcano (Figure 2).

The Quaternary volcanism of the Campania continental margin was controlled by three main rifting stages, allowing recognition of an ancient volcanic cycle, which was contemporaneous with the first rifting stage and of a young volcanic cycle, which acted simultaneously with the second and the third rifting stages [7]. This study established that the first rifting stage was Lower Pleistocene in age and was controlled by NE-SW trending normal faults. The second rifting stage, Middle Pleistocene in age, acted through NW-SE trending normal faults. The third rifting stage, Middle-Late Pleistocene in age, was controlled by the reactivation of previous fault systems and by N-S trending normal faults. The ancient volcanic cycle includes (among others) the V0 volcano [7] and the Parete volcano [39,40]. The volcanoes of the ancient cycle have a size comparable with that of the Somma-Vesuvius volcano and range in age between the Early Pleistocene and 0.4-0.1 My [7,41]. The Parete volcano is composed of thick basaltic and andesitic lavas, which were drilled by the "Villa Literno 2" and "Parete 2" wells [39,40,42] (Figure 3). The "Villa Literno 2" well has drilled the recent pyroclastic products, underlain by andesitic tuffs, having an average thickness of 150 m, underlain, in turn, by marine and transitional clastic deposits, thick about 650 m. These deposits overlie a thick volcanic sequence, composed of alternating lava and tuffs, thick about 2150 m, interpreted as the Villa Literno volcanic complex, recognized on the seismic profile Sister4_2 [39] and correlating with the "Parete 2" well (Figure 3). The "Parete 2" well has drilled a similar volcanic sequence, interpreted as the Parete volcano (Figure 3), and is overlain by recent pyroclastic products. The volcanoes of the young volcanic cycle include the lavas occurring in the subsurface of the Vesuvius volcanic complex ("Trecase 1" well) [35] (Figure 2), the Campanian Ignimbrite [11,13,43–51], and the eruptive sequence of the Vesuvius volcano (Figure 2).

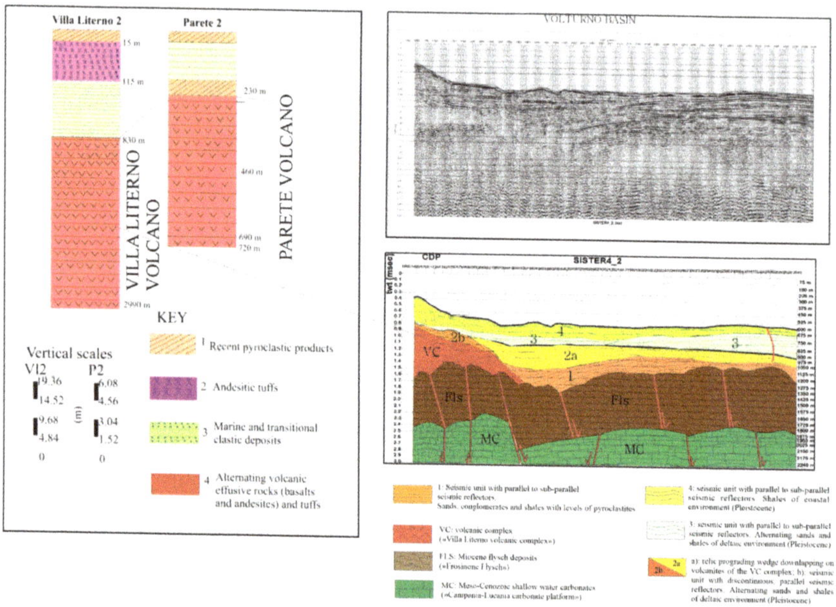

Figure 3. Sketch stratigraphy of the deep exploration wells "Villa Literno 2" and "Parete 2" (modified after [42]) showing the Villa Literno and Parete volcanic complexes and correlation of the Villa Literno volcanic complex with the seismic profile Sister4_2 (upper right) and corresponding geologic interpretation (lower right; modified after [40]).

In the northern Campania Volcanic Zone (CVZ), the stratigraphic data of more than 600 boreholes have allowed reconstruction of the stratigraphic architecture of the ignimbrite, lava flows, and marine deposits (Figure 4) [52]. The lowermost unit is composed of marine sediments (Tyrrhenian Layer M1), whose top was dated to 126 ky B.P. [53]. A volcanic unit overlies the marine deposits and is composed of massive ignimbrite deposits and lava flows. This unit is overlain by a succession of transitional and marine deposits, dated back to 55–50 ky (Tyrrhenian Unit M2) [54], overlain by the CI deposits [52,55–60]. The CI deposits are overlain by the Giugliano Ignimbrite (GI; Figure 4), while in the southern sector of the Campania Volcanic Zone, by grey Neapolitan Yellow Tuff (NYT) and by pyroclastic rocks younger than 10 ky.

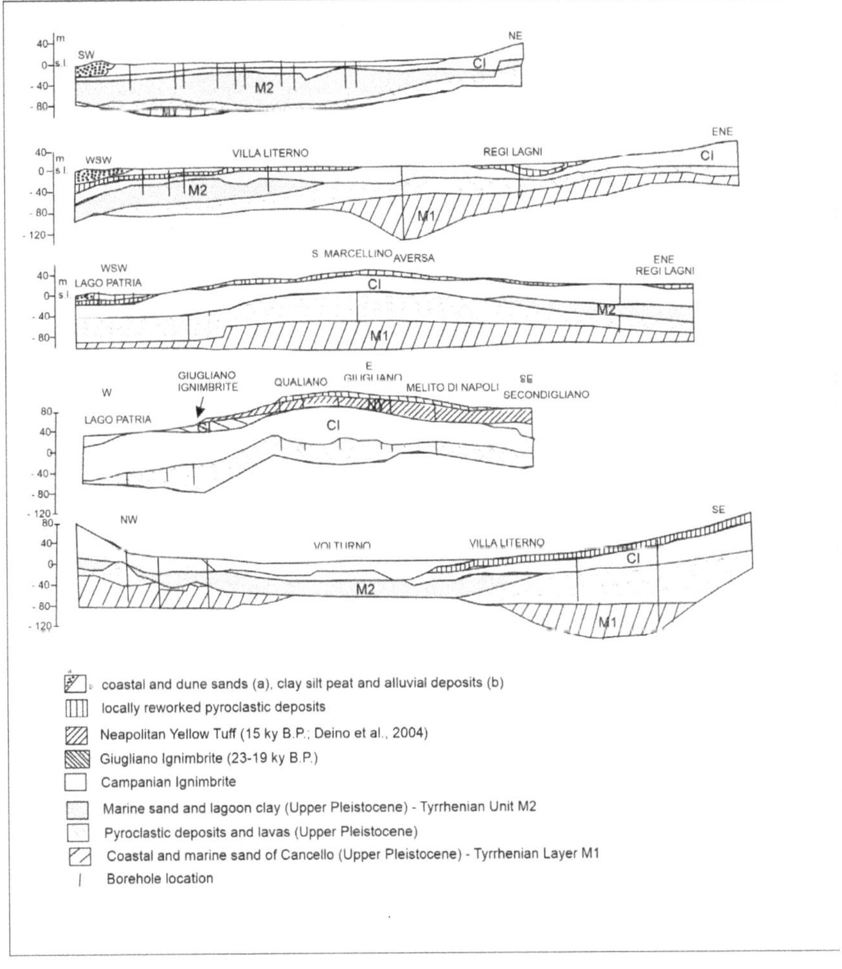

Figure 4. Geological sections and location of boreholes in the Northern Campania Volcanic Zone (N-CVZ; modified after [52]). Note that the Northern Campania Volcanic Zone (N-CVZ) is the broad area including the portion of the Campania Plain, which is characterized by intense volcanic activity, older than 12 ky B.P. [52]. The stratigraphic data of more than 600 boreholes were analyzed in order to construct the geological sections shown in the figure, showing the stratigraphic relationships between the volcanic units occurring in the subsurface of this sector of the Campania Plain.

The "Palazzo Reale borehole", whose stratigraphy is shown in Figure 2, has given key lithostratigraphic data on the Neapolitan Yellow Tuff (NYT) and on the Campanian Ignimbrite (CI) in the subsurface of Naples [37,52]. It represents a key lithostratigraphic column, taken from a borehole 465 m deep and drilled in 1845 in Naples, near the garden of Palazzo Reale (Figure 2) [37,52]. The stratigraphic column of the well has revealed that, below 16 m of reworked deposits of recent activity of Campi Flegrei, a sequence of Neapolitan Yellow Tuff, about 90 m thick, overlies a sequence of the Campanian Ignimbrite, about 40 m thick (Figure 2). The CI overlies a sequence of tuffs about 80 m thick. Such stratigraphy can be assumed as continuous in the eastern sector of the Bay of Naples. Previous seismo-stratigraphic studies have shown acoustically transparent thick ignimbrite sequences below the Bay of Naples, interpreted as the Campanian Ignimbrite and the Neapolitan Yellow Tuff [14,59].

Campi Flegrei is a complex volcanic area, characterized by monogenetic volcanoes, pyroclastic in nature (tuff cones, tuff rings, ash cones, cinder cones, huge calderas) and subordinately, by lava domes [8,10,11,42]. This volcanic area is located at the southwestern margin of the graben of the Campania Plain, between the Massico Mount and the Sebeto Plain [60–63]. This area represents the emerged northern margin of the Phlegrean caldera, which formed about 39 ky B.P. after the eruption of the Campanian Ignimbrite and the consequent down throwing of the Meso-Cenozoic basement [8–11,64–71].

The Campi Flegrei volcanic district includes numerous tuff rings [11]. One of the most important ones is the tuff ring of Averno, whose volcanology was studied in detail [72]. The tuff ring of Averno was dated at 3700 years B.P. and is a wide maar-type, lake-filled volcano [72]. Di Vito et al. [11] showed the characteristics of the volcanic units of the Campi Flegrei caldera younger than 12 ky B.P. Based on their interpretation, several tuff rings were distinguished, including the Porto Miseno volcano and the Archiaverno volcano. The Porto Miseno volcano is located above the volcanic deposits of the Bacoli volcano and below the deposits of the Capo Miseno volcano.

Somma-Vesuvius is a stratovolcano constituted by an older breached edifice (Mt. Somma) and an intra-caldera cone (Figure 5A; Mt. Vesuvius) [38,73–75]. The volcanic succession is composed of lava flows, interlayered with strombolian scoria fall deposits overlain by the deposits of four main Plinian eruptions [73]:

(i) 22,000 yr cal BP "Pomici di Base";
(ii) 8900 yr cal BP "Mercato Pumice";
(iii) 4300 yr cal BP "Avellino Pumice"; and
(iv) A.D. 79 "Pompeii Pumice".

The pyroclastic deposits corresponding with these four eruptions were interpreted as important stratigraphic markers (Figure 5B) [73]. These intervals, respectively, are:

(i) B-M inter-Plinian, between "Pomici di Base" and "Mercato Pumice";
(ii) M-A inter-Plinian, between "Mercato Pumice" and "Avellino Pumice";
(iii) A-P inter-Plinian, between "Avellino Pumice" and "Pompeii Pumice"; and
(iv) P-XX inter-Plinian, between the "Pompeii Pumice" and the last erupted products of the XXth century (Figure 5B) [73].

Moreover, a new geological map of the Somma-Vesuvius volcano at the 1:20,000 scale was recently constructed (Figure 5C) [75]. In this map, an interpretation of the volcanological evolution is shown, based on four main phases, including (Figure 5A): phase 1: the building of the Somma stratovolcano (39 ky–22 ky); phase 2: the caldera formation (22 ky–AD 79); phase 3: the post-caldera activity (AD 472–AD 1631); and phase 4: the growth of the Vesuvius cone (post-1631–1944). During the second phase, a multiphase caldera formation took place, following the emplacement of at least four Plinian eruptions (Pomici di Base, Mercato, Avellino, and Pompeii). During the third phase, the post-caldera activity took place, which was dominated by the deposits of two sub-Plinian eruptions: AD 472 Pollena (PPL) and AD 1631 (PMX). During the fourth phase, the Vesuvius cone grew (post-1631–1944), with the effusion of lavas and volcanic eruptions of a strombolian-type. The corresponding volcanic deposits are represented by wide lava flows, overlying the southwestern slopes of Vesuvius [75].

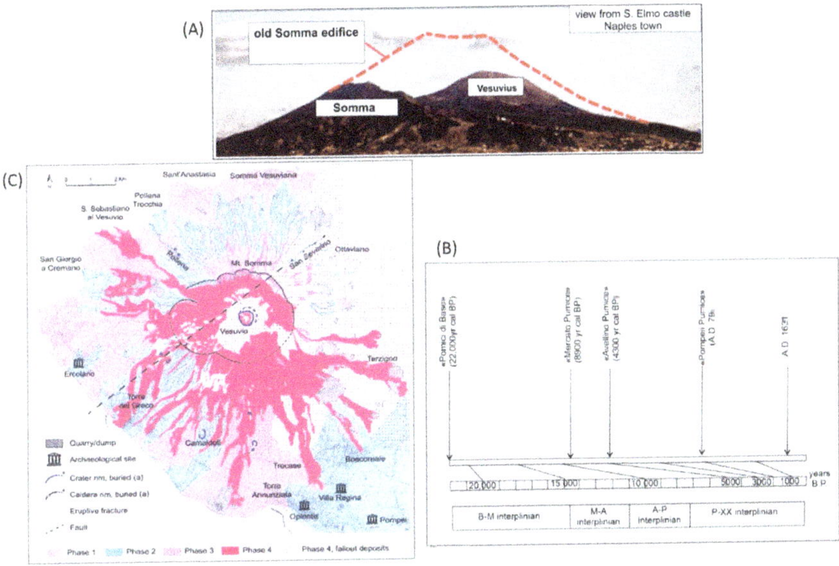

Figure 5. (**A**) Sketch reconstruction of the original profile of the old Somma volcano (modified after [74]); (**B**) Sketch diagram showing the eruptive history of the Somma-Vesuvius (calibrated ages; modified after [73]). (**C**) Geologic map of the Somma-Vesuvius (modified after [75]).

4. Results

4.1. Description of the Seismic Units

Seven seismic units (A, B, C, D, E, F, G) are identified on the seismic profiles of the eastern Bay of Naples based on the seismic stratigraphic and volcano-stratigraphic interpretation (Table 1, Figures 6–10). The G unit is recognized on the seismic profiles GRNA01 and GRNA03 (Figure 6). Its seismic facies is characterized by discontinuous, undulated, and wavy seismic reflectors. Moreover, the seismic unit shows thickness variations. The F unit is identified on the seismic profiles GRNA11, GRNA15, GRNA17, and GRNA21 (Figures 7–10). Its seismic facies is distinguished by parallel and continuous seismic reflectors, while the unit shows a wedge-shaped external geometry. The E unit is identified on the seismic sections GRNA11, GRNA15, and GRNA21 (Figures 7, 8 and 10). It is characterized by discontinuous and parallel seismic reflectors and shows thickness variations. The D unit is recognized on the seismic profiles GRNA17 and GRNA21 (Figures 9 and 10). Its seismic facies is acoustically-transparent and it is characterized by a wedge-shaped geometry. The C unit is identified on the seismic sections GRNA01, GRNA03, GRNA11, GRNA17, and GRNA21 (Figures 6, 7, 9 and 10). Its seismic facies shows two subfacies, the first one with parallel and continuous seismic reflectors and the second one with progradational reflectors. The B unit is an important volcanic seismic unit and is recognized in all the studied sections (Figures 6–10). Its seismic facies is acoustically transparent and the geometry is mounded-shaped. The A unit is another important volcanic seismic unit and is identified in all the studied sections (Figures 6–10). Its seismic facies is acoustically-transparent, while its shape is tabular (Table 1).

The 6 km long Grna01 seismic profile (Figure 6) is located offshore the Vesuvius volcano and crosses the easternmost sector of Naples Bay with an NW-SE trend (Figure 1). Four seismic units are recognized based on seismic facies and strata patterns: (G, C, B, A units; Figure 6). Based on the time-to-depth conversion of the seismic reflector corresponding to the top of the A seismic unit, this seismic reflector is located at a depth of about 412 m, which is in overall agreement with the depth of the Campanian Ignimbrite at the "Trecase 1" borehole (400 m; Figure 2). One mounded-shaped structure, composed of the B unit, is recognized. Proceeding from northwest to southeast, a significant thickness variation of the G seismic unit is observed, since the G unit increases in thickness (Figure 6).

The Grna03 seismic profile (Figure 6) has a length of about 8 km and crosses the eastern sector of Naples Bay with a NW-SE trend (Figure 1). The top of the A unit occurs at depths of 363 m (Figure 6), being in agreement with the depth of the top of the Campanian Ignimbrite unit at the "Trecase 1" borehole (400 m; Figure 2). Three mounded-shaped seismic structures, composed of the B unit, are recognized (Figure 8). The C unit drapes two of these three structures, while it onlaps the flank of the southeastward one (Figure 6). Proceeding from southeast to northwest, a thickness variation of the G unit is observed (Figure 6).

The Grna11 seismic profile (Figure 7) has a length of about 20 km and crosses the eastern Bay of Naples with a SE-NW trend (Figure 1).

Five seismic units are identified (A, B, C, E, F; Table 1; Figure 7). The E unit is located in depressions located between the seismic structures composed of the B unit or alternatively, at the top of the A unit (Figure 7).

The top of the A unit occurs at depths ranging between 330 m and 400 m (Figure 7), showing a good correlation with the top of the Campanian Ignimbrite (400 m; Figure 2).

Three broad seismic structures are recognized, composed of the B unit, respectively, having three mounds northwestwards, one mound at the center, and one mound southeastwards (Figure 7).

The C seismic unit has a progradational seismic facies and overlies the flanks of the mounded-shaped seismic structure composed of the B unit.

The E unit is extensive and in the northwest sector covers the volcanic edifices of the B unit, while in the southeast, it overlies the top of the A unit (Figure 7).

A part GRNA01 and GRNA03 (Figure 6), the F seismic unit is tabular in shape and was detected in all the seismic lines (Figure 7).

The Grna15 seismic profile (Figure 8) has a length of about 20 km and extends from Naples to the Sorrento offshore (Figure 1).

Five seismic units (A, B, C, E, F; are recognized) (Table 1; Figure 8).

The E unit is located in a depression between two seismic structures of the B unit, and on the top of the A unit (Figure 8).

The C unit is located in a depression on the B unit and on the top of the CI unit. Time-to-depth conversion yields a value of 462 m for the depth to the top of the A unit. This value is in agreement with the depth of the top of the Campanian Ignimbrite at the "Trecase 1" well (400 m; Figure 2).

The Grna17 profile (Figure 9) has a NW-SE orientation and a length of about 22 km, extending from the Naples offshore (Posillipo hill) to the Sorrento Peninsula offshore (Figure 1).

Table 1. Seismic units of the eastern Bay of Naples.

Seismic Profiles	Seismic Units	Seismic Facies External Geometry
GRNA01, GRNA03	G	Undulated and wavy seismic reflectors, discontinuous.
GRNA11, GRNA15, GRNA17, GRNA21	F	Parallel and continuous seismic reflectors and wedge-shaped external geometry
GRNA11, GRNA15, GRNA21	E	Discontinuous and sub-parallel seismic reflectors
GRNA17, GRNA21	D	Acoustically-transparent seismic facies, wedge-shaped unit
GRNA01, GRNA03, GRNA11, GRNA17, GRNA21	C	Parallel and continuous seismic reflectors. Prograding reflectors onlapping the top of a seismic structure composed of the B unit.
GRNA01, GRNA03, GRNA11, GRNA15, GRNA17, GRNA21	B	Acoustically-transparent seismic facies and mounded-shaped external geometry
GRNA01, GRNA03, GRNA11, GRNA15, GRNA17, GRNA21	A	Acoustically transparent seismic facies and tabular external geometry

Five seismic units are identified based on seismic interpretation (A, B, C, D, F).

The D unit is recognized offshore the city of Naples, unconformably overlying both the B and the C seismic units (Figure 9).

The C unit is deposited in a depression located at the top of the A unit, which is laterally bounded by the D and B seismic units.

The B unit includes a small seismic structure below the D unit and two wider seismic structures inter-layered in the A unit (Figure 9).

The Grna21 profile (Figure 10) crosses the Naples offshore from Posillipo to the Sorrento Peninsula (Figure 1). Six seismic units are recognized based on seismic interpretation (A, B, C, D, E, F; Table 1; Figure 10). D unit is deformed in the "Banco della Montagna" feature, a wide volcaniclastic field previously described in the Naples offshore [6,59,76,77]. The C unit is located in a depression, located at the top of the A unit and is laterally bounded by a seismic structure of the B unit. The E unit forms small deposits located at the top of B unit. The B unit includes three seismic structures (Figure 10). The A unit is wide and thick in this seismic line (Figure 10).

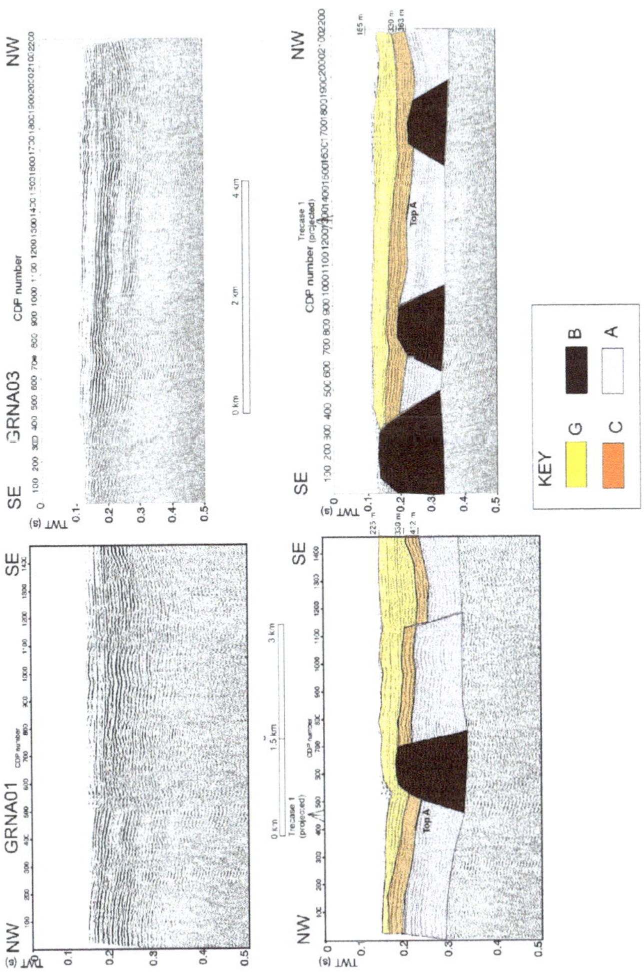

Figure 6. Seismic profile Grna01 (left side) and seismic profile Grna03 (right side), and corresponding geologic interpretation calibrated with the "Trecase 1" well (Figure 2).

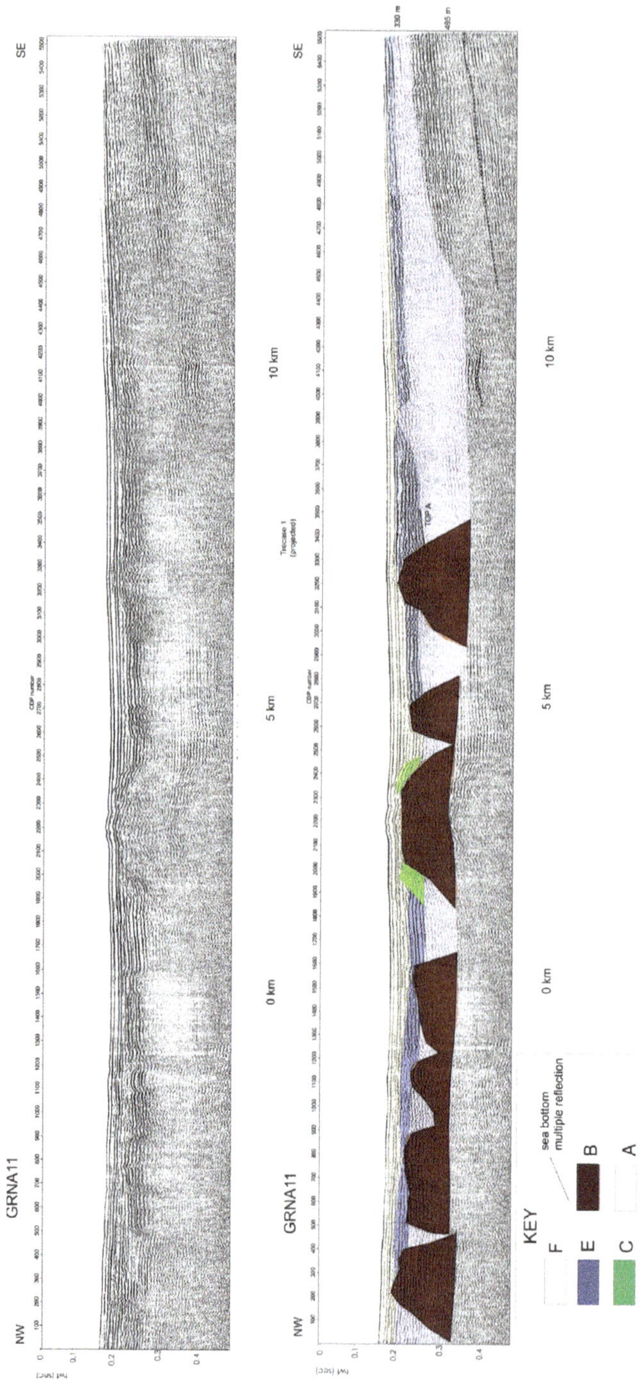

Figure 7. Seismic profile Grna11 and corresponding geologic interpretation, calibrated with the "Trecase 1" well; (Figure 2).

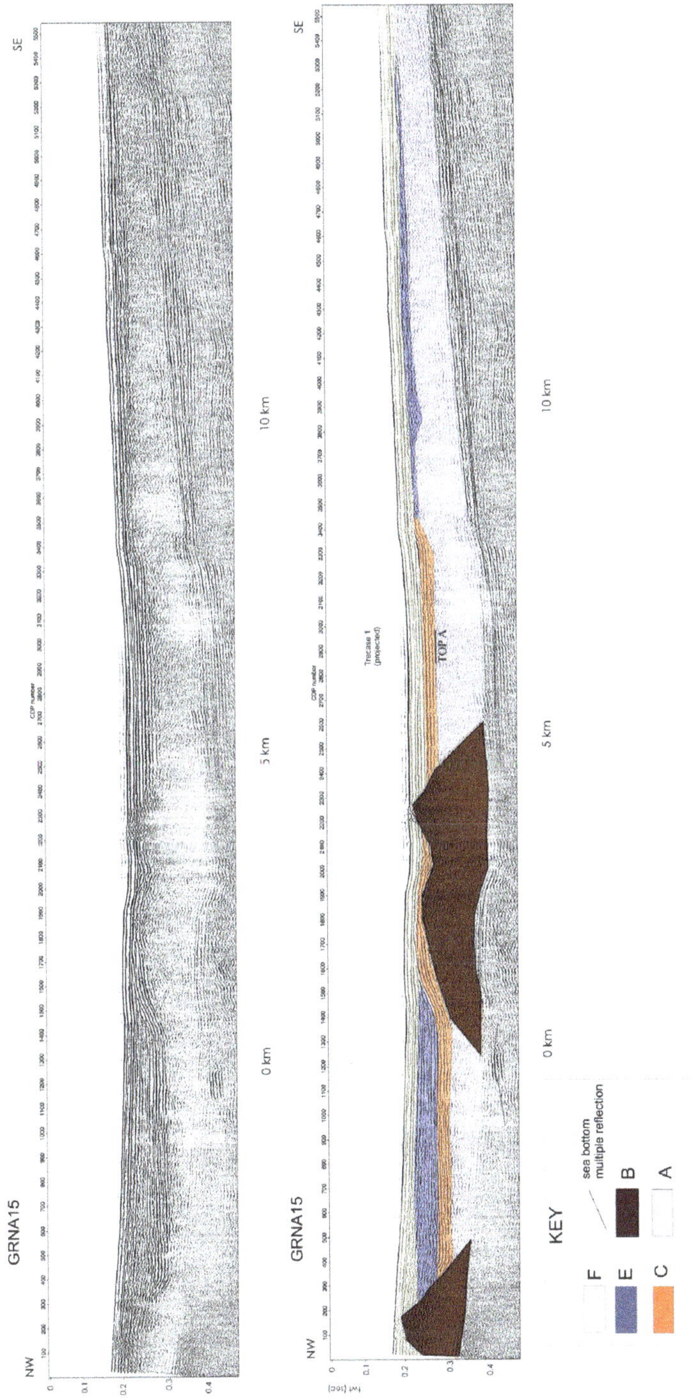

Figure 8. Seismic profile Grna15 and corresponding geologic interpretation, calibrated with the "Trecase 1" well (Figure 2).

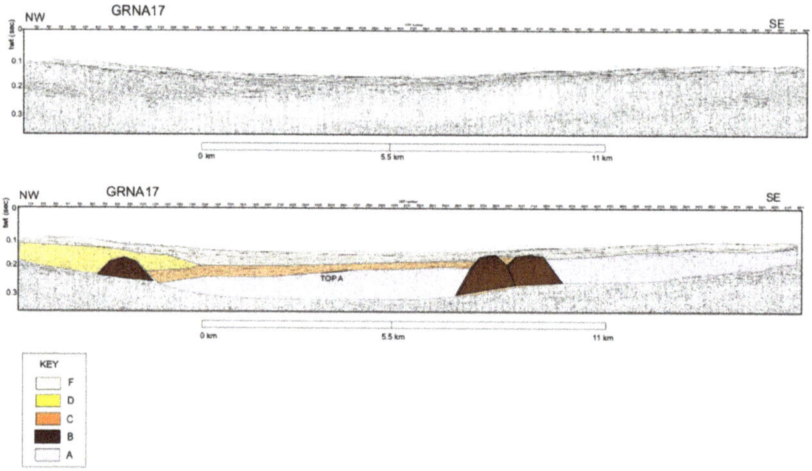

Figure 9. Seismic profile Grna17 and corresponding geologic interpretation.

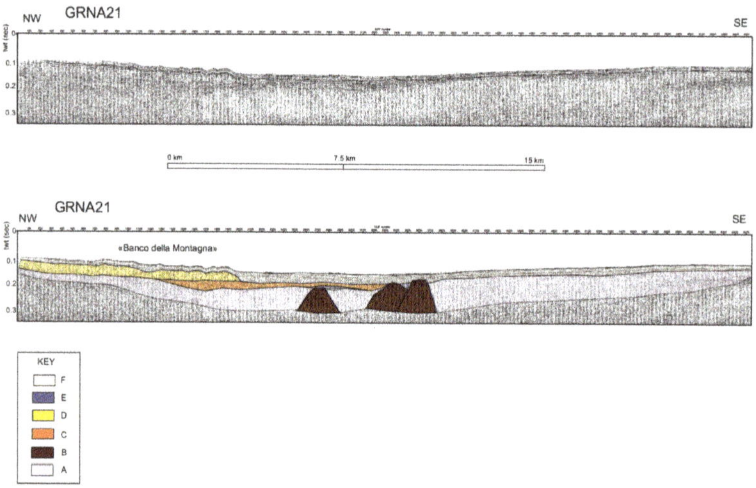

Figure 10. Seismic profile Grna21 and corresponding geologic interpretation.

4.2. Geological Interpretation of the Seismic Units

The G unit is identified on both the Grna01 and Grna03 seismic lines (Figure 6). The G unit could comprise volcanic deposits older than AD 79 and apparently capped by eruption deposits dated at their base.

The G unit probably represents pyroclastic flow deposits from Vesuvius, based on its location only offshore of Somma-Vesuvius and by the discontinuous and undulated character of the seismic reflectors. This interpretation is supported by the qualitative correlation with previous core sampling in the area, which have cored the fallout deposits of the AD 79 in the upper meters of the seismic unit [4], but the thickness of the seismic unit is much larger than the thickness of the AD 79 deposits.

Thus, the G unit is interpreted to mainly consist of PDCs of pre-AD 79 Plinian eruptions. Figure 11 shows the isopach maps of "Pomici di Avellino", "Pomici di Pompei", and "Pomici di Mercato" eruptions [25].

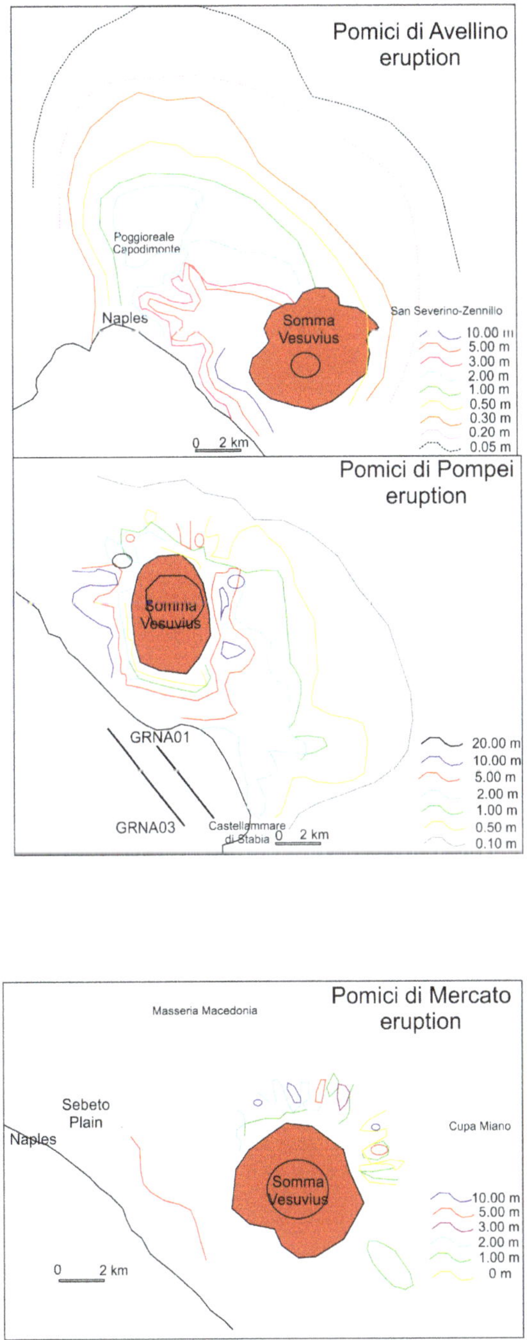

Figure 11. Isopach maps of the "Pomici di Avellino", "Pompei Pumice", and "Pomici di Mercato" eruptions (after [25]). These maps were used as geological constraints in order to support the interpretation of the G seismic unit as the PDCs older than the AD 79 eruption. The location of the GRNA01 and GRNA03 seismic profiles (Figure 6) is also shown.

In that paper, the authors discussed the dispersal, thickness, and extent of the PDC deposits, which were generated during seven plinian and sub-plinian eruptions (Pomici di Base, Greenish Pumice, Pomici di Mercato, Pomici di Avellino, Pompeii Pumice, AD 472 Pollena, and AD 1631 eruptions).

The isopach maps and the facies character reveal that the dispersal of the PDCs from Vesuvius was mainly controlled by the intensity of the eruption, by the location of the vent with respect to the Mt. Somma caldera and by the inherited topography. The most widely dispersed volcanic deposits are those of the Pomici di Avellino and Pomici di Pompei. In these maps, the isopachs of the volcanic deposits, reported in meters, indicate that they projected eastwards, not toward the Bay of Naples (Figure 11).

Based on the data of Gurioli et al. [25], the volcanic deposits of the Pomici di Mercato eruption are mainly located on the northern slopes of Mt. Somma, highlighting that the vent location was similar to that one of the present-day Vesuvius (Figure 11).

The Pomici di Avellino deposits reach their maximum thickness in the western sector of Vesuvius and are characterized by a strong facies variability. They are recognized as far as 25 km from the inferred vent, which was probably eccentric with respect to the volcano. Based on these data, the G unit is likely correlated with the volcanic deposits derived from the oldest plinian eruptions (Avellino, Mercato, and Pomici di Base).

The F and C units are recognized in the whole study area (Table 1; Figures 6–10) and are interpreted as highstand (HST) and lowstand (LST) system tract deposits from the Late Quaternary. The highstand deposits (HST) are younger than the phase of maximum marine ingression (about 4–5 ky B.P.) and were described in previous papers on the Bay of Naples [4]. The C unit, which is composed of lowstand deposits, was alsofound both in depressions between the buried tuff rings of the B unit and at the top of the A unit (Figures 6–10). The lowstand deposits (LST) were formed during the sea level lowstand during the last Quaternary glacial episode (isotopic stage 2) [54].

The D unit is identified on the seismic profiles Grna17 and Grna21 (Table 1; Figures 9 and 10). This unit is interpreted as the submarine extent of the Neapolitan Yellow Tuff (NYT) deposits, which underlie Naples and cropping out widely in the adjacent Posillipo hill. The NYT was emplaced during the second phreato-magmatic eruption from the Phlegraean caldera and crops out over an area of 1000 km^2 [8–11,16,18,67,68,70].

The NYT deposits were identified in the Gulf of Pozzuoli and in the northern Phlegraean Fields offshore [12,14,20,58,78–80]. Pyroclastic deposits represent main stratigraphic markers in the marine sedimentary record [81–83] and were documented on the Campania continental margin [84–88]. A continuous seismic reflector mapped in the northern Phlegraean offshore was correlated with the NYT deposits [21]. This horizon is inter-layered in transgressive deposits, which are thicker than those detected in other Campania outer shelf settings, as a consequence of a high input of pyroclastic and volcaniclastic deposits, related to the eruptive activity of the Campania Plain during the Late Pleistocene-Holocene.

The E unit (Table 1; Figures 7, 8 and 10) is found in depressions located between the buried tuff rings composed of the B unit or at the top of the A unit. Based on its location and on the seismic facies, characterized by discontinuous, sub-parallel seismic reflectors, the E unit is interpreted as a widely distributed debris flow unit, whose deposition is possibly related to eruptive processes in the Somma-Vesuvius offshore. Sulpizio et al. [89] have shown onshore stratigraphic data contrasting the occurrence of a large debris avalanche, which occurred just before the Avellino eruption of the Somma-Vesuvius volcano [90]. Indeed, volcaniclastic mass flows often occurred during the geological past and are characterized by rain-triggered debris flows and hyperconcentrated flows. These volcaniclastic deposits crop out widely along the cliffs of the western coastline and are separated by paleosols or erosional surfaces [89]. The volcaniclastic deposits ascribed to debris flows are coarse-grained and rich in lithic blocks, and are correlated with the pyroclastics with lava blocks of the Herculaneum log [90]. Based on these data, the E unit could represent a volcaniclastic mass flow, which occurred in the geological past, possibly after the eruption of the Campanian Ignimbrite (A unit) or after the emplacement of the B unit, interpreted as buried tuff rings [11,72].

The B seismic unit (Table 1; Figures 6–10) was previously interpreted as buried volcanic mounds [5] or cryptodomes [1] genetically related to Somma-Vesuvius eruptive activity. However, this interpretation needs be re-examined. For example, the height-to-width ratio of these bodies is very low (possibly 1:15–1:20), which seems more consistent with an interpreted of them as tuff rings, although they do not show any crater-related depressions, which should be evident in these edifices. This geological interpretation is supported by the width of the volcanic edifices (up to several km) and by comparison with similar volcanic edifices recognized in the Neapolitan area [11,72]. Figure 12 compares the structures of the B unit in seismic profiles GRNA11 and GRNA15 with the general types of volcanic edifices (lava dome, crater row, fissure vent, pyroclastic cone, tuff cone, tuff ring, maar). On the seismic profile GRNA11, two volcanic edifices, interpreted as tuff rings, are identified, both with a width of 6 km (tuff rings 1 and 2). On the seismic profile GRNA15, tuff ring 3 has a width of 5.6 km (Figure 12). The dimensions of the volcanic edifices, as measured on the seismic sections, are compatible with those of tuff rings recognized in the Neapolitan area and described in previous geological literature (Figure 12).

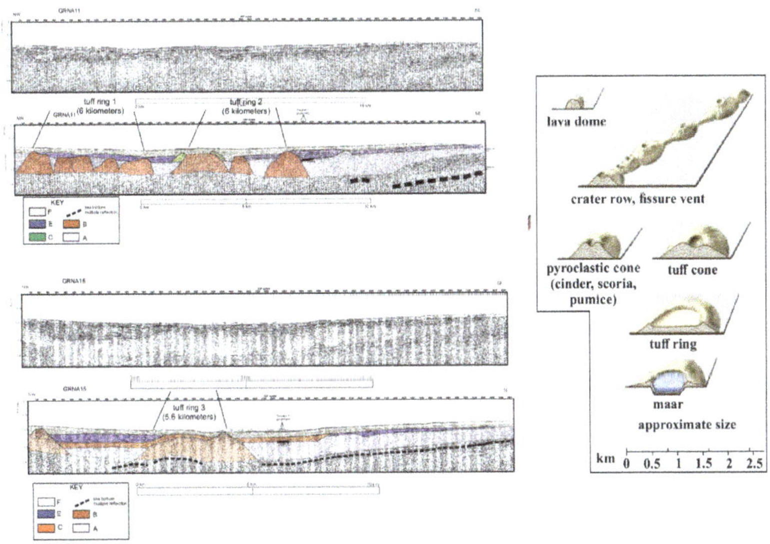

Figure 12. Sketch showing the geological constraints used for the interpretation of the B seismic unit. On the right side of the figure, several types of volcanic edifices were represented, based on the volcanological literature in order to show the shape and the approximate size of the tuff rings. On the left side of the figure, the GRNA11 and GRNA15 seismic profiles were reported (Figures 7 and 8), showing the interpretation of the volcanic edifices interpreted as tuff rings in the eastern sector of the Bay of Naples. Note that the dimensions of the volcanic edifices, measured on the seismic sections and ranging between 5.6 and 6 km, are compatible with those of the tuff rings recognized in literature in the Neapolitan area [11,72].

The A seismic unit (Table 1; Figures 6–10) is an important seismic unit in the stratigraphic architecture of the Bay of Naples. Due to its stratigraphic location and based on previous geological literature [7,59,91,92], it can be correlated with the Campanian Ignimbrite pyroclastic flow deposits. Based on literature data, this unit is important because it was identified in the whole subsurface of the eastern Bay of Naples [7,59,91,92]. The Campanian Ignimbrite is composed of thick pyroclastic deposits, trachytic and phonolitic in composition, and genetically related to a collapsed caldera [8–11,13,47,50,66–70,93,94].

5. Discussion and Conclusions

Seven seismic units were identified on the multichannel seismic lines presented in this paper (Figures 6–10), namely the G (pyroclastic flows of Vesuvius), F (highstand deposits), E (debris flows of Vesuvius), D (Neapolitan Yellow Tuff), C (lowstand deposits), B (buried tuff rings), and A (Campanian Ignimbrite) seismic units.

The seismo-stratigraphic data analyzed in this paper (Figures 6–10) can be interpreted as a succession of events for the eastern sector of the Naples Bay, referring, in particular, to the Somma-Vesuvius offshore area. The seismic profiles show a low seismic unit with an acoustically transparent seismic facies (A), corresponding to the 39 ky-old Campanian Ignimbrite and another acoustically-transparent seismic unit (B), located seawards that underlies the C unit (lowstand deposits) and is overlain by the F unit (highstand deposits; Figures 6–10).

The seismic unit B is interpreted as buried tuff rings. This interpretation is supported by the fact that tuff rings are abundant in the Neapolitan area [11,72]. For example, after the eruption of the Neapolitan Yellow Tuff (15 ky B.P.), the volcanic activity of Campi Flegrei volcanic complex was characterized by hydromagmatic phenomena, and by minor effusive activity forming lava domes. In the caldera of the Neapolitan Yellow Tuff, volcanism created numerous monogenetic vents, including tuff rings, tuff cones, cinder and spatter cones [11,72,95–98]. Among them, the tuff ring of Averno (3700 years BP) is interpreted as a maar-type, lake-filled volcano, formed during an explosive eruption of Campi Flegrei [72]. The Averno tuff ring is circular, with a rim diameter of 1.5 km, overlying an irregular topography composed of relicts of the earlier Archiaverno tuff ring [11]. Other significant tuff rings of the Neapolitan area were described by Di Vito et al. [11], discussing the characteristics of the main volcanic units of the Campi Flegrei caldera, younger than 12 ky B.P.

In proximal domains, the C unit is overlain by a chaotic seismic unit interpreted as formed by the volcanic deposits older than the AD 79 pyroclastic flow deposits (G; the pyroclastic deposits representing the base of this eruption; Figure 11) [25]. The G unit is an important seismic unit in the stratigraphic architecture of the eastern Naples Bay (Figure 6). The G unit is composed of some of the PDCs older than the AD 79 volcanic deposits. Previous studies on the geological setting of the Bay of Naples offshore the Sarno prodelta have documented the stratigraphic signature of the tephra deposit erupted by the Vesuvius volcano during the AD 79 volcanic event [4]. Moreover, the isopach maps of the PDCs constructed by Gurioli et al. [25] were used to check the correlation of this seismic unit with some of the PDCs older than AD 79 volcanic deposits (Figure 11) and have confirmed this interpretation. Figure 11 reports the isopach maps of Pomici di Avellino, Pompei, and Pomici di Mercato eruptions [25]. Based on the data of Gurioli et al. [25], five of the seven Vesuvius plinian and sub-plinian eruptions (Pomici di Base, Greenish Pumice, Pomici di Mercato, Pomici di Avellino, Pompeii Pumice, AD 472 Pollena, and AD 1631 eruptions) have dispersed PDCs radially and their dispersion was controlled by the location of the vent and by the palaeo-topography. The most dispersed deposits are those of the Pomici di Avellino and Pompeii Pumice, extending more than 20 km from the inferred vent [25]. The maximum thickness of Pomici di Mercato is 18 m (measured to the north), while the maximum thickness of Pomici di Avellino is 22 m (measured to the west). The isopach map of the Pomici di Mercato eruption has shown that these deposits are mainly located along the northern slopes of Mt. Somma, and that the corresponding vent was located in a location similar to that one of the Vesuvius. The isopach map of the Pomici di Avellino eruption has shown that they have their maximum thickness in the western sector of the volcano and are characterized by important lateral variations in lithofacies and thickness [99]. Sulpizio et al. [99] studied the sedimentology and the physical volcanology of the Pomici di Avellino eruption through field and laboratory studies, establishing five eruptive units (EUs), which emplaced during an opening, Plinian and final phreatomagmatic phase of this eruption. The PDCs of the Avellino eruption were erupted during EU1, EU2, EU3, EU4, and EU5 with different eruptive mechanisms.

Due to its seismic facies and to the stratigraphic relationships, the C unit is interpreted as the lowstand system tract (LST) deposits, i.e., the sediments that were deposited during the falling and lowstand of sea level. These deposits were disturbed by the eruption of the D unit, interpreted as the Neapolitan Yellow Tuff deposits (15 ky B.P.).

The seismo-stratigraphic analysis of multichannel profiles calibrated with gravity cores offshore the Vesuvius volcano and Posillipo hill has highlighted new implications for the Quaternary volcanism of the Campania continental margin. Multichannel seismic lines are interpreted to study the stratigraphic relationships between the Quaternary marine deposits deposited in this area and the volcanic acoustic substratum (A and D units). The volcano structure is controlled by NE-SW trending seismic structures, suggesting a volcano–structural correlation with the regional faults occurring in this area [100,101]. These structures correspond with tuff rings composed of the B unit, which underlie the lowstand deposits and seem to be younger than the A unit. The tuff rings seem to be younger than 37 ky B.P. (A unit) and older than 18 ky B.P. (approximate lower boundary of the lowstand deposits of the C unit).

The total magnetic field offshore the Somma-Vesuvius volcanic complex (Figure 13) shows that the shape of the anomaly is dipolar. There is no apparent effect caused by the occurrence of remnant magnetization with respect to that of the present-day main field. NE-SW trending seismic structures identified based on seismic data represent a preferential pathway of magma uprising for Vesuvius volcano. These structures mainly occur offshore of the Torre del Greco and the Torre Annunziata towns and are evident on the sketch magnetic map of Figure 13, which includes the tuff rings shown in this paper and the A and D seismic units. The geological structures identified through seismic interpretation are located in a complex magnetic anomaly area (Figure 13), which is made up of several anomalies, reaching a maximum intensity of 400 nT. This area represents the offshore prolongation of the Vesuvius volcano. The seismo-stratigraphic data suggest a close relationship between the seismic structures offshore the volcano and in the eastern sector of the gulf, and the structural setting of the Naples Bay, controlled by NE-SW trending (counter-Apenninic) regional faults (Figure 13).

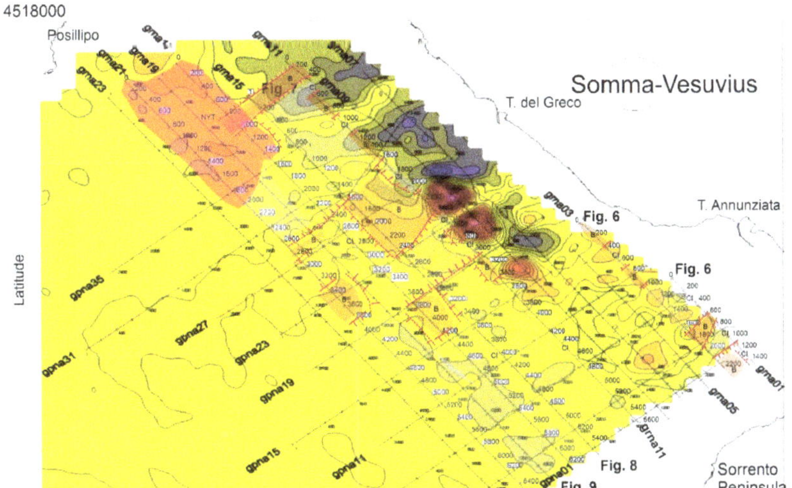

Figure 13. Magnetic map of the eastern Naples Bay and location of the seismic lines with superimposed A, B, and D seismic and volcanic structures based on this paper.

Funding: This research received no external funding.

Institutional Review Board Statement: Not applicable.

Informed Consent Statement: Not applicable.

Conflicts of Interest: The authors declare no conflict of interest.

References

1. Milia, A.; Mirabile, L.; Torrente, M.M.; Dvorak, J.J. Volcanism offshore of Vesuvius volcano in Naples Bay. *Bull. Volcanol.* **1998**, *59*, 404–413. [CrossRef]
2. Bruno, P.P.G.; Rapolla, A. Study of sub-surface structure of Somma-Vesuvius (Italy) by seismic reflection data: Implications for Campanian volcanism. *Tectonophysics* **1999**, *372*, 193–213. [CrossRef]
3. Milia, A.; Torrente, M.M.; Bellucci, F. A possible link between faulting, cryptodomes and lateral collapses at Vesuvius Volcano (Italy). *Glob. Planet. Change* **2012**, *90–91*, 121–134. [CrossRef]
4. Sacchi, M.; Insinga, D.D.; Milia, A.; Molisso, F.; Raspini, A.; Torrente, M.M.; Conforti, A. Stratigraphic signature of the Vesuvius 79 AD event off the Sarno prodelta system, Naples Bay. *Mar. Geol.* **2005**, *222–223*, 443–469. [CrossRef]
5. Aiello, G.; Angelino, A.; D'Argenio, B.; Marsella, E.; Pelosi, N.; Ruggieri, S.; Siniscalchi, A. Buried volcanic structures in the Gulf of Naples (Southern Tyrrhenian sea, Italy) resulting from high resolution magnetic survey and seismic profiling. *Ann. Geophys.* **2005**, *48*, 1–15.
6. Passaro, S.; Sacchi, M.; Tamburrino, S.; Ventura, G. Fluid Vents, Flank Instability, and Seafloor Processes along the Submarine Slopes of the Somma-Vesuvius Volcano, Eastern Tyrrhenian Margin. *Geosciences* **2018**, *8*, 60. [CrossRef]
7. Milia, A.; Torrente, M.M. Space-time evolution of an active volcanic field in an extensional region: The example of the Campania margin (eastern Tyrrhenian Sea). In *Vesuvius, Campi Flegrei, and Campanian Volcanism*, 1st ed.; De Vivo, B., Belkin, H., Rolandi, G., Eds.; Elsevier: New York, NY, USA, 2020; pp. 297–321.
8. Rosi, M.; Sbrana, A. *Phlegrean Fields*; Consiglio Nazionale delle Ricerche, Quaderni de La Ricerca Scientifica: Rome, Italy, 1987.
9. Scandone, R.; Bellucci, F.; Lirer, L.; Rolandi, G. The structure of the Campanian Plain and the activity of the Neapolitan volcanoes (Italy). *J. Volcanol. Geoth. Res.* **1991**, *48*, 1–31. [CrossRef]
10. Orsi, G.; De Vita, S.; Di Vito, M. The restless, resurgent Campi Flegrei nested caldera (Italy): Constraints on its evolution and configuration. *J. Volcanol. Geoth. Res.* **1996**, *74*, 179–214. [CrossRef]
11. Di Vito, M.A.; Isaia, R.; Orsi, G.; Southon, J.; De Vita, S.; D'Antonio, M.; Pappalardo, L.; Piochi, M. Volcanism and deformation since 12,000 years at the Campi Flegrei caldera (Italy). *J. Volcanol. Geoth. Res.* **1999**, *91*, 221–246. [CrossRef]
12. Aiello, G.; Marsella, E.; Di Fiore, V. New seismo-stratigraphic and marine magnetic data of the Gulf of Pozzuoli (Naples Bay, Tyrrhenian sea, Italy): Inferences for the tectonic and magmatic events of the Phlegrean Fields volcanic complex (Campania). *Mar. Geophys. Res.* **2012**, *33*, 93–125. [CrossRef]
13. Fitzsimmons, K.E.; Hambach, U.; Veres, D.; Iovita, R. The Campanian Ignimbrite Eruption: New Data on Volcanic Ash Dispersal and Its Potential Impact on Human Evolution. *PLoS ONE* **2013**, *8*, e65839. [CrossRef] [PubMed]
14. Aiello, G.; Giordano, L.; Giordano, F. High-resolution seismic stratigraphy of the Gulf of Pozzuoli (Naples Bay) and relationships with submarine volcanic setting of the Phlegrean Fields volcanic complex. *Rend. Lincei Sci. Fis. E Nat.* **2016**, *27*, 775–801. [CrossRef]
15. Giaccio, B.; Hajdas, I.; Isaia, R.; Deino, A.; Nomade, S. High-precision ^{14}C and ^{40}Ar/^{39}Ar dating of the Campanian Ignimbrite (Y-5) reconciles the time-scales of climatic-cultural processes at 40 ka. *Sci. Rep.* **2017**, *7*, 45940. [CrossRef] [PubMed]
16. Scarpati, C.; Cole, P.; Perrotta, A. The Neapolitan Yellow Tuff: A large volume multiphase eruption from Campi Flegrei, Southern Italy. *Bull. Volcanol.* **1993**, *55*, 343–356. [CrossRef]
17. Wohletz, K.; Orsi, G.; De Vita, S. Eruptive mechanism of the Neapolitan Yellow Tuff interpreted from stratigraphic, chemical and granulometric data. *J. Volcanol. Geoth. Res.* **1995**, *67*, 263–290. [CrossRef]
18. Deino, A.L.; Orsi, G.; Piochi, M.; De Vita, S. The age of the Neapolitan Yellow Tuff caldera forming eruption (Campi Flegrei caldera—Italy) assessed by ^{40}Ar/^{39}Ar dating method. *J. Volcanol. Geoth. Res.* **2004**, *133*, 157–170. [CrossRef]
19. Pabst, S.; Wörner, G.; Civetta, L.; Tesoro, R. Magma chamber evolution prior to the Campanian Ignimbrite and Neapolitan Yellow Tuff eruptions (Campi Flegrei, Italy). *Bull. Volcanol.* **2008**, *70*, 961–976. [CrossRef]
20. Sacchi, M.; Pepe, F.; Corradino, M.; Insinga, D.D.; Molisso, F.; Lubritto, C. The Neapolitan Yellow Tuff caldera offshore the Campi Flegrei: Stratal architecture and kinematic reconstruction during the last 15 ky. *Mar. Geol.* **2014**, *354*, 15–33. [CrossRef]
21. Aiello, G.; Insinga, D.D.; Iorio, M.; Meo, A.; Senatore, M.R. On the occurrence of the Neapolitan Yellow Tuff tephra in the Northern Phlegraean Fields offshore (Eastern Tyrrhenian margin; Italy). *Ital. J. Geosci.* **2017**, *136*, 263–274. [CrossRef]
22. Gurioli, L.; Cioni, R.; Sbrana, A.; Zanella, E. Transport and deposition of pyroclastic density currents over an inhabited area: The deposits of the AD 79 eruption of Vesuvius at Herculaneum (Italy). *Sedimentology* **2002**, *49*, 929–953. [CrossRef]
23. Cioni, R.; Gurioli, L.; Lanza, R.; Zanella, E. Temperatures of the A.D. 79 pyroclastic density current deposits (Vesuvius, Italy). *J. Geophys. Res.* **2004**, *109*, B02207. [CrossRef]
24. Gurioli, L.; Zanella, E.; Pareschi, M.T.; Lanza, R. Influences of urban fabric on pyroclastic density currents at Pompeii (Italy) I: Flow direction and deposition. *J. Geophys. Res.* **2007**, *112*, B05213. [CrossRef]

25. Gurioli, L.; Sulpizio, R.; Cioni, R.; Sbrana, A.; Santacroce, R.; Luperini, W.; Andronico, D. Pyroclastic flow hazard assessment at Somma-Vesuvius based on the geological record. *Bull. Volcanol.* **2010**, *72*, 1021–1038. [CrossRef]
26. Shea, T.; Gurioli, L.; Houghton, B.F.; Cioni, R.; Cashman, K.V. Column collapse and generation of pyroclastic density currents during the A.D. 79 eruption of Vesuvius: The role of pyroclastic density. *Geology* **2011**, *39*, 695–698. [CrossRef]
27. Cioni, R.; Tadini, A.; Gurioli, L.; Bertagnini, A.; Mulas, M.; Bevilacqua, A.; Neri, A. Estimating eruptive parameters and related uncertainties for pyroclastic density current deposits: Worked examples from Somma-Vesuvius (Italy). *Bull. Volcanol.* **2020**, *82*, 65. [CrossRef]
28. Tadini, A.; Bevilacqua, A.; Neri, A.; Cioni, R.; Biagioli, G.; Vitturi, M.; Esposti Ongaro, T. Reproducing pyroclastic density current deposits of the 79 CE eruption of the Somma-Vesuvius volcano using the box-model approach. *Solid Earth* **2021**, *12*, 119–139. [CrossRef]
29. Available online: http://www.isprambiente.gov.it/Media/carg/campania.html (accessed on 16 October 2022).
30. Colorado School of Mines. Seismic Unix, Center for Wave Phenomena. 2000. Available online: https://wiki.seismic-unix.org/doku.php (accessed on 16 October 2022).
31. Mitchum, R.M., Jr.; Vail, P.R.; Sangree, J.B. Stratigraphic interpretation of seismic reflection patterns in depositional sequences. In *Seismic Stratigraphy—Applications to Hydrocarbon Exploration*; Payton, C.E., Ed.; American Association of Petroleum Geologists: Tulsa, OH, USA, 1977; Volume 26, pp. 117–133.
32. Vail, P.R.; Mitchum, R.M.; Thompson, S. Relative changes of sea level from coastal onlap. In *Seismic Stratigraphy—Applications to Hydrocarbon Exploration*; Payton, C E., Ed.; American Association of Petroleum Geologists: Tulsa, OH, USA, 1977; Volume 26, pp. 83–97.
33. Vail, P.R.; Hardenbol, J.; Todd, R.G. Jurassic unconformities, chronostratigraphy and sea-level changes from seismic stratigraphy and biostratigraphy. In *Interregional Unconformities and Hydrocarbon Accumulation*; Schlee, J S., Ed.; American Association of Petroleum Geologists: Tulsa, OH, USA, 1984; Volume 36, pp. 129–144.
34. Planke, S.P.A.; Symonds, E.; Avelstad, J.; Skogseid, J. Seismic volcano stratigraphy of large-volume basaltic extrusive complexes on rifted margins. *J. Geophys Res* **2000**, *105*, 19335–19351.
35. Carlson, R.L.; Gangi, A.F.; Snow, K.L. Empirical reflection travel time versus depth and velocity versus depth functions for the deep sea sediments column. *J. Geophys. Res.* **1986**, *91*, 8249–8266. [CrossRef]
36. Brocchini, F.; Principe, C.; Castradori, D.; Laurenzi, M.A.; Gorla, L. Quaternary evolution of the southern sector of the Campanian Plain and early Somma-Vesuvius activity: Insights from the Trecase well. *Mineral. Petrol.* **2001**, *73*, 67–91. [CrossRef]
37. De Lorenzo, G. L'attività vulcanica dei Campi Flegrei. *Rend. Acc. Sci. Fis. E Mat. Napoli* **1904**, *10*, 203–221.
38. Santacroce, R. *Somma-Vesuvius*; CNR, Quaderni De La Ricerca Scientifica: Rome, Italy, 1987.
39. Ortolani, F.; Aprile, F. Nuovi dati sulla struttura profonda della Piana Campana a SE del Fiume Volturno. *Boll. Soc. Geol. Ital.* **1978**, *97*, 591–608.
40. Aiello, G.; Cicchella, A.G.; Di Fiore, V.; Marsella, E. New seismo-stratigraphic data of the Volturno Basin (northern Campania, Tyrrhenian margin, southern Italy): Implications for tectono-stratigraphy of the Campania and Latium sedimentary basins. *Ann. Geophys.* **2011**, *54*, 265–283.
41. Torrente, M.M.; Milia, A. Volcanism and faulting of the Campania margin (Eastern Tyrrhenian Sea, Italy): A three-dimensional visualization of a new volcanic field off Campi Flegrei. *Bull. Volcanol.* **2013**, *75*, 719. [CrossRef]
42. Ippolito, F.; Ortolani, F.; Russo, M. Struttura marginale tirrenica dell'Appennino Campano: Reinterpretazione di dati di antiche ricerche di idrocarburi. *Mem. Soc. Geol. Ital.* **1973**, *12*, 227–250.
43. Barberi, F.; Innocenti, F.; Lirer, L.; Munno, R.; Pescatore, T.; Santacroce, R. The Campanian Ignimbrite: A major prehistoric eruption in the Neapolitan area (Italy). *Bull. Volcanol.* **1978**, *41*, 10–31. [CrossRef]
44. Fisher, R.V.; Orsi, G.; Ort, M.; Heiken, G. Mobility of a large-volume pyroclastic flow—emplacement of the Campanian ignimbrite, Italy. *J. Volcanol. Geoth. Res.* **1993**, *56*, 205–220. [CrossRef]
45. Rosi, M.; Vezzoli, L.; Aleotti, P.; De Censi, M. Interaction between caldera collapse and eruptive dynamics during the Campanian Ignimbrite eruption, Phlegrean Fields, Italy. *Bull. Volcanol.* **1996**, *57*, 541–554. [CrossRef]
46. Civetta, L.; Orsi, G.; Pappalardo, L.; Fisher, R.V.; Heiken, G.; Ort, M. Geochemical zoning, mingling, eruptive dynamics and depositional processes—The Campanian Ignimbrite, Campi Flegrei caldera, Italy. *J. Volcanol. Geoth. Res.* **1997**, *75*, 183–219. [CrossRef]
47. Pappalardo, L.; Civetta, L.; D'Antonio, M.; Deino, A.; Di Vito, M.A.; Orsi, G.; Carandente, A.; De Vita, S.; Isaia, R.; Piochi, M. Chemical and Sr-isotopical evolution of the Phlegraean magmatic system before the Campanian Ignimbrite and the Neapolitan Yellow Tuff eruptions. *J. Volcanol. Geoth. Res.* **1999**, *91*, 141–166. [CrossRef]
48. Marianelli, P.; Sbrana, A.; Proto, M. Magma chamber of the Campi Flegrei supervolcano at the time of eruption of the Campanian Ignimbrite. *Geology* **2006**, *34*, 937–940. [CrossRef]
49. Fedele, F.G.; Giaccio, B.; Hajdas, I. Timescales and cultural process at 40,000 BP in the light of the Campanian Ignimbrite eruption, Western Eurasia. *J. Hum. Evol.* **2008**, *55*, 834–857. [CrossRef] [PubMed]
50. Costa, A.; Folch, A.; Macedonio, G.; Giaccio, B.; Isaia, R.; Smith, V.C. Quantifying volcanic ash dispersal and impact of the Campanian Ignimbrite super-eruption. *Geophys. Res. Lett.* **2012**, *39*, L10310. [CrossRef]
51. Marti, A.; Folch, A.; Costa, A.; Engwell, S. Recostructing the plinian and co-ignimbrite sources of large volcanic eruptions: A novel approach for the Campanian Ignimbrite. *Sci. Rep.* **2016**, *6*, 21220. [CrossRef] [PubMed]

52. Rolandi, G.; Bellucci, F.; Heizler, M.T.; Belkin, H.E.; De Vivo, B. Tectonic controls on the genesis of ignimbrites from the Campanian Volcanic Zone, southern Italy. *Mineral. Petrol.* **2003**, *79*, 3–31. [CrossRef]
53. Romano, P.; Santo, A.; Voltaggio, M. L'evoluzione geomorfologica della pianura del fiume Volturno (Campania) durante il tardo Quaternario (Pleistocene medio-superiore-Olocene). *Il Quat.* **1994**, *7*, 41–56.
54. Martinson, D.; Pisias, N.; Hays, J.; Imbrie, J.; Moore, T.; Shackleton, N. Age Dating and the Orbital Theory of the Ice Ages: Development of a High-Resolution 0 to 300,000-Year Chronostratigraphy. *Quat. Res.* **1987**, *27*, 1–29. [CrossRef]
55. De Vivo, B.; Rolandi, G.; Gans, P.B.; Calvert, A.; Bohrson, W.A.; Spera, F.J.; Belkin, H.E. New constraints on the pyroclastic eruptive history of the Campanian volcanic Plain (Italy). *Mineral. Petrol.* **2001**, *73*, 47–65. [CrossRef]
56. Webster, J.D.; Raia, F.; Tappen, C.; De Vivo, B. Pre-eruptive geochemistry of the ignimbrite-forming magmas of the Campanian Volcanic Zone, Southern Italy, determined from silicate melt inclusions. *Mineral. Petrol.* **2003**, *79*, 99–125. [CrossRef]
57. Bellucci, F.; Milia, A.; Rolandi, G.; Torrente, M.M. Structural control on the Upper Pleistocene ignimbrite eruptions in the Neapolitan area (Italy): Volcano-tectonic faults versus caldera faults. In *Vesuvius, Campi Flegrei, and Campanian Volcanism*, 1st ed.; De Vivo, B., Belkin, H., Rolandi, G., Eds.; Elsevier: New York, NY, USA, 2020; pp. 163–180.
58. Zanchetta, G.; Sulpizio, R.; Giaccio, B.; Siani, G.; Paterne, M.; Wulf, S.; D'Orazio, M. The Y-3 tephra: A Last Glacial stratigraphic marker for the central Mediterranean basin. *J. Volcanol. Geoth. Res.* **2008**, *177*, 145–154. [CrossRef]
59. Aiello, G.; Iorio, M.; Molisso, F.; Sacchi, M. Integrated Morpho-Bathymetric, Seismic-Stratigraphic, and Sedimentological Data on the Dohrn Canyon (Naples Bay, Southern Tyrrhenian Sea): Relationships with Volcanism and Tectonics. *Geosciences* **2020**, *10*, 319. [CrossRef]
60. Rolandi, G.; De Natale, G.; Kilburn, C.R.J.; Troise, C.; Somma, R.; Di Lascio, M.; Fedele, A.; Rolandi, R. The 39 ka Campanian Ignimbrite eruption: New data on source area in the Campanian Plain. In *Vesuvius, Campi Flegrei, and Campanian Volcanism*, 1st ed.; De Vivo, B., Belkin, H., Rolandi, G., Eds.; Elsevier: New York, NY, USA, 2020; pp. 175–205.
61. D'Argenio, B.; Pescatore, T.; Scandone, P. Schema geologico dell'Appennino meridionale (Campania e Lucania). In *Proceedings of the Conference Moderne Vedute sulla Geologia dell'Appennino, Roma*; Accademia Nazionale dei Lincei; Rome, Italy, 1973; pp. 1–183.
62. Bigi, G.; Bonardi, G.; Catalano, R.; Cosentino, D.; Lentini, F.; Parotto, M.; Sartori, R.; Scandone, P.; Turco, E. *Structural Model of Italy, 1:500.000*; Progetto Finalizzato Geodinamica, Consiglio Nazionale delle Ricerche: Rome, Italy, 1992.
63. Cinque, A.; Aucelli, P.P.C.; Brancaccio, L.; Mele, R.; Milia, A.; Robustelli, G.; Romano, P.; Russo, F.; Russo, M.; Santangelo, N.; et al. Volcanism, tectonics and recent geomorphological change in the Bay of Napoli. *Suppl. Geogr. Fis. Dinam. Quat.* **1997**, *3*, 123–141.
64. Vitale, S.; Ciarcia, S. Tectono-stratigraphic setting of the Campania region (southern Italy). *J. Maps* **2018**, *14*, 9–21. [CrossRef]
65. Perrotta, A.; Scarpati, C.; Luongo, G.; Morra, V. The Campi Flegrei caldera boundary in the city of Naples. In *Vesuvius, Campi Flegrei, and Campanian Volcanism*, 1st ed.; De Vivo, B., Belkin, H., Rolandi, G., Eds.; Elsevier: New York, NY, USA, 2020; pp. 85–96.
66. Acocella, V.; Funiciello, R.; Marotta, E.; Orsi, G.; De Vita, S. The role of extensional structures on experimental calderas and resurgence. *J. Volcanol. Geoth. Res.* **2004**, *129*, 199–217. [CrossRef]
67. Acocella, V. Activating and reactivating pairs of nested collapses during caldera forming eruptions: Campi Flegrei (Italy). *Geophys. Res. Lett.* **2008**, *35*, L17304. [CrossRef]
68. Acocella, V. Evaluating fracture patterns within a resurgent caldera: Campi Flegrei, Italy. *Bull. Volcanol.* **2010**, *72*, 623–638. [CrossRef]
69. Capuano, P.; Russo, G.; Civetta, L.; Orsi, G.; D'Antonio, M.; Moretti, R. The active portion of the Campi Flegrei caldera structure imaged by 3-D inversion of gravity data. *Geochem. Geophys. Geosystems* **2013**, *14*, 276. [CrossRef]
70. Vitale, S.; Isaia, R. Fractures and faults in volcanic rocks (Campi Flegrei, southern Italy): Insight into volcano-tectonic processes. *Int. J. Earth Sci.* **2014**, *103*, 801–819. [CrossRef]
71. Ascione, A.; Aucelli, P.P.C.; Cinque, A.; Di Paola, G.; Mattei, G.; Ruello, M.R.; Russo Ermolli, E.; Santangelo, N.; Valente, E. Geomorphology of Naples and the Campi Flegrei: Human and natural landscapes in a restless land. *J. Maps* **2021**, *17*, 18–28. [CrossRef]
72. Mastrolorenzo, G. Averno tuff ring in Campi Flegrei (south Italy). *Bull. Volcanol.* **1994**, *56*, 561–572. [CrossRef]
73. Santacroce, R.; Cioni, R.; Marianelli, P.; Sbrana, A.; Sulpizio, R.; Zanchetta, G.; Donahue, D.J.; Joron, J.L. Age and whole rock-glass compositions of proximal pyroclastics from the major explosive eruptions of Somma-Vesuvius. A review as a tool for distal tephrostratigraphy. *J. Volcanol. Geoth. Res.* **2008**, *177*, 1–18. [CrossRef]
74. Cioni, R.; Bertagnini, A.; Santacroce, R.; Andronico, D. Explosive activity and eruption scenarios at Somma-Vesuvius (Italy). Towards a new classification scheme. *J. Volcanol. Geoth. Res.* **2009**, *178*, 331–346. [CrossRef]
75. Sbrana, A.; Cioni, R.; Marianelli, P.; Sulpizio, R.; Andronico, D.; Pasquini, G. Volcanic Evolution of the Somma-Vesuvius Complex (Italy). *J. Maps* **2020**, *16*, 137–147. [CrossRef]
76. Sacchi, M.; D'Argenio, B.; Morra, V.; Petrazzuoli, S.; Aiello, G.; Budillon, F.; Sarnacchiaro, G.; Tonielli, R. Pyroclastic lumps: Quick diapiric structures off the Naples Bay, Italy. In Proceedings of the European Geophysical Society (EGS), 25th General Assembly, Millenium Conference on Earth, Planetary & Solar Systems, Nice, France, 25–29 April 2000.
77. Aiello, G.; Budillon, F.; Cristofalo, G.; D'Argenio, B.; De Alteriis, G.; De Lauro, M.; Ferraro, L.; Marsella, E.; Pelosi, N.; Sacchi, M.; et al. Marine geology and morpho-bathymetry in the Bay of Naples (South eastern Tyrrhenian sea, Italy). In *Mediterranean Ecosystems: Structures and Processes*; Faranda, F.M., Guglielmo, L., Spezie, G., Eds.; Springer: Milano, Italy, 2001; pp. 1–8.
78. Steinmann, L.; Spiess, V.; Sacchi, M. The Campi Flegrei caldera (Italy): Formation and evolution in interplay with sea-level variations since the Campanian Ignimbrite eruption at 39 ka. *J. Volcanol. Geoth. Res.* **2016**, *327*, 361–374. [CrossRef]

79. Steinmann, L.; Spiess, V.; Sacchi, M. Post-collapse evolution of a coastal caldera system: Insights from a 3-D multichannel seismic survey from the Campi Flegrei caldera (Italy). *J. Volcanol. Geoth. Res.* **2018**, *349*, 83–98. [CrossRef]
80. Natale, J.; Ferranti, L.; Isaia, R.; Marino, C.; Sacchi, M.; Spiess, V.; Steinmann, L.; Vitale, S. Integrated on-land-offshore stratigraphy of the Campi Flegrei caldera: New insights into the volcano-tectonic evolution in the last 15 kyr. *Basin Res.* **2021**, *34*, 1–28. [CrossRef]
81. Lowe, D.J. Tephrochronology and its application: A review. *Quat. Geochronol.* **2011**, *6*, 107–153. [CrossRef]
82. Zanchetta, G.; Sulpizio, R.; Roberts, N.; Cioni, R.; Eastwood, W.J.; Siani, G.; Caron, B.; Paterne, M.; Santacroce, R. Tephrostratigraphy, chronology and climatic events of the Mediterranean basin during the Holocene: An overview. *Holocene* **2011**, *21*, 33–52. [CrossRef]
83. Insinga, D.D.; Tamburrino, S.; Lirer, F.; Vezzoli, L.; Barra, M.; De Lange, G.J.; Tiepolo, M.; Vallefuoco, M.; Mazzola, S.; Sprovieri, M. Tephrochronology of the astronomically-tuned KC01B deep-sea core, Ionian Sea: Insights into the explosive activity of the Central Mediterranean area during the last 200 ka. *Quat. Sci. Rev.* **2014**, *85*, 63–84. [CrossRef]
84. Buccheri, G.; Capretto, G.; Di Donato, V.; Esposito, P.; Ferruzza, G.; Pescatore, T.S.; Russo Ermolli, E.; Senatore, M.R.; Sprovieri, M.; Bertoldo, M.; et al. A high resolution record of the last deglaciation in the southern Tyrrhenian sea: Environmental and climatic evolution. *Mar. Geol.* **2002**, *186*, 447–470. [CrossRef]
85. Iorio, M.; Sagnotti, L.; Angelino, A.; Budillon, F.; D'Argenio, B.; Dinares-Turell, J.; Macrì, P.; Marsella, E. High resolution petrophysical and palaeomagnetic study of Late Holocene shelf sediments, Salerno Gulf, Tyrrhenian Sea. *Holocene* **2004**, *14*, 425–426. [CrossRef]
86. Insinga, D.D.; Molisso, F.; Lubritto, C.; Sacchi, M.; Passariello, L.; Morra, V. The proximal marine record of Somma-Vesuvius volcanic activity in the Naples and Salerno bays, Eastern Tyrrhenian sea, during the last 3 kyrs. *J. Volcanol. Geoth. Res.* **2008**, *177*, 170–186. [CrossRef]
87. Lirer, F.; Sprovieri, M.; Ferraro, L.; Vallefuoco, M.; Capotondi, L.; Cascella, A.; Petrosino, P.; Insinga, D.D.; Pelosi, N.; Tamburrino, S.; et al. Integrated stratigraphy in the eastern Tyrrhenian sea. *Quat. Int.* **2013**, *292*, 71–85. [CrossRef]
88. Margaritelli, G.; Vallefuoco, M.; Di Rita, F.; Capotondi, L.; Bellucci, L.G.; Insinga, D.D.; Petrosino, P.; Bonomo, S.; Cacho, I.; Cascella, A.; et al. Marine response to climate changes during the last five millennia in the central Mediterranean Sea. *Glob. Planet. Change* **2016**, *142*, 53–72. [CrossRef]
89. Sulpizio, R.; Cioni, R.; Di Vito, M.A.; Santacroce, R.; Sbrana, A.; Zanchetta, G. Comment on: "The dark nature of Somma-Vesuvius volcano: Evidence from the 3.5 ka B P Avellino eruption" by Milia, A., Raspini, A., Torrente, M.M. *Quat. Int.* **2008**, *192*, 102–109. [CrossRef]
90. Milia, A.; Raspini, A.; Torrente, M.M. The dark nature of Somma-Vesuvius volcano: Evidence from the 3.5 ka B.P. Avellino eruption. *Quat. Int.* **2007**, *173–174*, 57–66. [CrossRef]
91. De Vivo, B.; Belkin, H.E.; Rolandi, G. *Introduction to Vesuvius, Campi Flegrei, and Campanian Volcanism, In Vesuvius, Campi Flegrei, and Campanian Volcanism*, 1st ed.; De Vivo, B., Belkin, H., Rolandi, G., Eds.; Elsevier: New York, NY, USA, 2020; pp. 1–8.
92. Silleni, A.; Giordano, G.; Isaia, R.; Ort, M. Magnitude of the 39.8 ka Campanian Ignimbrite, Italy: Insights from an ignimbrite isopach map. *Earth Space Sci. Open Arch.* **2020**, *40*, 502011. [CrossRef]
93. Pyle, D.M.; Ricketts, D.G.; Margari, V.; Van Andel, T.H.; Sinitsyin, A.A.; Praslov, N.D.; Lisitsyn, S. Wide dispersal and deposition of distal tephra during the Pleistocene Campanian Ignimbrite/Y5 eruption, Italy. *Quat. Sci. Rev.* **2006**, *25*, 2713–2728. [CrossRef]
94. Giaccio, B.; Isaia, R.; Fedele, F.G.; Di Canzio, E.; Hoffecker, J.; Ronchitelli, A.; Sinitsyn, A.A.; Anikovich, M.; Lisitsyn, S.N.; Popov, V.V. The Campanian Ignimbrite and Codola tephra layers: Two temporal/stratigraphic markers for the Early Upper Palaeolithic in southern Italy and eastern Europe. *J. Volcanol. Geoth. Res.* **2008**, *177*, 208–226. [CrossRef]
95. De Vita, S.; Orsi, G.; Civetta, L.; Carendente, A.; D'Antonio, M.; Di Cesare, T.; Di Vito, M.A.; Fisher, R.V.; Isaia, R.; Marotta, E.; et al. The Agnano-Monte Spina eruption (4.1 ka) in the resurgent, nested Campi Flegrei caldera (Italy). *J. Volcanol. Geoth. Res.* **1999**, *91*, 269–301. [CrossRef]
96. Di Renzo, V.; Arienzo, I.; Civetta, L.; D'Antonio, M.; Tonarini, S.; Di Vito, M.A.; Orsi, G. The magmatic feeding system of the Campi Flegrei caldera: Architecture and temporal evolution. *Chem. Geol.* **2011**, *281*, 227–241. [CrossRef]
97. Smith, V.C.; Isaia, R.; Pearce, N.J.G. Tephrostratigraphy and glass compositions of post-15 kyr Campi Flegrei eruptions: Implications for eruption history and chronostratigraphic markers. *Quat. Sci. Rev.* **2011**, *30*, 3638–3660. [CrossRef]
98. Isaia, R.; Vitale, S.; Di Giuseppe, M.G.; Iannuzzi, E.; D'Assisi Tramparulo, F.; Troiano, A. Stratigraphy, structure, and volcano-tectonic evolution of Solfatara maar-diatreme (Campi Flegrei, Italy). *GSA Bull.* **2015**, *127*, 1485–1504. [CrossRef]
99. Sulpizio, R.; Bonasia, R.; Dellino, P.; Mele, D.; Di Vito, M.A.; La Volpe, L. The Pomici di Avellino eruption of Somma-Vesuvius (3.9 ka BP), part II: Sedimentology and physical volcanology of pyroclastic density current deposits. *Bull. Volcanol.* **2009**, *72*, 559–577. [CrossRef]
100. Milia, A.; Torrente, M.M.; Russo, M.; Zuppetta, A. Tectonics and crustal structure of the Campania continental margin: Relationships with volcanism. *Mineral. Petrol.* **2003**, *79*, 33–47. [CrossRef]
101. Acocella, V.; Funiciello, R. Transverse systems along the extensional Tyrrhenian margin of central Italy and their influence on volcanism. *Tectonics* **2006**, *25*, TC2003. [CrossRef]

Journal of
Marine Science and Engineering

MDPI

Article

Formation Mechanism of NW-Trending Faults and Their Significance on Basin Evolution in Zhu III Depression of the Pearl River Mouth Basin, SE China

Pengfei Zhu [1,2,†], Lintao Zhao [3,4,†], Jiantai Zhang [5], Dunling Mu [1,2,*], Yichun Chen [1,2] and Pengfei Rong [1,2]

1 Institute of Digital Geology and Energy, Linyi University, Linyi 276000, China; zzpf4377zz@163.com (P.Z.); chunzhi233233@hotmail.com (Y.C.); rpf15954643405@163.com (P.R.)
2 Shandong Provincial Key Laboratory of Water and Soil Conservation and Environmental Protection, School of Resource and Environment (College of Carbon Neutrality), Linyi University, Linyi 276000, China
3 Key Laboratory of Tropical Marine Ecosystem and Bioresource, Fourth Institute of Oceanography, Ministry of Natural Resources, Beihai 536000, China; zhaolintao@4io.org.cn
4 Guangxi Key Laboratory of Beibu Gulf Marine Resources, Environment and Sustainable Development, Fourth Institute of Oceanography, Ministry of Natural Resources, Beihai 536000, China
5 The Seventh Geological Brigade of Shandong Provincial Bureau of Geology and Mineral Exploration and Development, Linyi 276000, China; jiantai26@126.com
* Correspondence: 17505324322@163.com
† These authors contributed equally to this work.

Abstract: The Zhu III Depression, situated in the northern Pearl River Mouth Basin, features a complex fault system composed of NE–SW-, nearly E–W-, and NW–SE-oriented faults. However, there is limited research on NW-trending faults, especially regarding their formation mechanisms. Through structural analysis of 3D seismic profiles, we have revealed the geometric and kinematic characteristics of NW-trending faults and categorized them into three types based on their formation mechanisms: extensional fault, dextral transtensional fault, and sinistral strike–slip fault. The extensional faults predominantly developed as boundary faults during the rifting I period, caused by tectonic inversion of the NW–NWW-trending basement faults since early Eocene. The transtensional fault resulted from the dextral strike–slip motion of the NE-trending basin-controlling faults since late Eocene, under the regional dextral extension stress setting. The sinistral strike–slip faults have been dominant during the post-rifting period since early Oligocene. This is due to the sinistral shearing action related to the southeastward lateral extrusion of the Indochina Block and slab pull southward by subduction of the proto-SCS. The NW-trending faults controlled the development of local tectonics and structures, the depocenter migration during the rifting period, and the trapping, migration, and preservation of oil and gas.

Keywords: strike–slip fault; formation mechanism; seismic interpretation; Zhu III Depression; Pearl River Mouth Basin

1. Introduction

Faults are the main deformation style of basin tectonic activity, not only controlling the evolution of the basin, but also serving as the main migration channel for oil and gas in the basin, and further controlling the formation and distribution of oil and gas reservoirs [1–6]. In the Cenozoic, the East Asian Continental Margin developed intense rifting, producing massive rift basins and marginal seas. Among these, the South China Sea (SCS) and adjacent continental margin have attracted attention due to the obscure opening mechanism and strong lithospheric thinning [7–9]. As a representative of the epicontinental rift basin in the northern margin of the South China Sea (NSCS), the Pearl River Mouth Basin (PRMB) is an ideal study area for a better understanding of the tectonic evolution of the South China Block and the time limit of the opening of the SCS. The Zhu III Depression, a secondary structural unit located in the northwest of the PRMB, is an important region for oil and gas exploration in the west of the SCS. In recent years, extensive research has been conducted on the complex fault system and its relationship with sedimentary evolution and oil–gas distribution in the basin [10–16]. The tectonic units of the Zhu III Depression (even the PRMB) and main faults in the rifting period mostly extend in the NE striking, thus research on the fault system of the Zhu III Depression has mainly focused on the NE-trending faults, while relatively less attention has been paid to the NW–NWW-trending faults. However, there are several large basin-controlling fault zones trending NW, such as the Yangjiang–Yitong Fault. Additionally, some NW-trending secondary faults related to the basement faults are also found in depressions [17–19]. Especially in the post-rifting stage, the NW-trending faults increased and became the dominant faults [20–22], which induced local uplift activity and controlled the trapping, migration, and preservation of oil and gas [23,24]. It is thus clear that the NW-trending fault is very important for the evolution of the basin. However, there are still differing views on the formation mechanisms of NW-trending faults, including extension and strike–slip motion [25–27].

A detailed study of faults is crucial for understanding the process of tectonic evolution and the dynamic mechanisms of the basin, revealing patterns of hydrocarbon accumulation and guiding the implementation of oil and gas exploration. The continuous acquisition of 3D seismic and well data by the China National Offshore Oil Corporation (CNOOC) Shenzhen Branch and associated cooperative services companies over the past decades has provided a sufficient basis for our research. Therefore, building on previous studies, this paper conducts a detailed interpretation and analysis of the NW-trending (including NWW-trending) faults in the Zhu III Depression using 3D seismic data. Our aim is to reveal the geometric and kinematic characteristics of the faults, discuss their formation mechanisms, and explore their impact on the tectonic evolution of the basin and hydrocarbon accumulation. This study will provide reference for the formation mechanism of the rifting basins in the northern continental margin of the South China Sea.

2. Geological Setting

The South China Sea (SCS) is at the convergence of the Pacific Plate (Philippines Sea Plate), the Indo–Australian Plate, and the Eurasian Plate. It records the interaction among multiple surrounding plates and has been considered a natural laboratory for studying the tectonic evolution of marginal seas [28–32]. However, the origin of the SCS remains obscure, and four principal tectonic models have been proposed for SCS opening: (1) slab pull from subduction of a proto-South China Sea; (2) extrusion tectonics from the India–Asia collision; (3) basal drag from a mantle plume; and (4) backarc rifting [33–38]. The PRMB (Figure 1a) located on the NSCS is a Cenozoic petroliferous basin developed above the Mesozoic tectonic basement [18,39]. Previous studies have revealed that the Mesozoic basement structure of the PRMB is mainly controlled by a NE- and NW-trending conjugate fault system [40–42]. During the Cenozoic, the basin experienced a rifting period in Paleocene–early Oligocene and a depression period since late Oligocene (Figure 2). The rifting of the basin is controlled by the Zhu–Qiong Movement. The first episode of the Zhu–Qiong Movement occurred in the Early to Middle Eocene, corresponding to the seismic reflection interface T90, which is the bottom interface of the Wenchang Formation. The seismic wave features above the interface are distinctly stratified, while those below are more chaotic, representing the beginning of large-scale rifting and faulting. The second episode of the Zhu–Qiong Movement is a significant tectonic event that occurred in the Late Eocene, represented on the seismic profile as the T80 unconformity, which is the bottom interface of the Enping Formation. This event, distinguished from the first episode of the Zhu–Qiong Movement, caused regional uplift in the basin, accompanied by faulting and magmatic activity, although it has been considered a continuation of the first episode. Thus, the rifting period has been further divided into the rifting I period and the rifting II period, with the T80 interface as the boundary [43,44]. In addition, the Transition and Neotectonic period can be classified as the post-rifting thermal subsidence period (depression) [45,46].

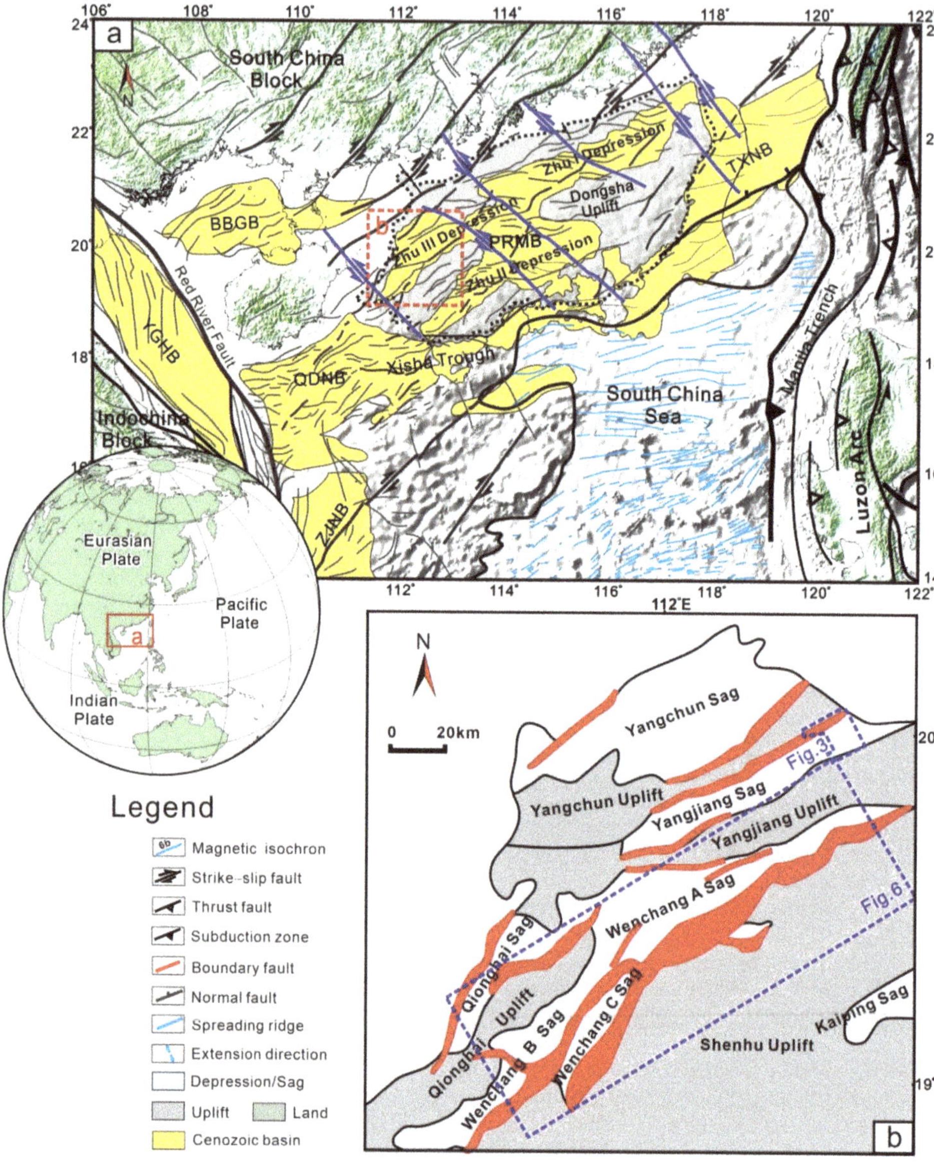

Figure 1. Regional structural map of PRMB (**a**) and the distribution of structural units and fault system of Zhu III Depression (**b**). The schematic diagram in the lower left shows the location of the PRMB and the NSCS. The fault pattern is modified after [8,47]. The magnetic isochrons are adopted from [48]. BBGB: Beibu Gulf Basin; QDNB: Qiongdongnan Basin; PRMB: Pearl River Mouth Basin; TXNB: Taixinan Basin; YGHB: Yinggehai Basin; ZJNB: Zhongjiannan Basin.

As a result, a tectonic framework characterized by lower faulting and upper depression has been established. From north to south, the basin can be divided into five first-order structural units: the northern uplift zone, the northern depression zone, the central uplift zone, the southern depression zone, and the southern uplift zone (Figure 1a).

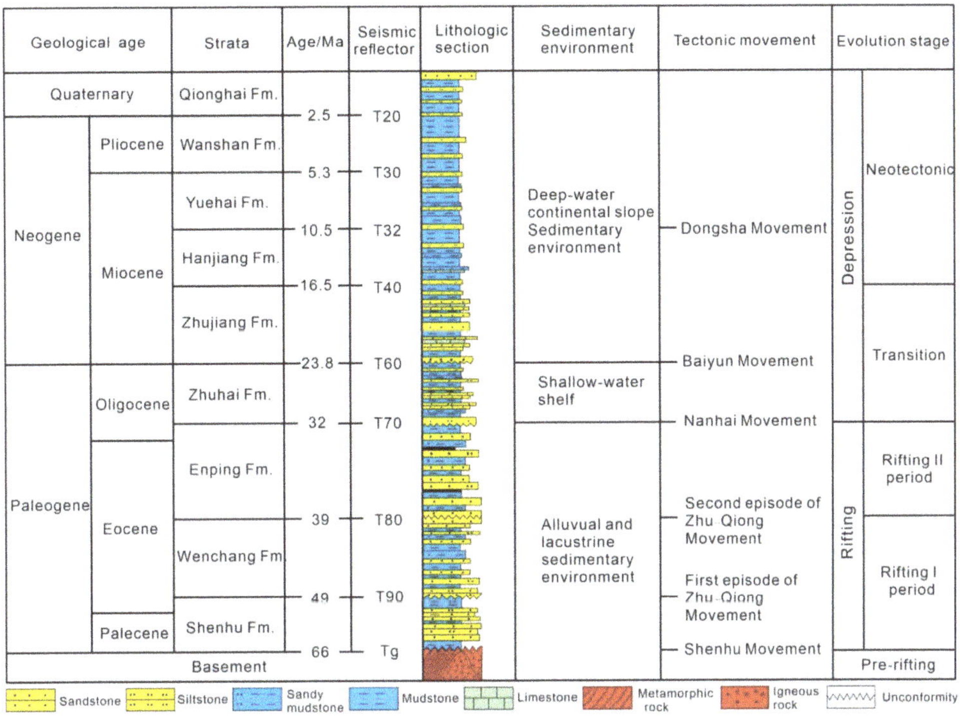

Figure 2. Comprehensive sequence columns of the PRMB [2,21].

The Zhu III Depression is a secondary structural unit in the west of the northern depression zone. It is separated from the Zhu I Depression by the Yangjiang–Yitong fault zone to the east and is adjacent to the Shenhu uplift to the south. From north to south, the Zhu III Depression is composed of the Yangchun Sag, Yangchun Uplift, Yangjiang Sag, Yangjiang Uplift, Qionghai Sag, Qionghai Uplift, and Wenchang Sag. These secondary structural units mainly appear in a NE direction, characterized in belt distribution (Figure 1b). From bottom to top, the sedimentary strata developed in the depression include the Paleogene Shenhu Formation, Wenchang Formation, Enping Formation, Zhuhai Formation, Neogene Zhujiang Formation, Hanjiang Formation, Yuehai Formation, Wanshan Formation, and Quaternary Qionghai Formation (Figure 2). Exploration practices have revealed that the Yangjiang Sag and Wenchang Sag are significant hydrocarbon-generating sags in the Zhu III Depression, and two sets of important hydrocarbon source rocks are developed in the Paleogene Wenchang and Enping Formation [24,49,50].

The Zhu III Depression features a complex fault system containing NE-, nearly E–W-, and NW-striking faults. It is revealed that the formation of multi-striking faults is closely related to both the pre-existing basement structures and the continuous dextral variation of the regional extensional tectonic stress field since the Cenozoic [19,42,51,52]. There is a basin-controlling fault zone trending NW in the east of the Zhu III Depression called the Yangjiang–Yitong fault zone. The northern segment of the fault zone separates the Zhu III and Zhu I depressions and was active from Mesozoic to Cenozoic, which has a great impact on the development of the Zhu III Depression [41,53], especially for the adjacent Yangjiang Sag [27,54]. Additionally, a NE-striking depression-controlling fault, known as the Zhu III South Fault, governed the formation of the Zhu III Depression as a half-graben, with faulting in the south and overlapping in the north. Given the development of en echelon folds, horsetail faults, and oblique steps in the Wenchang Sag, it is revealed that the

Zhu III South Fault has undergone dextral strike–slip deformation [10,15,25,26,32]. Except the boundary fault, it has also been revealed that strike–slip activity occurred along the faults within the depression. Based on the tectonic analysis of the Wenchang Sag, the fault system is further categorized into early NE-trending extensional faults, middle E–W-trending extensional–dextral strike–slip faults, and late NW-trending extensional–sinistral strike–slip faults [11,55]. Structure studies show that after the initial rifting, the Yangjiang Sag successively experienced two stages of strike–slip pull apart movement in NE and NW directions, respectively [8,21,54,56]. This provides a basis for establishing the strike–slip model for basin formation rather than a single extension model.

3. Data and Methods

The 3D seismic and well data used in our fault interpretation were provided by the China National Offshore Oil Corporation (CNOOC) Shenzhen Branch and associated cooperative service companies. The seismic profiles utilized in this study offer comprehensive coverage of the Yangjiang Sag, spanning an area of over 5600 km^2. The bin size of the survey is 12.5 m $\times$ 12.5 m. The inline seismic profiles are oriented NW, which is approximately perpendicular to the trend of the basin. We adhered to standard seismic interpretation workflows for tectonic–stratigraphic analysis and identification of magmatic bodies,. We established a robust stratigraphic framework by integrating 3D seismic and log data with vertical seismic profile (VSP) information from select typical wells. Exploration wells from the Yangjiang Sag and Enping Sag have also been utilized for depth conversion by (1). Given the relatively complete stratigraphic sequence, petroleum exploration boreholes served as a calibration tool for the Cenozoic strata. This approach ensures that our interpretations are rooted in well–understood geological structures and sequences, thereby enhancing the reliability and validity of our findings.

$$D = 0.00027451t2 + 0.72410996t + 31.78776360, R2 = 0.99991635 \tag{1}$$

where D represents the depth, t represents the time, and R2 represents the goodness of fit.

We employed traditional manual interpretation of the 3D seismic data to unveil the geometric shape, combination style, and distribution pattern of faults in plan-section views. This was further validated and constrained using methods such as coherence analysis of the 3D seismic data. Our workflow comprised the following steps: (1) identification and correlation of seismic horizons; (2) generation of time structural maps, seismic volume slices, and fault distribution maps during the different seismic reflectors; (3) observation of strike–slip structures; (4) investigation of changes of depocenters; (5) calculation of fault activity rates; (6) analysis of the mechanism and dynamic process of fault formation; (7) exploration of the significance of faults on basin evolution. Among them, the interpretation of strike–slip faults is relatively difficult. Based on previous research [57,58], we mainly applied the following approaches for identification: (1) determination of fault location based on seismic attributes. We analyzed seismic wavegroup characteristics, maximum amplitude attributes, and data volume slices to identify fault locations. (2) Identification of strike–slip structures based on fault combination relationships. This was comprehensively identified from two aspects: plan and section views. (a) On the plan view, we focused on linearly extended or band-distributed structures, pull-apart structures, en echelon faults or en echelon folds, horsetail structures, plumosestructures, and horizontal offset of geological boundaries on both sides of the faults. (b) On the section view, based on the fault surface attitude (steep and penetrating the basement), we identified flower-like structures, "ribbon effect", "dolphin effect", and discordance in the thickness, sedimentary facies, and stratum attitude of the same stratigraphic unit on both sides of the fault.

4. Results

4.1. Fault Patterns of the Yangjiang Sag

The Yangjiang Sag is located in the east of the Zhu III Depression, with the Yangjiang–Yitong fault zone crossing its eastern part. NW-trending faults have developed in the Yangjiang Sag, even though the fault system was dominated by NE-direction faults during the rifting I period during early–middle Eocene (Figure 3). In the east Yangjiang Sag, the F20 fault with a westward-dipping fault surface and the F19 fault with an east-dipping fault surface were formed trending NNW, becoming important boundary faults in the east and west of the Enping 20 subsag. The two faults cut the Wenchang Formation with significant fault throws at about 800 m, and there are obvious differences in strata thickness between the two sides of the fault. Therefore, they were synsedimentary faults and controlled the deposition of Wenchang Formation. Contemporaneously, the NWW-trending, south-dipping, sag-controlling F7 fault also formed, of which the maximum dip–slip active rate has reached 130 m/Ma (Figure 4b), which controlled the intense initial rifting process of the Enping 27 subsag. Additionally, some secondary faults, trending NW–NWW and mostly curved in shape, have formed within the Yangjiang Sag. These faults are relatively small in size with short lengths. These faults have partially influenced the control of rifting and sedimentation.

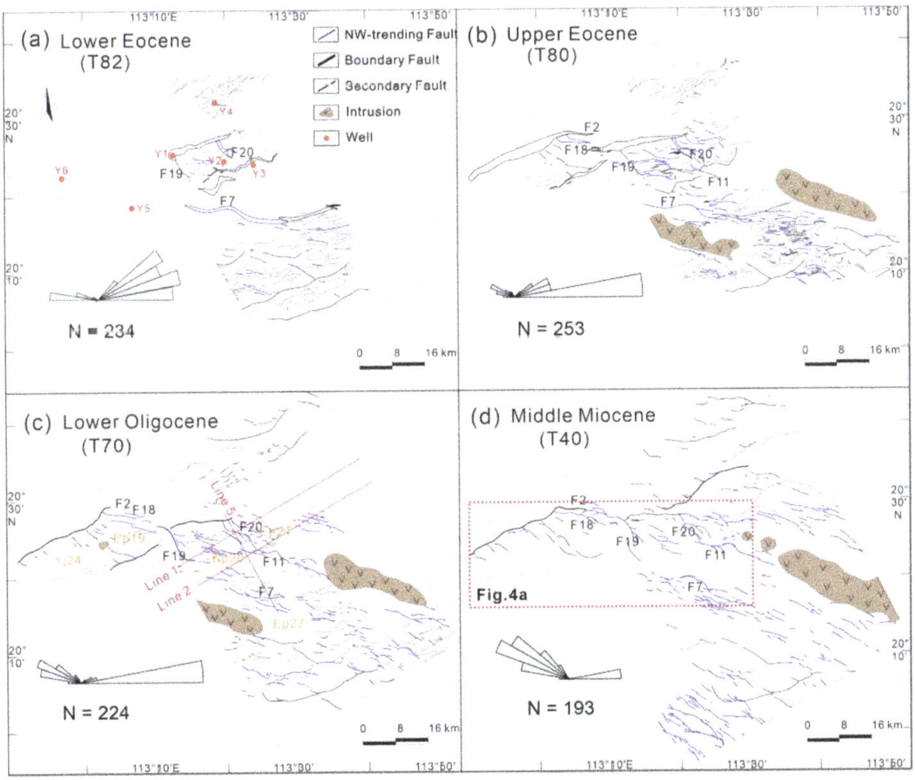

Figure 3. Fault patterns and rose diagram in different seismic reflectors of the Yangjiang Sag. The rose diagram shows stripes and strikes of the faults in seismic reflectors. The faults were drawn using more than 100 seismic lines for the Yangjiang Sag, and the strike data of the fault are from [27]. Yj24—Yangjiang 24 subsag; Ep19—Enping 19 subsag; Ep 20—Enping 20 subsag; Ep 21—Enping 21 subsag; Ep27—Enping 27 subsag.

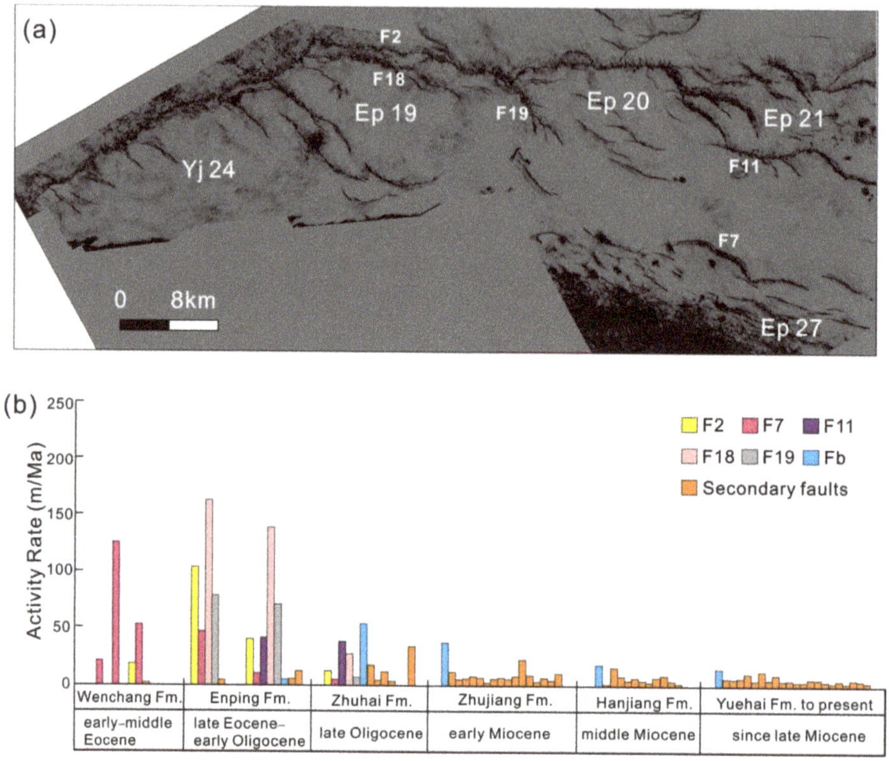

Figure 4. Fault patterns of East Yangjiang Sag shown in time slice at 1500ms (**a**) (see Figure 3 for locations), and the dip–slip active rate of major NW-trending faults in the Zhu III Depression (**b**). The data of the secondary faults are from [22].

During the late Eocene–early Oligocene rifting II period of the Yangjiang Sag, the number of NW-direction faults significantly increased. Among these, pre-existing major faults, such as the NW-trending F7, F19, and F20 faults, were characterized by inherited activity. However, their activity intensity changed during this stage, as represented by the reduced intensity of F7 activity. In addition, a few new major faults formed at this stage. The F18 was formed and was active as a branch fault of F1 at the highest rate of ~170 m/Ma. This led to a transition of the Enping 19 subsag from a single half-graben to a composite half-graben in the north–south double-fault model. Additionally, another southward dipping major fault F11 formed in a NNW trend, and its strong activity caused the rifting of the Enping 20 subsag to move southwards continuously. On the sides of the active major faults in both the NE and NW trends, some newly generated faults with NWW trending were developed and combined into en echelon structures and horsetail structures horizontally. The newly generated faults cut the Enping Formation with a thickness difference between the two sides of the faults, indicating the extensional characteristics. In addition, several secondary faults in a NW trend formed cross the sag diagonally. In a cross-section, the faults were identified as forming step-shaped, Y-shaped, and negative flower structural styles (Figure 5), manifesting strike–slip characteristics.

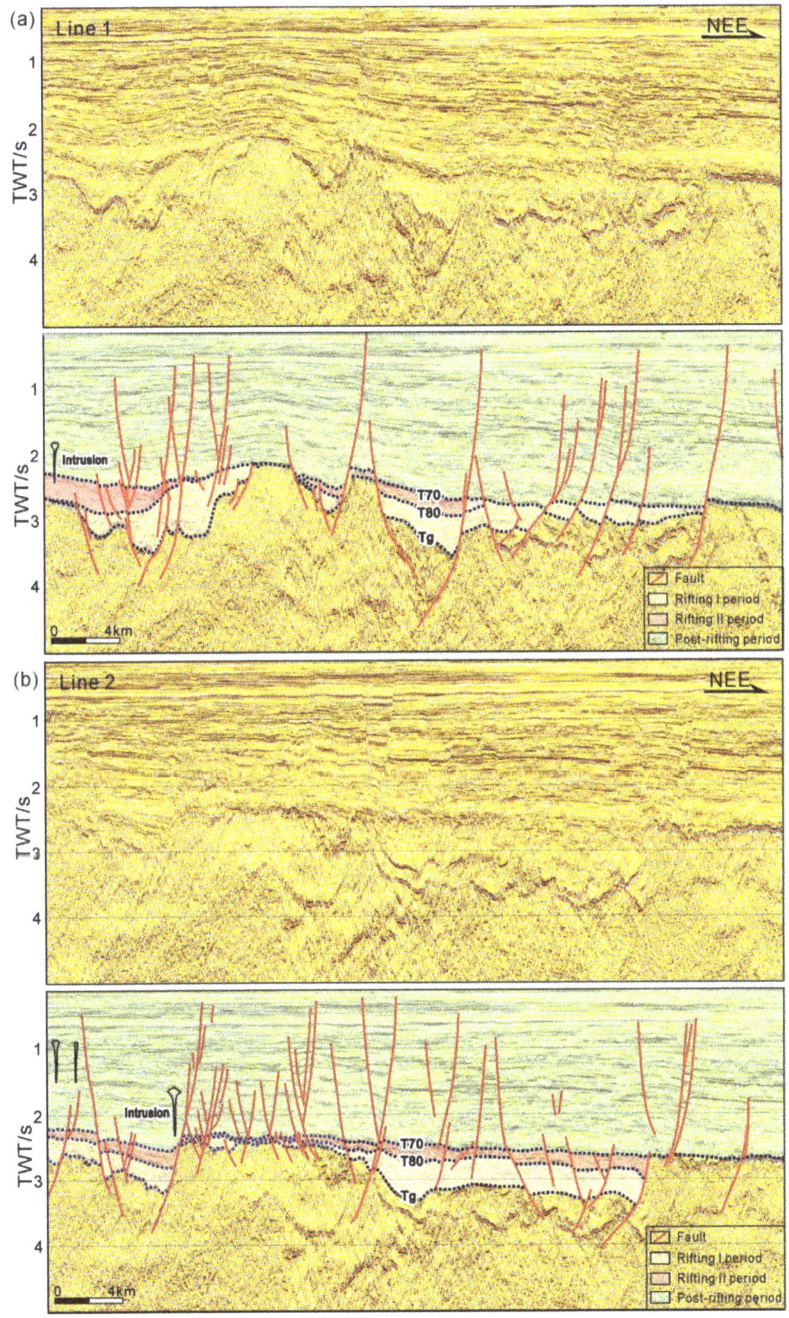

Figure 5. Fault system and structural style showed in the seismic profiles of the Yangjiang Sag, with corresponding unexplained seismic profiles (see Figure 3 for locations). The (**a**,**b**) corresponds the profile in Lines 1, 2. The rifting I period corresponds to the First Zhu–Qiong Movement in early–middle Eocene and the sedimentation of the Wenchang Formation in the basin, while the rifting II period corresponds to the Second Zhu–Qiong Movement in late Eocene when the Enping Formation developed.

In the post-rifting period since late Oligocene, there was a significant decrease in the number of NE- and nearly E–W-oriented faults, while the NW-oriented fault system emerged as the dominant feature within the sag (Figure 4a). The formation of this fault system can be attributed to the continuous expansion and linkage of the initially formed short NW-oriented faults from the rifting stage. Additionally, some new NW-oriented faults developed around pre-existing ones, merging to form Y-shaped patterns and negative flower structures, as shown in Figure 5. These newly formed faults, located in the shallow layer, exhibit minor fault displacement, with virtually no variation in the formation thickness between the two sides. This means that the extensional intensity of the fault activity was obviously weakened, while the strike–slip component dominated. During this stage, on the Yangchun Uplift in the north of the Yangjiang Sag, some steep and erect NW-trending faults developed, which often cut down to the basement. Negative flower structures were also identified in this section, and the fault throws of these faults are relatively small, showing the characteristics of strike–slip motion.

4.2. Fault Patterns of the Wenchang Sag

Under the influence of the Zhu III South Fault's activity, the Wenchang Sag commenced its rifting process during the Paleocene, marking it as the earliest sag formation within the Zhu III Depression. During the rifting I period of the sag, the dominant fault was the NE-oriented Zhu III South Fault (Fa). This fault, characterized by segmented activity, governed the development of secondary sags, namely, Wenchang A, B, and C sags. Simultaneously, the NW-oriented No. 3 fault (Fb), which emerged within the Wenchang B sag, served as an eastward-dipping extensional fault, thereby segmenting the sag's internal structure (Figure 6).

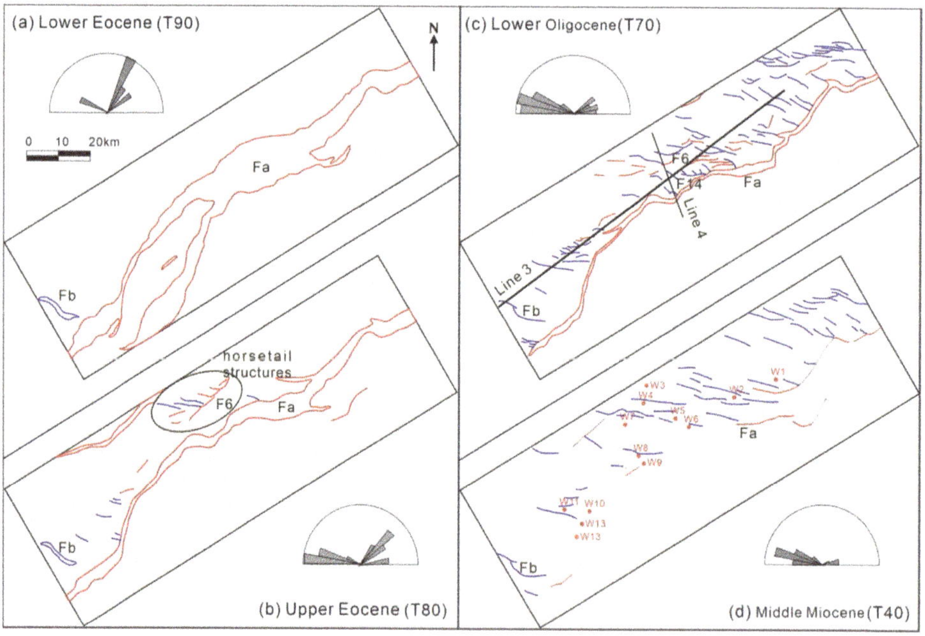

Figure 6. Fault patterns and rose diagram in different seismic reflectors of the Wenchang Sag. The rose diagram is drawn based on statistical analysis of the fault strikes. The fault data are from [37], and the well data are from [59]. The red lines represent the NE- to EW-trending faults, and the blue lines represent the NW-oriented faults. Fa represents the Zhu III south fault, and Fb represents the No. 3 fault in the Wenchang B Sag.

During the rifting II period of the Wenchang Sag, faults with nearly E–W and NWW orientations were not only generated but also increased in number. These faults are curved in shape and relatively short in length, most of which show left-stepping en echelon assemblage in plan view. Furthermore, some of the NWW-oriented faults laterally connect with the NE-oriented faults, forming horsetail structures. These structures are particularly well-developed near the F6 fault belt in the Wenchang A sag. In the seismic section, it is observed that the fault surfaces mostly tilt northward and are distributed in a stepped combination. Some faults can extend downwards and link to the NE-trending major fault or sag-controlling fault, forming Y-shaped patterns (Figure 7). It is implied that the dextral strike–slip motion might take place along the NE-trending fault by the geometric and kinematic analysis, which is similar to that in the Yangjiang Sag. Additionally, the NW-oriented faults, which cut through the Enping Formation, function as synsedimentary normal faults and contribute to the evolution of the depocenter.

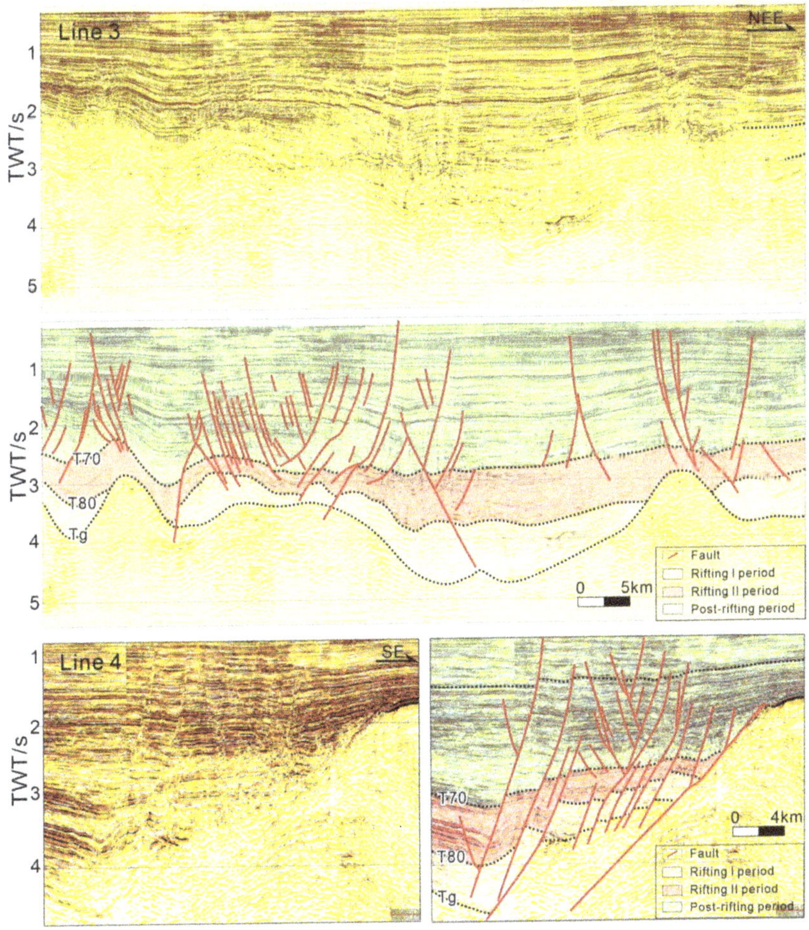

Figure 7. Fault system and structural style shown in the seismic profiles of the Wenchang Sag, with corresponding unexplained seismic profiles (see Figure 6 for locations; the seismic data are from [22,32]).

In the post-rifting period, the NW-oriented faults became dominant within the depression, coinciding with the cessation of activity in the early NE- and nearly E–W-oriented faults. Although the number of active faults in this stage significantly decreased, the scale and length of the faults expanded compared to that in the previous stage. The NW-trending faults mostly appear in straight rather than curved shapes in plan view. The newly formed NW-trending faults are mostly arranged in en echelon and combine with the early faults in Y-shaped and negative flower structures on the profile. It can also be seen that the NW-trending faults have smaller fault throws, meaning that the activity of the fault weakens, and the depression enters the stage of thermal subsidence as a whole.

5. Discussion

5.1. Development Stages of NW-Trending Faults

As shown in the rose diagram of the fault strikes (Figures 3 and 6), the orientations of the Cenozoic faults in the Zhu III Depression underwent a transition from NE (T80) to nearly E–W (T70), and subsequently to NWW (T40). This change reflects a clockwise rotation of the regional extension direction from NW–SE to N–S to NNE–SSW. Under the dextral transtensional stress setting, the direction of primary rotational main extensional stress intersects obliquely with the strike of pre-existing basement faults or pre-formed main controlling faults. This intersection can induce oblique extension or strike–slip activity of early faults in the basin [55]. Consequently, the NW-oriented faults, which started forming during the basin's initial rifting stage and continued to develop and dominate in the post-rifting stage, could exhibit varying activity characteristics under the regional dextral transtensional stress conditions.

Given the geometric and kinematic characteristics of faults, it becomes apparent that NW-oriented faults exhibit significant variations across different structural layers. The NW-oriented faults, which developed in the deep layers (Shenhu and Wenchang formations) were major sag-controlling faults during the rifting I period. They are small in number and exhibit strong continuity, large fault displacement, and a shovel-shaped fault surface, indicative of synsedimentary boundary faults. At this stage, however, the dominant faults are mainly developed in the NE orientation, especially reflected in the boundary faults. The number of faults in the Enping Formation has increased, and the faults are distributed in belts in plan view, arranged in en echelon, and connected laterally to the early NE-trending major faults. Vertically, the newly generated faults form a Y-shaped combination with the major faults, and the negative flower structures also developed, all of which reveal that the NW-trending faults were characterized by strike–slip motion in the rifting II period. A fault displacements analysis reveals that the activity intensity of the newly formed major faults was greater than that in the rifting I period. Furthermore, secondary faults played a significant role in controlling the sedimentation of the Enping Formation. It can be seen that main motion of the faults developed in this stage is extension with dextral slip component.

Within the shallow structural layers of the post-rifting stage, the newly formed faults primarily exhibit a NW orientation and are predominantly arranged in an en echelon pattern. The fault surfaces present as nearly vertical, high-angle structures with minimal fault displacement, as observed on the profile. Additionally, negative flower structures were developed. It is reasonable to infer that the NW-trending faults in this stage are mainly characterized by strike–slip motion. Taking the changing dynamics of the regional stress field into account, the evolution of the NW-oriented faults in the Zhu III Depression can be segmented into three stages: the extension stage during rifting I period, the extension–strike–slip stage during rifting II period, and the strike–slip stage during the post-rifting period. The newly formed faults in different stages correspond to different genetic types and mechanisms.

5.2. Genetic Types and Formation Mechanism

According to the structural analysis on seismic profiles, two groups of pre-Cenozoic faults were identified in the basement, which manifested as conjugate pre-existing fault systems trending in the WNW and ENE directions (Figure 8). The conjugate fault systems appear as imbricate and flat-ramp fault structures on the section, suggesting the development of thrust fault systems during the pre-Cenozoic. The ENE-trending thrust faults cuts the WNW-trending thrust faults on the seismic reflection profile, indicating it formed at a relatively later stage and preserved more completely. The thrust faults trending in ENE and WNW have been also suggested to be widespread in the basement of the Zhu I and Zhu II depressions, based on the borehole-constrained high-quality 3D seismic data [18,19,27,60,61]. Additionally, it was also discovered that the boundary faults and some major normal faults within the Cenozoic basin cut through the Tg and converged with the pre-existing faults, similar to that in the Enping and Baiyun sags. It is concluded that the pre-rifting structure played a key role in controlling the architecture and rifting of the PRMB.

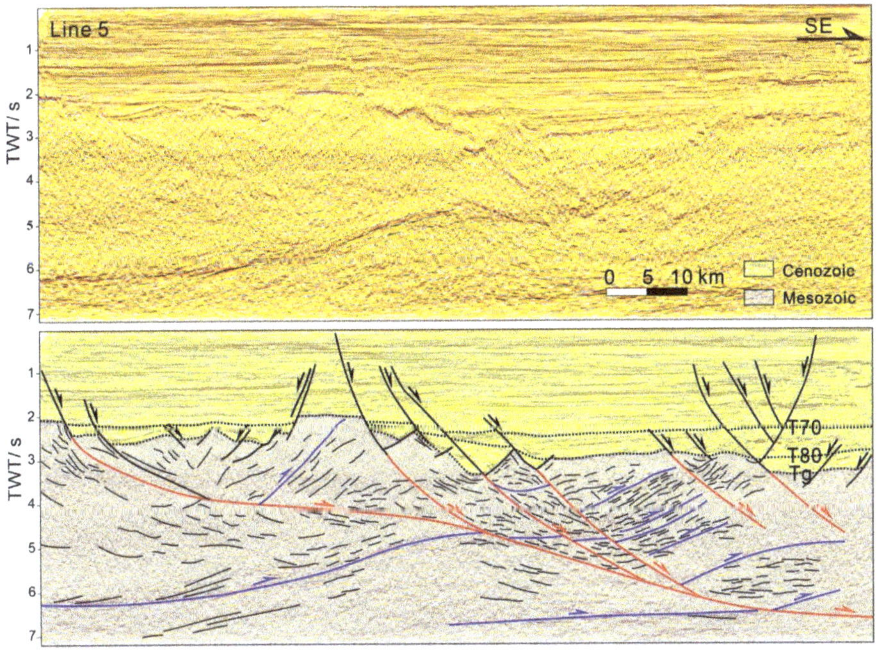

Figure 8. The distribution of pre-existing structures in the basement of the Yangjiang Sag and their relationship with the development of Cenozoic faults in the basin (see Figure 3 for location). The red lines represent the ENE-trending normal faults in the basement, which were reactivated and inverted from Mesozoic thrust faults during the Cenozoic. The blue lines represent the WNW-trending Mesozoic thrust faults, while the black lines represent the Cenozoic faults in the basin.

As previously mentioned, during the initial rifting of the Zhu III Depression, the primary extensional stress direction in the region was NW–SE. This led to the formation of a fault system primarily composed of NE-oriented faults (Figures 3 and 6). Concurrently, there are also a few conjugate NW–NWW-trending faults formed within sags as sag-controlling boundary faults. These faults manifest as curved structures in plan view and exhibit substantial fault displacement in cross-section, reflecting the characteristics of extensional faults. Moreover, they extend downwards and connect to the pre-existing basement faults. It is plausible to infer that the formation of NW-oriented faults during the rifting I period was closely associated with the reactivation of pre-existing NW-oriented faults in the basement. That is to say, under the background of NW–SE regional extension, the NE oriented pre-existing faults in the basement will be preferentially activated, forming boundary faults in a NE trend that controlled the development of the initial depression. Meanwhile, the conjugate NW-trending faults, as weak structural zones in the basement, will also partially activate to form several NW–NWW-trending boundary faults and some secondary faults within the depression (Figure 9a,b). Consequently, the formation of NW-oriented extension faults during the Zhu III Depression's rifting I period can likely be attributed to the reactivation of the basement faults.

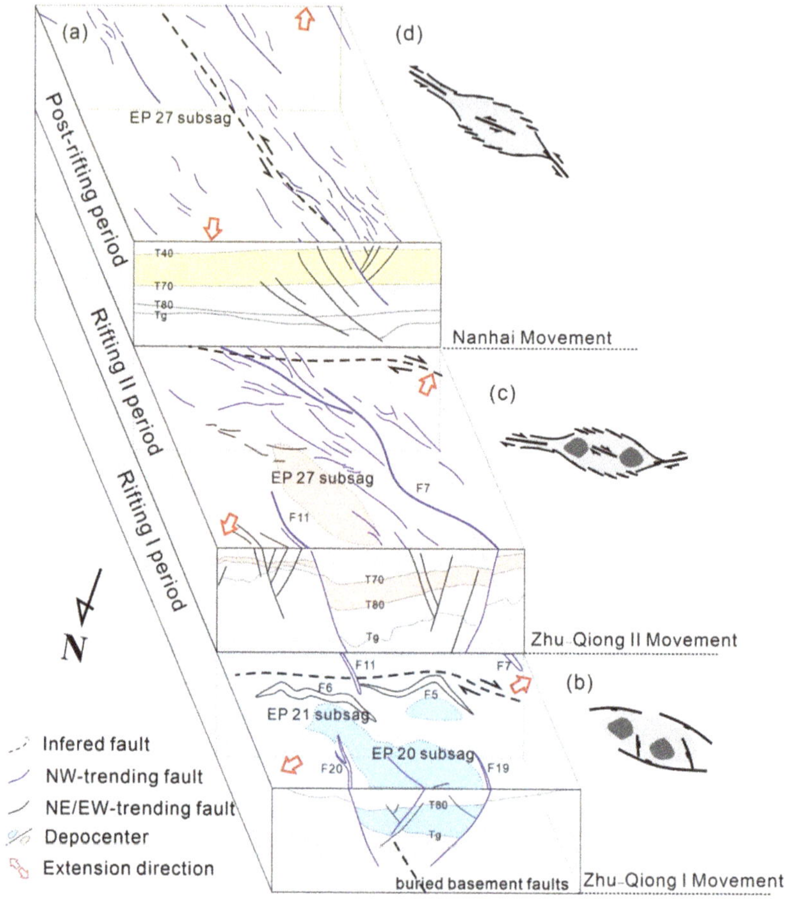

Figure 9. Simplified model showing three-stage evolution process of the NW-trending faults. Subfigures (**a**) displays the three evolutionary stages in the form of spatial superposition to reveal the vertical correlation of the faults and the southward migration of sedimentary center. The ideal model on the right side of the map show the tectonic evolution mechanism of the extensional basin (**b**), transtensional basin (**c**), and strike–slip basin (**d**), and the connections between faults also can be explained.

During the rifting II period, the regional extension direction experienced a dextral shift, transitioning to nearly an S–N orientation. Meanwhile, the NE-trending strike–slip fault systems have been revealed to develop in the marginal basin on the NSCS, according to a structural analysis of field and seismic data and numerical modeling [35,56,62,63]. Accompanied by the faults in the nearly E–W direction dominant at this period, the number of NWW-oriented faults increased. These NWW-oriented faults displayed left-stepping en echelon arrangements and were predominantly connected to boundary faults, forming horsetail structures. This suggests that the newly formed faults likely resulted from dextral strike–slip motion along the major NE-oriented faults. Moreover, the faults also showed a stepped combination on the profile and formed Y-shaped and negative flower-shaped structures with the boundary fault, all of which could prove the development of strike–slip structures in the depression. In particular, within the East Yangjiang Sag, the NE-oriented boundary faults give rise to dextral pull-apart structures [24]. Consequently, a secondary NW-oriented strike–slip fault zone is established along the diagonal of the rhombic-shaped Enping 20 subsag. Similarly, in the Wenchang Sag, the combination styles of the nearly E–W- and NWW-trending faults also revealed that the NE-trending major faults, represented by the Zhu III South Fault, experienced dextral strike–slip motion. The newly formed NWW-oriented synsedimentary faults penetrate downwards through the T80 reflector, suggesting that strike–slip activity commenced during the rifting II period (Enping Formation period). This contrasts with previous assertions that dextral transtensional faults emerged in the late Oligocene to early Miocene [26,55]. Due to the clockwise rotation of the regional extensional stress direction, the strike of boundary faults formed in the early stage, mainly oriented in the NE direction (basement structural line direction), has an oblique intersection with the direction of regional extensional stress. As the included angle of the oblique intersection continuously decreases, it prompts a gradual amplification of the dextral shear component produced by the active NE-oriented faults, thereby inducing dextral strike–slip activity. As a result, faults in E–W and NWW directions were derived within the depression, forming broom structures with the major faults, or left-stepping en echelon assemblages by lateral connection, which is consistent with the model of dextral transtensional basin (Figure 9c). Hence, the NWW-oriented faults that developed during this stage can likely be classified as strike–slip-derived types along the NE-oriented faults, characterized by extension with dextral strike–slip action (Figure 9a).

During the post-rifting stage, the basin entered the stage of thermal subsidence. There is weak fault activity from late Oligocene to early Miocene, which was known as a transition period [43,45]. The regional extensional direction shifted to NNE–SSW, leading to the formation of a fault system primarily composed of NW-oriented faults. The activity intensity of the NE- and nearly E–W-oriented faults formed during the rifting period significantly decreased or even ceased, while some NWW-oriented faults demonstrated continued activity. However, the activity intensity also weakened with the basin entering the depression period. During the Neotectonic period, there was an obvious rise in activity intensity of the NW-trending faults, which resulted in complex fault block activity in the region (Figure 4b). The Y-shaped and negative flower structures formed by the newly developed NWW-oriented faults and earlier faults in the profile, along with the right-stepping en echelon arrangement, suggest the evolution of the NW-oriented sinistral strike–slip faults. The NW-trending strike–slip activity was also highly developed in the uplift areas, such as the Qionghai Uplift, manifested as a single fault with a high angle and upright plane, cutting through the Hanjiang Formation upwards, even shallower strata, down to the basement, generally [11,59]. Furthermore, the Yangjiang–Yitong fault zone has been recognized as an active sinistral strike–slip fault zone since the Early Oligocene [17,64–66]. This supports the assertion that the NW-oriented faults that developed during this stage were associated with strike–slip motion, trending NW under the NNE–SSW extensional setting. The faulting in this stage is most notably characterized by sinistral strike–slip, which is consistent with model of the strike–slip basin (Figure 9d). However, within the Zhu III Depression, there may not be a continuous main strike–slip zone. Instead, the strike–slip faults could be the product of strike–slip motion along the adjacent Yangjiang–Yitong fault zone.

5.3. Geodynamic Process

The PRMB is situated at the northern continental margin of the SCS, where the convergence of the Pacific Plate (Philippines Sea Plate), the Indo–Australian Plate, and the Eurasian Plate is located (Figure 1). Generally, the sequence of stress superposition across different temporal domains within the same spatial domain, or the ratio of the intensities of various superimposed stresses within the same temporal domain, can lead to substantial variations in the ultimate structural characteristics [67]. Hence, study of the PRMB's tectonic evolution dynamics should consider the two major geodynamic processes: the Pacific–Eurasian subduction and the India–Eurasian collision. This is essential for gaining a deep understanding of the primary dynamic sources driving the various stages of basin evolution.

In the early Eocene, the Indian Plate underwent rapid subduction and collision with the Eurasian Plate [68–70]. Concurrently, the subduction rate of the Pacific Plate, trending NNW, decreased, initiating the retreat of the subduction zone [71]. This led to a transition of the tectonic regime in the NSCS from compression to NW–SW extension. Under the extensional background, the Zhu III Depression in the PRMB began to rift, and major NE-oriented faults developed due to the preferential reactivation of pre-existing NE-oriented faults in the basement. Meanwhile, a few NW-trending pre-existing faults in the basement also reversed and reactivated [18,56] to form the first period of NW-trending extensional faults by tectonic inversion. Since the late Eocene, the subduction direction of the Pacific Plate shifted from NNW to NWW toward the southeastern edge of the Eurasian Plate [71–73]. The ongoing India–Eurasia collision and the oblique subduction between the Pacific and Eurasian plates have led to the formation of a dextral transtensional tectonic setting in the NSCS [74,75]. Dextral strike–slip motion occurred along the NE-oriented major faults within the basin. This triggered the development of NW-oriented secondary faults, characterized by both extension and strike–slip movements. In the Early Oligocene, the continuous collision of the India–Eurasian plate caused lateral extrusion of the Indochina Block southeastwards. Meanwhile, slab pull due to southward subduction of the proto-SCS caused breakup of the NSCS and spreading of the SCS [8,68]. A widespread transition from rifting to post-rifting thermal subsidence occurred in the NSCS where the post-rift fault activities are widely distributed [64,76]. The Zhu III Depression in the PRMB entered the depression period, accompanied by the weakening fault activity. The activity of NW-trending faults became the dominant faults, exhibiting sinistral strike–slip motion. In middle Miocene, the arc–continent collision ended the spreading of the SCS, and the sustained subduction of Philippine Sea Plate towards the Eurasian Plate resulted in compressive uplift and erosion in the PRMB.

5.4. Significance on Basin Evolution

The inversion-related extensional faults trending NW formed during the rifting I period have a relatively limited distribution within the Zhu III Depression. They primarily play the role of boundary faults that control the development of sags or subsags. The F19 and F20 faults, acting as synsedimentary faults, bound the Enping 20 subsag of the Yangjiang Sag. These faults cut through the Tg interface with significant fault throws, playing a crucial role in the deposition processes of the Wenchang Formation and restricting the spatial distribution and morphological development of the sag. The development of the F18 fault, the boundary fault in the north of the Enping 19 subsag, has transformed the subsag from an initial half-graben with northern-faulted and southern-overlapped patterns to a double-faulted graben. In addition, the F7 fault located in the southern Yangjiang Sag is the main controlling fault of the Enping 27 subsag. The fault was formed in the upper Wenchang Formation period, causing the depocenter of the Yangjiang Sag to migrate southwards. Similarly, the No. 3 fault located in the Wenchang B sag is a segmented fault of the Wenchang depression zone, controlling the development of the internal tectonic and sedimentation. It is evident, therefore, that the NW-oriented faults formed during the rifting I period exert a significant influence on the division of internal structural units, the evolution of local tectonics, and the development of the depression's depocenter.

The NW-oriented faults, resulting from dextral strike–slip movements, developed during the rifting II period. Despite their wide distribution within the depression, the size of individual faults is generally relatively small. As a product of the strike–slip activity of

NE–NEE-trending boundary faults/major faults, they regulated the continuous dextral extensional stress at the beginning of strike–slip motion, and derived extension–strike–slip faults in the overlapping area of the main strike–slip faults. This increased the complexity of the local tectonic within the depression. Meanwhile, the NW-trending sag-controlling faults formed in the rifting I period continued to be active, causing the southward migration of the depocenter. Moreover, during this stage, the long-axis direction of deposition shifted from ENE to WNW, which is a response to the dextral transition of the regional stress field [21,27]. With the migration of the rifting and deposition, tectonic uplift took place in local areas. For example, with the steady strengthening of F11 fault activity, the Enping 21 subsag located in the footwall continuously tilted and uplifted, leading to the formation of a faulted nose structure and the erosion in the northern ramp of the Yangjiang Sag. In addition, the NW-oriented faults that formed during this stage intersected with earlier faults, leading to the creation of block traps. These became a significant type of oil and gas trap during the rifting period.

The NW-oriented strike–slip faults formed during the post-rifting stage predominantly developed within the Neogene sedimentary layer. These faults, defined by their short developmental periods and modest sizes, typically extend in length from hundreds to thousands of meters. Some of the faults could cut down to the basement with a high angle, complicating local structures in the depression. There are also differences in the activity intensity of NW-trending faults during this stage. During the early post-rifting stage, a period marked by tectonic tranquility, fault activity was generally subdued. In the late post-rifting, the activity of the fault enhanced (Figure 4) while the basin entered the tectonic activation period, which affected the activity of fault blocks in the depression and the formation of local structures. Additionally, fault activity during this stage significantly influenced the trapping, migration, and preservation of oil and gas. In the overlap region of the sinistral left-stepping faults in the Yangjiang Sag, a series of faulted nose and fault block traps were formed. Based on the drilling results, it has also been confirmed that the NW-trending fault is of good lateral seal ability and became a kind of trap-controlling fault in the oil fields of the Enping 20 subsag [23,24]. Furthermore, the NW-oriented faults can also facilitate the vertical migration of oil and gas to the Neogene Zhujiang Formation, thereby becoming a major conduit for oil and gas transportation within the Yangjiang Sag. Within the Wenchang Sag, the strengthening NW-oriented faults, in conjunction with the NE-oriented major faults, led to the formation of S-shaped, dextral left-stepping, and skew-shaped stress-increased traps [26]. These traps play a controlling role in the trapping and accumulation of hydrocarbons.

As mentioned above, the faults formed during the initial rifting stage of the basin are often basin-controlling or sag-controlling faults. The development of these faults determines the location, shape, and basic structural units of the early depression. The faults formed during the rifting II period played a controlling role in the migration of the basin's rifting and sedimentary centers, making the internal structure of the basin more complex. The faults in the post-rifting stage exhibit a general decrease in activity, with their influence on the basin structure being less pronounced. They induce fault block activity within certain uplifted areas. Moreover, the faults have a significant impact on oil and gas reservoirs. Early rifting faults control the distribution of source rocks and play a crucial role in providing channels for oil and gas transportation, especially the long-active faults, which facilitate the migration of oil and gas from deep to shallow layers. Some secondary faults are crucial for the lateral sealing of oil and gas, developing important structural traps. In addition, the fault activities in the post-rifting stage also have a significant impact on the trapping and migration of oil and gas reservoirs. However, it is worth noting that the impact has both advantages and disadvantages. While forming effective traps, it may also destroy existing oil and gas reservoirs.

In summary, the seismic data used in this study, underpinned by well-log correlations, constitute high-precision 3D seismic data. These data are instrumental in elucidating the sedimentary–structural characteristics of the Zhu III Depression. However, it is important to note that the data distribution is confined to the Yangjiang and Wenchang Sags and does not extend to include other areas such as the Yangchun and Qionghai Sags. Moreover, our structural analysis is primarily grounded in the results derived from seismic section interpretation, though different time slices also prove extremely useful in unveiling strike–slip structures. Concurrently, a broader array of methods, including physical simulation, can be employed to examine the formation and evolution of strike–slip structures. These methodologies will continue to be integrated into our ongoing research efforts.

6. Conclusions

Based on a detailed structural analysis of the Zhu III Depression of the Pearl River Mouth Basin, we arrive at the following conclusions.

Considering their geometric and kinematic characteristics, as well as the regional dextral extensional stress setting, the NW-oriented faults can be classified into three evolutionary stages: extension in Palecene to middle Eocene, dextral transtension during late Eocene to early Oligocene, and sinistral strike–slip since late Oligocene.

The NW-trending extensional fault represents a type of tectonic inversion, triggered by the reactivation of basement faults. The transtensional faults are strike–slip-derived faults, resulting from dextral strike–slip motion of the NE-trending major faults under the regional dextral extensional stress setting. The strike–slip fault is caused by the sinistral shearing action related to the lateral extrusion of the Indochina Block and slab pull of the proto-SCS subduction.

The NW-oriented extensional faults in the rifting I period influenced the division of sag structures and depocenter distribution. The late Eocene–early Oligocene transtensional faults controlled the local tectonic development within the depression and the formation of oil and gas structural traps, as well as depocenter migration. The sinistral strike–slip faults affected the activity of fault blocks, along with the trapping, migration, and preservation of oil and gas since late Oligocene.

In our future work, more time slices will be applied to explore strike–slip faults. Concurrently, we aim to establish relationships between major and minor faults through simulation experiments. We will also focus on the study of strike–slip faults' influence on the migration and storage of oil and gas and formulate a strike–slip controlled reservoir model.

Author Contributions: Conceptualization, L.Z. and D.M.; Methodology, P.Z., L.Z., J.Z., D.M. and Y.C.; Software, P.Z. and L.Z.; Validation, L.Z. and Y.C.; Formal analysis, P.Z., L.Z., J.Z., D.M., Y.C. and P.R.; Investigation, J.Z. and D.M.; Resources, L.Z. and J.Z.; Data curation, P.Z., J.Z., D.M., Y.C. and P.R.; Writing—original draft, P.Z. and L.Z.; Writing—review & editing, P.Z., L.Z., J.Z., D.M., Y.C. and P.R.; Visualization, P.Z., Y.C. and P.R.; Supervision, D.M. and P.R.; Funding acquisition, L.Z. and D.M. All authors have read and agreed to the published version of the manuscript.

Funding: This work was financially supported by Natural Science Foundation of Shandong Province of China (No. ZR2022QD079), the Project of Introducing and Cultivating Young Talent in the Universities of Shandong Province (No. LUJIAOKEHAN–2021–51, granted to L. Yu), and the Scientific Research Start Funds Project of the Fourth Institute of Oceanography, Ministry of Natural Resources (No. 202106).

Institutional Review Board Statement: Not applicable.

Informed Consent Statement: Not applicable.

Data Availability Statement: Data are contained within the article.

Conflicts of Interest: The authors declare no conflict of interest.

References

1. Wu, Z.P.; Li, W.; Ren, C.J.; Li, C.S. Basin Evolution in the Mesozoic and Superposition of Cenozoic Basin in the area of the Jiyang Depression. *Acta Geol. Sin.* **2003**, *77*, 280–286. (In Chinese with English Abstract)
2. Quan, Y.B.; Liu, J.Z.; Zhao, D.J.; Hao, F.; Wang, Z.F.; Tian, J.Q. The origin and distribution of crude oil in Zhu III sub-basin, Pearl River Mouth Basin, China. *Mar. Pet. Geol.* **2015**, *66*, 732–747. [CrossRef]

3. Liu, L.; Sun, Y.H.; Chen, C.; Lou, R.; Wang, Q. Fault reactivation in No.4 structural zone and its control on oil and gas accumulation in Nanpu sag, Bohai Bay Basin, China. *Pet. Explor. Dev.* **2022**, *49*, 824–836. [CrossRef]
4. Allgaier, F.; Busch, B.; Hilgers, C. Fault leakage and reservoir charging in the Upper Rhine Graben, Germany—Assessment of the Leopoldshafen fault bend. *Mar. Pet. Geol.* **2023**, *156*, 106428. [CrossRef]
5. Sobhy, H.; Moustafa, A.R. Impact of structural geometry of tilted fault blocks on hydrocarbon entrapment and deposition of syn–rift clastic reservoirs: Belayim Marine field (Gulf of Suez rift). *Mar. Pet. Geol.* **2024**, *160*, 106631. [CrossRef]
6. Bashmagh, N.M.; Lin, W.R.; Radwan, A.E.; Manshad, A.K. Comprehensive analysis of stress magnitude and orientations and natural fractures in complex structural regimes oil reservoir: Implications for tectonic and oil field development in the Zagros suture zone. *Mar. Pet. Geol.* **2024**, *160*, 106615. [CrossRef]
7. Sun, Z.; Zhong, Z.H.; Keep, M.; Zhou, D.; Cai, D.S.; Li, X.S.; Wu, S.M. 3D analogue modeling of the South China Sea: A discussion on breakup pattern. *J. Asian Earth Sci.* **2009**, *34*, 544–556. [CrossRef]
8. Wang, P.C.; Suo, Y.H.; Peng, G.R.; Li, S.Z.; Du, X.D.; Chao, X.Z.; Zhou, J.; Wang, G.Z.; Santosh, M.; Jiang, S.H.; et al. Three-stage extension in the Cenozoic Pearl River Mouth Basin triggering onset of the South China Sea spreading. *Gondwana Res.* **2023**, *120*, 31–46. [CrossRef]
9. Nguyen, H.H.; Carter, A.; Hoang, L.V.; Fox, M.; Pham, S.N.; Vinh, H.B. Evolution of the continental margin of South to Central Vietnam and its relationship to opening of the South China Sea (East Vietnam Sea). *Tectonics* **2022**, *41*, e2021TC006971. [CrossRef]
10. Lei, B.H.; Zheng, Q.G.; Li, J.L.; Liu, H.; Wang, H.R. Formation and evolution of Zhu-3 south fault and its control on the depocenter shift in Zhu-3 depression, Pearl River Mouth Basin. *Acta Geol. Sin.* **2012**, *33*, 807–813. (In Chinese with English Abstract)
11. Li, H.; Zhang, Y.Z.; Gan, J.; Lu, J.; Zhan, Z.P. The style distribution and hydrocarbon accumulation of inverted structures in Zhu-3 depression. *J. Oil Gas Technol.* **2014**, *36*, 1–6.
12. Quan, Y.B.; Liu, J.Z.; Hao, F.; Bao, X.H.; Xu, S.; Teng, C.Y.; Wang, Z.F. Geochemical characteristics and origins of natural gas in the Zhu III sub-basin, Pearl River Mouth Basin, China. *Mar. Pet. Geol.* **2018**, *101*, 117–131. [CrossRef]
13. Heinemann, N.; Haszeldine, R.S.; Shu, Y.T.; Stewart, R.J.; Scott, V.; Wilkinson, M. CO_2 sequestration with limited sealing capability: A new injection and storage strategy in the Pearl River Mouth Basin (China). *Int. J. Greenh. Gas Control* **2018**, *68*, 230–235. [CrossRef]
14. Fu, X.Y.; Chen, S.J.; You, J.J.; Li, H.; Lei, M.Z. Geochemical characteristics and sources of crude oil in the Wenchang B depression and the Western Qionghai uplift of the Zhu-3 sub-basin, Pearl River Mouth Basin, South China Sea. *J. Pet. Sci. Eng.* **2022**, *219*, 111091. [CrossRef]
15. Zheng, Q.G.; Li, J.L.; Lei, B.H.; Song, P.; Li, Q.; Shi, D.F.; Liu, H.; Lin, C.S. Differential tectonic evolution and formation mechanism of three subsags in Wenchang Sag of Pearl River Mouth Basin, South China Sea. *Pet. Sci.* **2023**, *20*, 1379–1394. [CrossRef]
16. Ge, J.W.; Dong, Y.L.; Tan, M.X.; Chen, H.H.; Sun, L.P.; Shun, L.P.; Li, S.L.; Zhao, X.M. Tecono stratigraphy of Paleogene Zhu-3 depression of the Pearl River Mouth Basin, South China Sea: Implications for syn-rift architecture in multiphase rifts. *Mar. Pet Geol.* **2023**, *155*, 106389. [CrossRef]
17. Wang, J.H.; Pang, X.; Tang, D.Q.; Liu, B.J.; Xu, D.H. Transtensional tectonism and its effects on the distribution of sandbodies in the Paleogene Baiyun Sag, Pearl River Mouth Basin, China. *Mar. Geophys. Res.* **2013**, *34*, 195–207. [CrossRef]
18. Ye, Q.; Mei, L.F.; Shi, H.S.; Camanni, G.; Shu, Y.; Wu, J.; Yu, L.; Deng, P.; Li, G. The Late Cretaceous tectonic evolution of the South China Sea area, An overview, and new perspectives from 3D seismic reflection data. *Earth Sci. Rev.* **2018**, *187*, 186–204. [CrossRef]
19. Suo, Y.H.; Li, S.Z.; Peng, G.R.; Du, X.D.; Zhou, J.; Wang, P.C.; Wang, G.Z.; Somerville, I.; Diao, Y.X.; Liu, Z.Q.; et al. Cenozoic basement-involved rifting of the northern South China Sea margin. *Gondwana Res.* **2022**, *120*, 20–30. [CrossRef]
20. Li, G.; Mei, L.F.; Pang, X.; Zheng, J.Y.; Ye, Q.; Hao, S.H. Magmatism within the northern margin of the South China Sea during the post-rift stage: An overview, and new insights into the geodynamics. *Earth-Sci. Rev.* **2022**, *225*, 103917. [CrossRef]
21. Mu, D.L.; Peng, G.R.; Zhu, D.W.; Li, S.Z.; Suo, Y.H.; Zhan, H.W.; Zhao, L.T. Structure and formation mechanism of the Pearl River Mouth basin: Insights from multi-phase strike-slip motions in the Yangjiang Sag, SE China. *J. Asian Earth Sci.* **2022**, *226*, 105081. [CrossRef]
22. Wu, W.Q.; Li, W.; Fan, C.W.; Li, H.; Li, M.; Li, J.; Zhao, Y.L.; Meng, M.F. Control of the Cenozoic transformation in regional extension direction on the development and evolution of fault system in Zhusan Depression. *Mar. Geol. Front.* **2023**, *39*, 52–65. (In Chinese with English Abstract)
23. Du, X.D.; Peng, G.R.; Wu, J.; Zhang, Z.W.; Xu, X.M.; Zhu, D.W. Faults and Its Impacts on Petroleum Accumulation in Eastern Yangjiang Sag, Pearl River Mouth Basin. *Xinjiang Pet. Geol.* **2020**, *41*, 414–421. (In Chinese with English Abstract)
24. Liu, J.; Peng, G.R.; Zhu, D.W.; Wu, J.; Zhang, Z.W.; Du, X.D.; Wang, X.M.; Liu, Q.Y.; Li, S.Z.; Suo, Y.H. Fault-controlled Hydrocarbon Accumulation in the Eastern Yangjiang Sag, Pearl River Mouth Basin. *Geotecton. Metallog.* **2021**, *45*, 123–130. (In Chinese with English Abstract)
25. Zhang, Y.C.; Gan, J.; Li, H.; Yuan, B.; Deng, G.J.; Zheng, R.F.; Wu, Y.Y.; Cao, Z. Strike-slip deformation mechanism and its petroleum geology significance along south fault in Zhu III depression under extensional tectonic setting. *China Offshore Oil Gas* **2013**, *25*, 9–15. (In Chinese with English Abstract)
26. Jiang, R.F.; Zhou, J.X.; Yang, X.B.; You, J.J.; Li, S.S.; Chen, L.; Zhang, X. The different extension and strike-slip mechanism and its accumulation controlling effect in Wenchang B sag. *Acta Geol. Sin.* **2020**, *94*, 2422–2432. (In Chinese with English Abstract)

27. Zhan, H.W.; Wang, G.Z.; Peng, G.R.; Suo, Y.H.; Wang, P.C.; Du, X.D.; Zhou, J.; Li, S.Z.; Zhu, D.W. Cenozoic evolution of the Yangjiang-Yitong'ansha fault zone in the northern South China Sea: Evidence from 3D seismic data. *Front. Earth Sci.* **2023**, *10*, 1070004. [CrossRef]
28. Clift, P.; Lin, J. Preferential mantle lithospheric extension under the South China margin. *Mar. Pet. Geol.* **2001**, *18*, 929–945. [CrossRef]
29. Xie, X.N.; Ren, J.Y.; Wang, Z.F.; Li, X.S.; Lei, C. Difference of tectonic evolution of continental marginal basins of South China Sea and relationship with SCS spreading. *Earth Sci. Front.* **2015**, *22*, 77–87.
30. Barckhausen, U.; Engels, M.; Franke, D.; Ladage, S.; Pubellier, M. Reply to chang et al. 2014, evolution of the South China Sea: Revised ages for breakup and seafloor spreading. *Mar. Pet. Geol.* **2015**, *59*, 679–681. [CrossRef]
31. Wen, Y.L.; Li, C.F.; Wang, L.J.; Liu, Y.T.; Peng, X.; Yao, Z.W.; Yao, Y.J. The onset of seafloor spreading at the northeastern continent-ocean boundary of the South China Sea. *Mar. Pet. Geol.* **2021**, *133*, 105255. [CrossRef]
32. Wang, Z.Z.; Hu, L.; Wang, S.C.; Lei, M.Z.; Li, M.; Hu, Q.M.; Liu, K. Structural characteristics and reservoir-control mechanism of Wenchang 9-7 transfer slope zone in Zhu III Depression, Pearl River Mouth Basin. *Mar. Orig. Pet. Geol.* **2023**, *28*, 83–93. (In Chinese with English Abstract)
33. Molnar, P.; Tapponnier, P. Relation of the tectonics of eastern China to the India-Eurasia collision: Application of slip-line field theory to large-scale continental tectonics. *Geology* **1977**, *5*, 212–216. [CrossRef]
34. Yin, A. Cenozoic tectonic evolution of Asia: A preliminary synthesis. *Tectonophysics* **2010**, *488*, 293–325. [CrossRef]
35. Xu, J.Y.; Ben-Avraham, Z.; Kelty, T.; Yu, H.S. Origin of marginal basins of the NW Pacific and their plate tectonic reconstructions. *Earth Sci. Rev.* **2014**, *130*, 154–196. [CrossRef]
36. Jolivet, L.; Faccenna, C.; Becker, T.; Tesauro, M.; Sternai, P.; Bouilhol, P. Mantle Flow and Deforming Continents: From India-Asia Convergence to Pacific Subduction. *Tectonics* **2018**, *37*, 2887–2914. [CrossRef] [PubMed]
37. Wang, P.C.; Li, S.Z.; Suo, Y.H.; Guo, L.L.; Santosh, M.; Li, X.Y.; Wang, G.Z.; Jiang, Z.X.; Liu, B.; Zhou, J.; et al. Structural and kinematic analysis of Cenozoic rift basins in South China Sea: A synthesis. *Earth Sci. Rev.* **2021**, *216*, 103522. [CrossRef]
38. Burton-Johnson, A.; Cullen, A.B. Continental rifting in the South China Sea through extension and high heat flow: An extended history. *Gondwana Res.* **2023**, *120*, 235–263. [CrossRef]
39. Li, S.B.; Wang, Y.J.; Wu, S.M. Meso-Cenozoic tectonothermal pattern of the Pearl River Mouth Basin: Constraints from zircon and apatite fission track data. *Earth Sci. Front.* **2018**, *25*, 95–107.
40. Wang, J.L.; Zhang, X.B.; Wu, J.S.; Chen, B.; Zhong, H.Z.; Hao, H.J.; Li, P.L.; Su, N.E. Integrated geophysical researches on base texture of Zhujiang River Mouth Basin. *J. Trop. Oceanogr.* **2002**, *21*, 13–22. (In Chinese with English Abstract)
41. Chen, H.Z.; Wu, X.J.; Zhou, D.; Wang, W.Y.; He, H.J. Meso-Cenozoic faults in Zhujiang River Mouth Basin and their geodynamics background. *J. Trop. Oceanogr.* **2005**, *24*, 52–61. (In Chinese with English Abstract)
42. Sun, Z.; Zhou, D.; Sun, L.T.; Chen, C.M.; Pang, X.; Jiang, J.Q.; Fan, H. Dynamic Analysis on Rifting Stage of Pearl River Mouth Basin through Analogue Modeling. *J. Earth Sci.* **2010**, *21*, 439–454. [CrossRef]
43. Li, P.L. Cenozoic tectonic movement in the Pearl River Mouth Basin. *China Offshore Oil Gas* **1993**, *7*, 11–17. (In Chinese with English Abstract)
44. Cai, Z.R.; Liu, W.L.; Wan, Z.F.; Guo, F. Determination of Cenozoic tectonic movement in the northern South China Sea and the relationship between oil-gas reservoir and tectonic movement. *Mar. Sci. Bull.* **2010**, *29*, 161–165. (In Chinese with English Abstract)
45. Jiang, H.; Wang, H.; Xiao, J.; Chen, S.P.; Lin, Z.L. Control of Paleomorphology to Sedimentary Filling in Marginal Sea Basin—By Taking Zhu III Depression for Example. *J. Oil Gas Technol.* **2008**, *30*, 10–16. (In Chinese with English Abstract)
46. Gong, Y.; Lin, C.S.; Zhang, Z.T.; Zhang, B.; Shu, L.F.; Feng, X.; Hong, F.H.; Xing, Z.C.; Liu, H.Y.; Su, E.Y. Breakup unconformities at the end of the early Oligocene in the Pearl River Mouth Basin, South China Sea: Significance for the evolution of basin dynamics and tectonic geography during rift-drift transition. *Mar. Geophys. Res.* **2019**, *40*, 371–384. [CrossRef]
47. Morley, C.K. Major unconformities/termination of extension events and associated surfaces in the South China Seas: Review and implications for tectonic development. *J. Asian Earth Sci.* **2016**, *120*, 62–86. [CrossRef]
48. Sibuet, J.; Yeh, Y.; Lee, C. Geodynamics of the South China Sea. *Tectonophysics* **2016**, *692*, 98–119. [CrossRef]
49. Niu, Z.C.; Liu, G.D.; Ge, J.W.; Zhang, X.T.; Cao, Z.; Lei, Y.C.; Yuan, A.; Zhang, M.Y. Geochemical characteristics and depositional environment of Paleogene lacustrine source rocks in the Lufeng sag, Pearl River Mouth basin, South China Sea. *J. Asian Earth Sci.* **2018**, *171*, 60–77. [CrossRef]
50. Peng, G.R.; Zhang, X.T.; Xu, X.M.; Bai, H.J.; Cai, G.F.; Zhao, C.; Zhang, Z.W. Important discoveries and understandings of oil and gas exploration in Yangjiang sag of the Pearl River Mouth Basin, northern South China Sea. *China Pet. Explor.* **2019**, *24*, 267–279. (In Chinese with English Abstract)
51. Nanni, U.; Pubellier, M.; Chan, L.S.; Senell, R.J. Rifting and reactivation of a Cretaceous structural belt at the northern margin of the South China Sea. *J. Asian Earth Sci.* **2017**, *136*, 110–123. [CrossRef]
52. Tang, X.; Yu, Y.X.; Zhang, X.T.; Peng, G.R.; Niu, S.L.; Qiu, X.W.; Lu, M.S.; He, Y.B. Multiphase faults activation in the southwest Huizhou Sag, Pearl River Mouth basin: Insights from 3D seismic data. *Mar. Pet. Geol.* **2023**, *152*, 106257. [CrossRef]
53. Li, S.Z.; Cao, X.Z.; Wang, G.Z.; Liu, B.; Li, X.Y.; Suo, Y.H.; Jiang, Z.X.; Guo, L.L.; Zhou, J.; Wang, P.C.; et al. Meso-Cenozoic tectonic evolution and plate reconstruction of the Pacific Plate. *J. Geomech.* **2019**, *25*, 642–677. (In Chinese with English Abstract)

54. Zhan, H.W.; Cai, G.F.; Zhang, Z.W.; Wang, G.Z.; Li, Y.W.; Suo, Y.H.; Wang, P.C.; Jiang, S.H.; Liu, B.; Guo, L.L.; et al. Paleogene fault activity and basin controlling characteristics in the northern South China Sea margin—A case study of the eastern Yangjiang sag. *Geotecton. Metallog.* **2021**, *45*, 20–39. (In Chinese with English Abstract)
55. Li, J.L.; Lei, B.H.; Zheng, Q.G.; Duan, L.; Yan, Y. Stress Field Evolution and its Controls on Oil Accumulation in the Wenchang Sag. *Geotecton. Metallog.* **2015**, *39*, 601–609. (In Chinese with English Abstract)
56. Zhou, J.; Li, S.Z.; Suo, Y.H.; Zhang, L.; Du, X.D.; Cao, X.Z.; Wang, G.Z.; Li, F.K.; Liu, Z.; Liu, J.; et al. NE-Trending transtensional faulting in the Pearl River Mouth basin of the northern South China Sea margin. *Gondwana Res.* **2022**, *120*, 4–19. [CrossRef]
57. Zalán, P.V. Identification of strike-slip faults in seismic sections. In *SEG Technical Program Expanded Abstracts*; Society of Exploration Geophysicists: Houston, TX, USA, 1987; pp. 116–118.
58. Xia, Y.P.; Liu, W.H.; Xu, L.G.; Zheng, L.H. Identification of strike-slip fault and its petroleum geology significance. *Pet. Geol.* **2007**, *1*, 17–23. (In Chinese with English Abstract)
59. Xie, G.J.; Chen, D.X.; Chang, L.; Li, J.H.; Yin, Z.J. Migration and accumulation of crude oils in the Qionghai Uplift, Pearl River Mouth Basin, Offshore South China Sea. *J. Pet. Sci. Eng.* **2021**, *205*, 108943. [CrossRef]
60. Zhao, F.; Alves, T.M.; Xia, S.H.; Li, W.; Wang, L.; Mi, L.J.; Wu, S.G.; Cao, J.H.; Fan, C.Y. Along strike segmentation of the South China Sea margin imposed by inherited pre-rift basement structures. *Earth Planet. Sci. Lett.* **2020**, *530*, 115862. [CrossRef]
61. He, Z.Y.; Bai, Z.Z.; Wang, W.Y.; Li, L.Z.; Zhang, Y.M.; Chen, Y.; He, T.; Ma, R.Y. Tectonic framework research in Zhujiang River Mouth Basin based on gravity and magnetic data. *Haiyang Xuebao* **2023**, *45*, 25–43. (In Chinese with English Abstract)
62. Xu, J.Y.; Zhang, L.Y. Genesis of Cenozoic basins in northwest Pacific margin (2): Linked dextral pull-apart basin system. *Oil Gas Geol.* **2000**, *21*, 185–190. (In Chinese with English Abstract)
63. Liu, Z.; Li, S.Z.; Suo, Y.H.; Bukhari, S.W.H.; Ding, X.; Zhou, J.; Wang, P.; Cheng, H.; Somerville, I. Evolution of pull-apart basins with overlapping NE-trending strike-slip fault systems in the northern South China Sea margin: Insight from numerical modeling. *Tectonophysics* **2023**, *846*, 229679. [CrossRef]
64. Sun, Z.; Xu, Z.Y.; Sun, L.T.; Pang, X.; Yan, C.Z.; Li, Y.P.; Zhao, Z.X.; Wang, Z.W.; Zhang, C.M. The mechanism of post-rift fault activities in Baiyun sag, Pearl River Mouth basin. *J. Asian Earth Sci.* **2014**, *89*, 76–87. [CrossRef]
65. Cai, G.F.; Zhang, X.T.; Peng, G.R.; Wu, J.; Liu, B.J., Bai, H.J., Li, Z.S., Ma, X.N.; Li, S.Z.; Suo, Y.H. Neogene volcanism and tectonics along the Yangjiang-Yitong'ansha Fault Zone in the Northern South China Sea margin. *Geotecton. Metallog.* **2021**, *45*, 40–52. (In Chinese with English Abstract)
66. Zhang, G.C.; Qu, H.J.; Jia, Q.J.; Zhang, L.G.; Yang, B.; Chen, S.; Ji, M.; Sun, R.; Guan, L.M.; Hayat, K. Passive continental margin segmentation of the marginal seas and its effect on hydrocarbon accumulation: A case study of the northern continental margin in South China Sea. *Mar. Pet. Geol.* **2021**, *123*, 104741. [CrossRef]
67. Hu, Z.W.; Xu, C.G.; Wang, D.Y.; Ren, J.; Liu, Y.F.; Xiao, S.G.; Zhou, X. Superimposed characteristics and genetic mechanism of strike–slip faults in the Bohai Sea, China. *Pet. Explor. Dev.* **2019**, *46*, 254–267. (In Chinese with English Abstract) [CrossRef]
68. Hall, R. Cenozoic geological and plate tectonic evolution of SE Asia and the SW Pacific: Computer-based reconstructions and animations. *J. Asian Earth Sci.* **2002**, *20*, 353–434. [CrossRef]
69. Molnar, P., Stock, J.M. Slowing of India's convergence with Eurasia since 20 Ma and its implications for Tibetan mantle dynamics. *Tectonics* **2009**, *28*, 1–11. [CrossRef]
70. Copley, A.; Avouac, J.-P.; Royer, J.-Y. India-Asia collision and the Cenozoic slowdown of the Indian plate: Implications for the forces driving plate motions. *J. Geophys. Res. Solid Earth* **2010**, *115*, 1–14. [CrossRef]
71. Li, Y.H.; Zhu, R.W.; Liu, H.L.; Qiu, X.L.; Huang, H.B. The Cenozoic activities of Yangjiang-Yitongdong Fault: Insights from analysis of the tectonic characteristics and evolution processes in western Zhujiang (Pearl) River Mouth Basin. *Acta Oceanol. Sin.* **2019**, *38*, 87–101. [CrossRef]
72. Northrup, C.J.; Royden, L.H.; Burchfiel, B.C. Motion of the Pacific Plate relative to Eurasia and its potential relation to Cenozoic extension along the eastern margin of Eurasia. *Geology* **1995**, *23*, 719–722.
73. Müller, R.D.; Seton, M.; Zahirovic, S.; Williams, S.E.; Cannon, J. Ocean Basin Evolution and Global-Scale Plate Reorganization Events Since Pangea Breakup. *Annu. Rev. Earth Planet. Sci.* **2016**, *44*, 107–138. [CrossRef]
74. Ren, J.Y.; Tamaki, K.; Li, S.T.; Zhang, J.X. Late Mesozoic and Cenozoic rifting and its dynamic setting in Eastern China and adjacent areas. *Tectonophysics* **2002**, *344*, 175–205. [CrossRef]
75. Li, S.Z.; Suo, Y.H.; Li, X.Y.; Wang, Y.M.; Cao, X.Z.; Wang, P.C.; Guo, L.L.; Yu, S.Y.; Lan, H.Y.; Li, S.J.; et al. Mesozoic plate subduction in West Pacific and tectono-magmatic response in the East Asian ocean-continent connection zone. *Chin. Sci. Bull.* **2018**, *63*, 1550–1593. [CrossRef]
76. Ye, Q.; Shi, H.S.; Mei, L.F.; Shu, Y.; Liu, H.L.; Wei, T.; Yan, H. Post-Rift Faulting Migration, Transition and Dynamics in Zhu I Depression, Pearl River Mouth Basin. *Earth Sci.* **2017**, *42*, 105–118. (In Chinese with English Abstract)

Journal of
Marine Science and Engineering

MDPI

Article

A Novel Method for Estimating the Undrained Shear Strength of Marine Soil Based on CPTU Tests

Sai Fu [1], Yanghai Shen [2], Xianlin Jia [1], Zhiqing Zhang [3,*] and Xibin Li [2,*]

1 PowerChina Huadong Engineering Corporation Limited, Hangzhou 311122, China; fu_s@hdec.com (S.F.); jia_xl@hdec.com (X.J.)
2 College of Landscape Architecture, Zhejiang A & F University, Hangzhou 311300, China; shenyanghai2021@163.com
3 College of Architecture and Energy Engineering, Wenzhou University of Technology, Wenzhou 325035, China
* Correspondence: zhangzhiqing2000@163.com (Z.Z.); ytulxb@zafu.edu.cn (X.L.)

Abstract: The undrained shear strength is an essential parameter in the foundation design of marine structures. Due to the complex marine environment and technical limitations, it is difficult and costly to obtain offshore samples. Piezocone penetration tests (CPTU) are relatively low-cost compared to drilling and sampling methods. Therefore, based on the soil behavior type index (I_c) derived from CPTU results, a model for estimating cone factors (N_{kt}, N_{ke}) is proposed to improve the accuracy of estimation of undrained shear strength. The result shows that the soil behavior type index (I_c) and cone factors take on a negatively correlated exponential relation. Incorporating a cone factor that varies with the soil behavior type index (I_c) significantly enhances the accuracy of undrained shear strength predictions compared to the conventional method of using a constant cone factor. This approach reduces the root mean square error (*RMSE*) for N_{kt} (N_{ke}) from 0.124 (0.126) MPa to 0.056 (0.06) MPa, and the mean absolute error (*MAE*) from 0.0154 (0.016) MPa to 0.0032 (0.0036) MPa. The method was validated at an additional location and the predictions were in high agreement with the results of the consolidated quick direct shear test. The developed method can serve as an effective tool used in the design of foundations of marine structures.

Keywords: piezocone penetration test; soil behavior type index; undrained shear strength; consolidated quick direct shear test

1. Introduction

The utilization of offshore wind energy resources is a crucial driver for economic development [1,2] and an essential pathway towards achieving net-zero carbon goals. In recent years, offshore wind power construction has been gradually expanded from offshore areas to deeper parts of the sea, which further causes more difficulties and costs in soil sampling of marine geotechnical surveys [3]. As a result, in situ testing techniques, such as standard penetration tests (SPT) [4], cone penetration tests (CPT) [5], and piezocone penetration tests (CPTU) [6], have been applied in marine soil investigations. Compared to the traditional SPT, CPTUs have significant advantages, such as providing continuous testing data and suitability for more complex environmental conditions; as a result, they are extensively utilized in marine geotechnical surveys. However, existing empirical formulae relating CPTU results to soil parameters are primarily based on terrestrial environments [7,8] and are unsuitable for marine conditions [9]. Therefore, it is of great importance to conduct in-depth research on the empirical formulae connecting CPTU results and marine soil parameters.

In the context of pile foundation design for marine structures, the undrained shear strength of cohesive soil emerges as a critical parameter [10]. Consequently, the accurate prediction of undrained shear strength assumes paramount importance as a central focus in exploring the correlation between CPTU data and soil parameters. The penetration mechanism of CPTU in cohesive soils can be classified into theoretical and empirical methods. The theoretical method commonly includes bearing capacity theory [11], cavity expansion theory [12], the strain path method [13], and moving point dislocation theory [14]. Due to idealized assumptions in these theoretical methods, significant deviations often occur. Hence, scholars began to explore suitable empirical methods to improve the reliability of the prediction with various empirical methods being developed, such as the "total" cone tip resistance method [15], the effective cone resistance method [12], and the excess pore water pressure method [16]. However, the cone factors are still not fixed and vary greatly. To obtain more accurate cone factors, researchers further considered soil characteristic factors to improve the estimating precision of the undrained shear strength from CPTU. When considering the plasticity index of the soil [17], the estimated undrained shear strength values obtained from CPTU are usually very close to laboratory experiment results. Nevertheless, the correlation between cone factors and the plasticity index is only valid for specific soil conditions in certain regions [18,19]. In addition, cone factors are also related to the over-consolidation ratio [20], natural moisture content [21], and fines content [22] of the soil. To obtain the plasticity index, the over-consolidation ratio of soils, etc., soil sampling is required, which unavoidably increases the engineering costs. In order to solve this problem, Bol et al. [23,24] classified soils by means of a soil behavior type index and then gave empirical coefficients for the cone coefficients corresponding to the different soil classes. Nevertheless, the quantitative relation between I_c and cone factors was not provided in Bol's study. With the development of computer science, machine learning algorithms have been employed to estimate the undrained shear strength with high accuracy [25–27]. Moreover, these algorithms are gradually overcoming the limitation of not being able to provide deterministic functional relationships.

To facilitate practical engineering applications and reduce costs associated with marine geotechnical exploration, this study categorized soil into distinct intervals based on the soil behavior type index. Utilizing data obtained from indoor consolidated quick direct shear tests, it established a robust regression model to elucidate the explicit functional relationship between the soil behavior type index and the cone factor. The predictive performance of this model was rigorously assessed through comprehensive evaluation metrics, including the coefficient of determination (R^2), root mean square error ($RMSE$), and mean absolute error (MAE). Furthermore, validation exercises were conducted in nearby marine project sites to confirm the applicability of the proposed methodology.

2. Methods for Calculating Undrained Shear Strength and Soil Behavior Type Index

The "total" cone tip resistance q_t method and the effective cone resistance q_e method are common methods for estimating undrained shear strength S_u from CPTU results. In the following part, we also discuss the undrained shear strength from a consolidated quick direct shear test and the computation of the soil behavior type index I_c.

2.1. Estimation of S_u by "Total" Cone Tip Resistance q_t

Following Lunne et al. [15], the unified formula for calculating the undrained shear strength of clayey soils can be written as:

$$S_u = \frac{q_t - \sigma_v}{N_{kt}} \tag{1}$$

where S_u is the undrained shear strength; q_t is the total cone tip resistance; σ_v is the total vertical stress of the overlying soil; N_{kt} is the empirical cone factor.

2.2. Estimation of S_u by Effective Cone Tip Resistance q_e

Considering the corrected total cone tip resistance q_t and the difference in pore pressure u_2, Senneset et al. [12] proposed an expression for the undrained shear strength S_u of clayey soils as listed below:

$$S_u = \frac{q_e}{N_{ke}} = \frac{q_t - u_2}{N_{ke}} \tag{2}$$

where N_{ke} is the effective cone factor.

2.3. Calculation of Undrained Shear Strength S_u

The consolidated quick direct shear test expeditiously assesses the shear strength of soil by subjecting a sample to shearing within a confined shear box under applied pressure. Despite potential limitations in accuracy, notably concerning sensitive marine soils, its widespread applicability stems from its simplicity, expediency, and cost-effectiveness. This method finds significant utility, particularly in marine geotechnical investigations, where its deployment often exceeds that of triaxial compression tests in terms of quantity. The undrained shear strength of soil at different depths can be estimated by the cohesive strength C_u, angle of internal friction φ_u, and unit weight of soil γ obtained from the consolidated quick direct shear test [28]. The average values of cohesive strength and angle of internal friction are generally used to calculate the indoor undrained shear strength. Due to the difficulties and high cost of offshore sampling [29], only a few holes are drilled for soil sampling. Hence, we adopt a combined method of indoor experiments and empirical formulae from CPTU to obtain the unit weight of the soil, which is a common practice in current practical engineering [30,31].

The consolidation stress σ_3 and the undrained shear strength S_u can be written as:

$$\sigma_3 = \frac{\sum \gamma_i h_i (1 + 2(1 - \sin\varphi_u))}{3} \tag{3}$$

$$S_u = c_u + \sigma_3 \tan\varphi_u \tag{4}$$

where h_i and γ_i are the thickness and weight of layer i, respectively. Note that the first layer is actually seawater.

2.4. Calculation of Soil Behavior Type Index I_c

The normalized soil behavior type chart based on CPT proposed by Robertson (1990) [32] classifies soils. Jeffries and Davies [33] discovered the soil behavior type index, which defines the boundaries of soil types without the need for soil sampling. According to the generalized definition provided by Robertson [34], the soil behavior type index can be further described as follows:

$$I_c = \sqrt{[3.47 - \log_{10} Q]^2 + [1.22 + \log_{10} F]^2} \tag{5}$$

$$Q = \frac{q_t - \sigma_v}{p_a} \left(\frac{p_a}{\sigma_v'}\right)^n \tag{6}$$

$$F = \frac{f_s}{q_t - \sigma_v} \tag{7}$$

where Q is the normalized tip resistance; P_a is the atmospheric pressure; F is the normalized friction ratio; f_s is the side friction resistance; σ_v' is the effective covering total stress; n is the stress index, and its value is set to 1. The revised classification of soil behavior types index [35] is listed in Table 1.

Table 1. The corrected I_c classification.

Range of I_c	Classification of the Soil in China
$I_c > 3.45$	Muck and mucky soil
$3.00 < I_c < 3.45$	Clay
$2.80 < I_c < 3.00$	Silty clay, clay
$2.60 < I_c < 2.80$	Silt, silty clay
$2.40 < I_c < 2.60$	Silt
$2.10 < I_c < 2.40$	Silty sand, silt
$1.87 < I_c < 2.10$	Fine sand
$I_c < 1.87$	Medium sand

3. CPTU Results

As depicted in Figure 1, the experimental sites A and B are located in the eastern sea area of Cangnan and the southeastern sea area of Xiangshan in Zhejiang Province, China, respectively. Figures 2 and 3 illustrate the distribution of drilling holes at sites A and B. Throughout the CPTU drilling process, parameters such as "total" cone tip resistance (q_t), sleeve friction (f_s), and pore water pressure at the cone shoulder (u_2) were continuously measured with a data acquisition interval not exceeding 100 mm. At site A, there were 39 CPTU boreholes and 10 soil sampling holes, from which 202 soil samples were collected, and 68 sets of laboratory consolidated quick direct shear tests were conducted. Figure 4 displays the soil profile data of the CPTU borehole log section No. JT-1 along the depth direction at site A. The upper sediment at site A primarily consists of Quaternary (Q_4) shallow marine deposits, including mud, muddy silty clay, silt, sandy silt, and silty clay. The lower sediment comprises Late Pleistocene ($Q_3{}^2$) deposits from terrestrial and coastal environments, mainly composed of silt, silty clay, and sandy silt within the depth range of the exploration. At site B, there were 19 CPTU test holes and 28 soil sampling holes, from which 484 soil samples were collected, and 68 sets of laboratory consolidated quick direct shear tests were performed. The upper sediment at site B primarily consists of Quaternary (Q_4) shallow marine deposits, including mud, muddy silty clay, sandy silt, silt, silty clay, and gravel. The lower sediment consists of Late Pleistocene ($Q_3{}^2$) deposits from estuarine and coastal environments, characterized by sandy silt and silty clay within the depth range of the exploration. In summary, the geological conditions of experimental sites A and B are relatively similar. We first proposed an undrained shear strength formula based on CPTU results derived from the data of experimental site A and then verified it using the data from experimental site B. Table 2 presents the types and quantities of soil layers studied at experimental sites A and B.

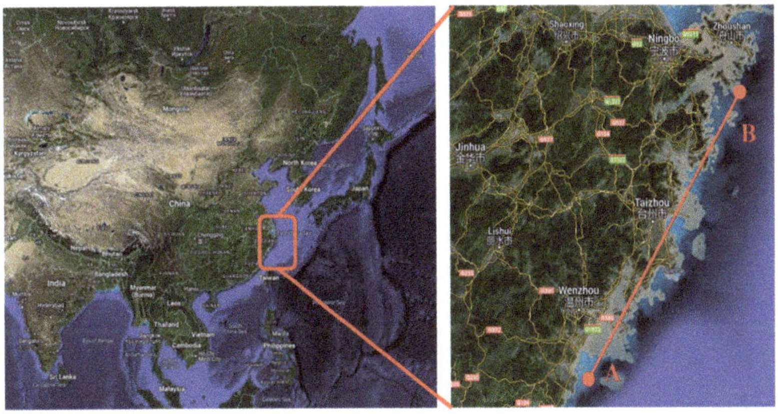

Figure 1. Relative location of the experiment at sites A and B.

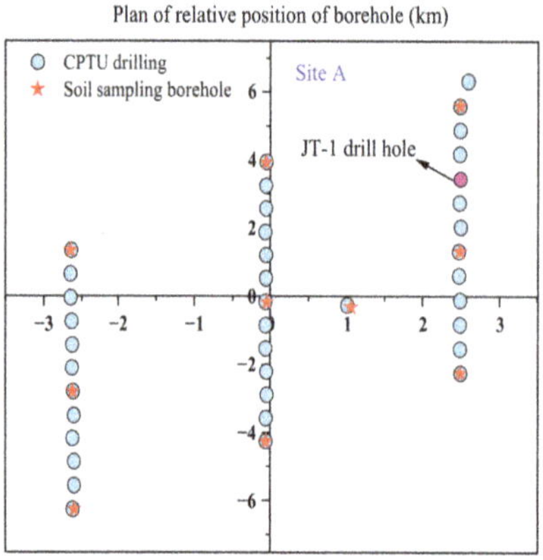

Figure 2. Distribution of boreholes at site A.

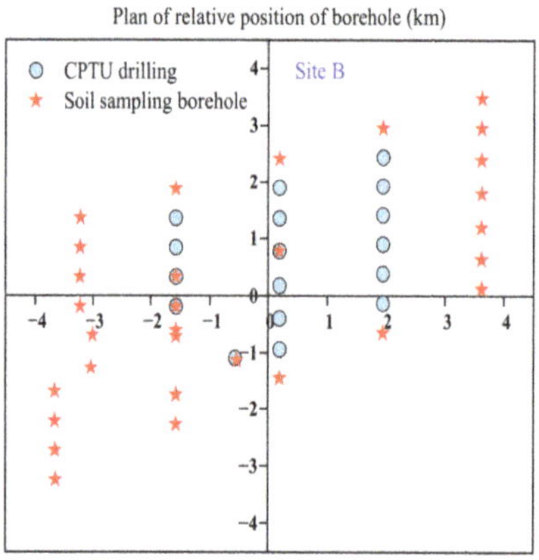

Figure 3. Distribution of boreholes at site B.

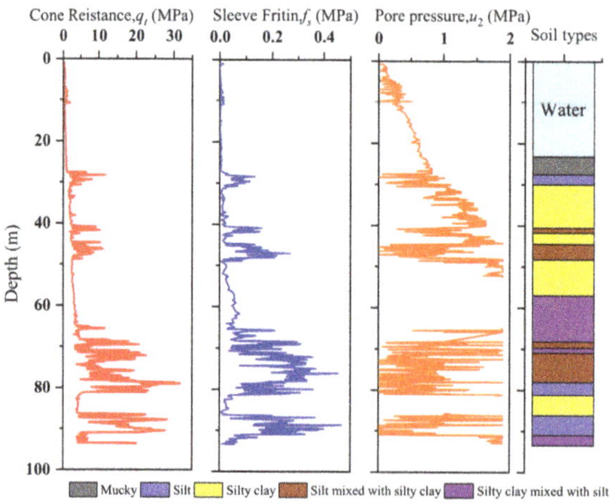

Figure 4. Soil profile data of the CPTU borehole log section No. JT-1 at site A.

Table 2. The information of soil layers.

Experiment in A		Experiment in B	
Soil Layers	**Statistic**	**Soil Layers**	**Statistic**
Silty clay	116	Muddy silty clay	27
Silt mixed with silty clay	67	Silty clay	26
Silty clay mixed with silt	53	Muddy clay	16
Muddy silty clay	18	Silty clay mixed with silt	9
Silty clay mixed with silt	13	Clayey silt mixed with silty clay	4
Silty clay with silty sand	7		
Silt mixed with silty clay	6		

Data extraction was first performed on the test results from 39 CPTU boreholes at experimental site A. After eliminating data from sand layers and surface mucky layers, 344 soil layer data were obtained. The histogram in Figure 5 displays the distribution of I_c data. Figure 6 shows the results of the selection of 280 soil layer data. To minimize random errors, this study focused on the range [1.8, 2.8], which has a dense distribution of I_c data. Figure 6 shows that the data points are relatively discrete and cannot form a functional relationship.

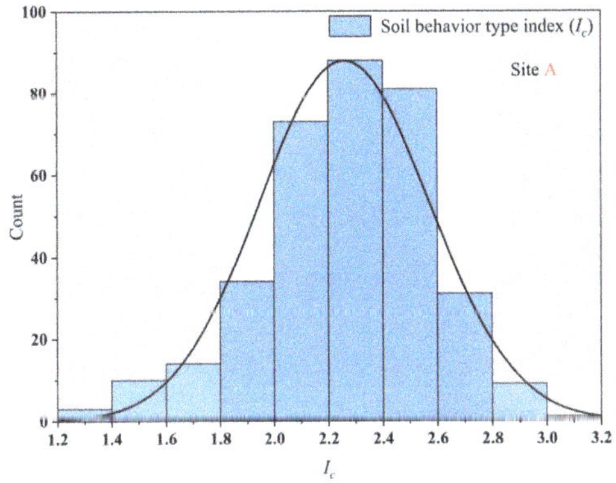

Figure 5. Histogram of I_c at site A.

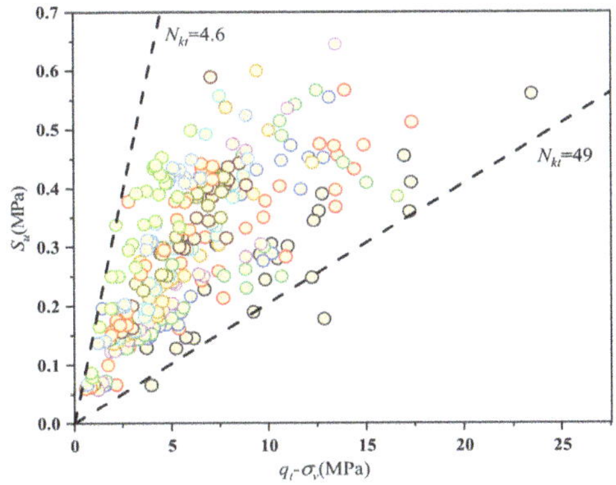

Figure 6. Empirical cone factor N_{kt} vs. S_u.

To investigate the functional relationship between the related variables (q_t-σ_v, S_u), 280 soil samples were divided into 12 groups based on the magnitude of I_c. Each group contained 23 soil samples, except for the last one, as shown in Figure 7. Two clear outliers were excluded based on observation. The results indicate a linear relationship between these variables, as demonstrated by the coefficient of determination R^2 in Table 3. All results, except for I_c ranging from 2.12 to 2.22, indicate good fitting effectiveness, with values larger than 0.95. Additionally, using the method described above, the relationship between q_t-u_2 and S_u can be determined (see Figure 8 and Table 3).

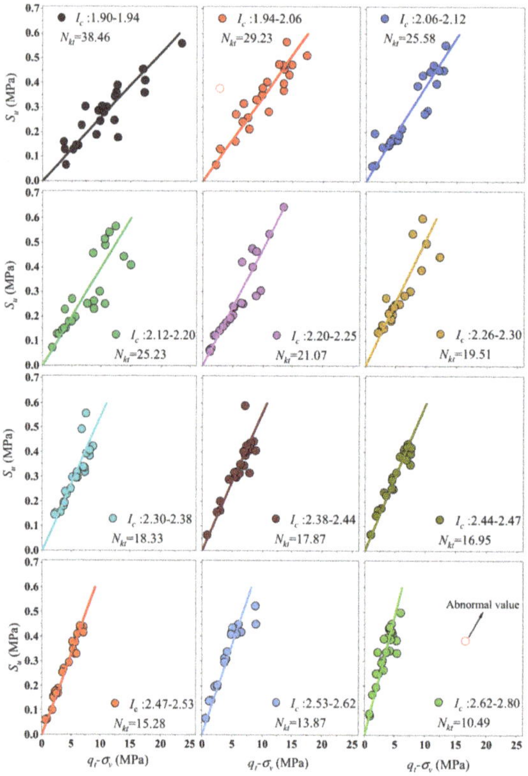

Figure 7. N_{kt} corresponding to different intervals of I_c at site A.

Table 3. N_{kt} and N_{ke} of I_c in different intervals at site A.

Depth Range of the Soil Layer Bottom (m)	I_c	I_c (Mean Value)	N_{kt}	R^2	N_{ke}	R^2
[43–54]	[1.80–1.94]	1.88202	38.46	0.96	38.43	0.96
[52–65]	[1.94–2.06]	2.01657	29.23	0.97	29.36	0.97
[47–59]	[2.06–2.12]	2.09938	25.58	0.97	25.95	0.96
[46–63]	[2.12–2.20]	2.15491	25.23	0.93	25.75	0.93
[46–61]	[2.20–2.25]	2.22081	21.07	0.96	21.43	0.96
[50–65]	[2.26–2.30]	2.27726	19.51	0.95	19.69	0.94
[56–72]	[2.30–2.38]	2.34376	18.33	0.97	18.90	0.96
[67–83]	[2.38–2.44]	2.41203	17.87	0.97	18.76	0.97
[59–78]	[2.44–2.47]	2.45625	16.95	0.98	17.59	0.98
[54–74]	[2.47–2.53]	2.49873	15.28	0.99	15.86	0.98
[62–82]	[2.53–2.62]	2.57546	13.87	0.97	13.90	0.97
[67–84]	[2.62–2.80]	2.69623	10.49	0.97	10.89	0.96

Note: The water depth within this region ranges from approximately 20 m to 26 m, with the exclusion of surface mucky layers from the research scope.

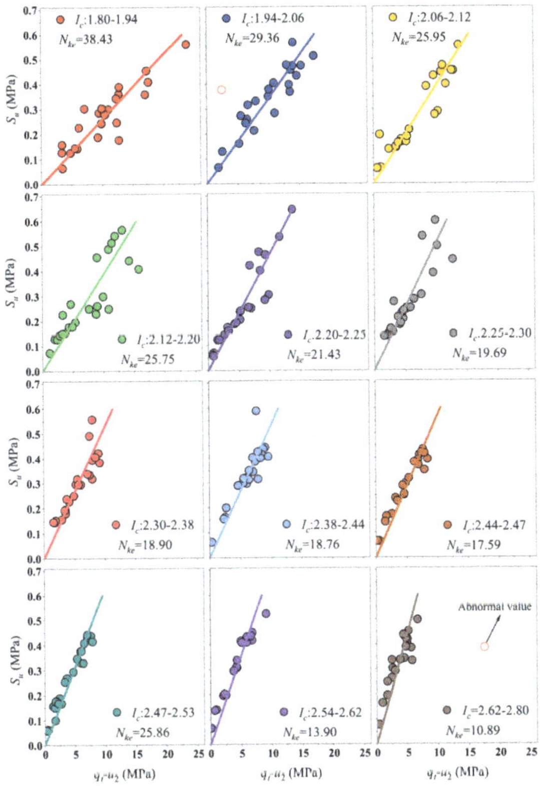

Figure 8. N_{ke} corresponding to different intervals of I_c at site A.

To further probe the functional relation between I_c and N_{kt}/N_{ke}, the regression analysis was performed on the average values of I_c and the empirically obtained N_{kt} and N_{ke} within each interval. It can be observed from Figure 9 that R^2 is greater than 0.97 and a significant nonlinear relation exists between I_c and N_{kt}/N_{ke}.

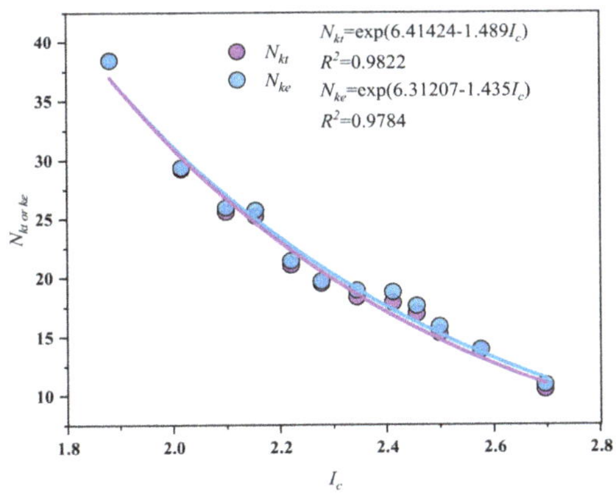

Figure 9. Relation between N_{kt}/N_{ke} and I_c at site A.

As presented in Figure 9, a new computational formula for estimating cone factors (N_{kt}, N_{ke}) can be provided as:

$$S_u = \frac{q_t - \sigma_v}{N_{kt}} \tag{8}$$

in which:

$$N_{kt} = \exp(6.41424 - 1.489 I_c), \quad 1.8 \le I_c \le 2.8 \tag{9}$$

$$S_u = \frac{q_t - u_2}{N_{ke}} \tag{10}$$

in which:

$$N_{ke} = \exp(6.31207 - 1.435 I_c), \quad 1.8 \le I_c \le 2.8 \tag{11}$$

According to Equation (9), the average value of N_{kt} in the soil behavior type index I_c range of 1.8 to 2.8 is 21.76, which is close to the statistical average value of 17.9 obtained by the consolidated quick direct shear test specified in the Technical Code for Static Penetration Test in Water Transportation Engineering of China [21]. According to Table 2, the study involved various types and numbers of soil layers. However, the lack of clay layers may result in an overestimation of the calculated value of N_{kt} in the formula. Furthermore, research conducted in the marine areas near Hangzhou Bay, Hangzhou City, Zhejiang Province, China, indicates that the consolidated quick direct shear tests conducted in this region yielded an overall average value of 20.25 for the cone tip factor (N_{kt}) in terms of undrained shear strength [36]. This further corroborates the reliability of the formula in the marine areas adjacent to Zhejiang Province, China.

Three statistical criteria (coefficient of determination R^2, root mean square error *RMSE*, and mean absolute error *MAE*) are utilized to evaluate the proposed predictive model. The calculation formulas are as follows:

$$R^2 = 1 - \frac{\sum_{i=1}^{n} (L_i - E_i)}{\sum_{i=1}^{n} (L_i - \overline{E})^2} \tag{12}$$

$$RMSE = \sqrt{\frac{\sum_{i=1}^{n} (L_i - E_i)^2}{n}} \tag{13}$$

$$MAE = \frac{\sum_{i=1}^{n} (L_i - E_i)^2}{n} \tag{14}$$

where L_i represents the undrained shear strength obtained from consolidated quick direct shear test; E_i represents the undrained shear strength predicted from CPTU results. The closer R^2 is to 1, the better the model's predictive performance. A smaller *RMSE* or *MAE* indicates better predictive performance of the model.

The cone factor values are not fixed and exhibit significant variation, making it challenging to obtain precise values for the undrained shear strength of soil. In practical engineering, traditional approaches often involve using maximum and minimum cone values of N_{kt} and N_{ke}, or directly employing empirical values specific to the region, to calculate the undrained shear strength of soil. To compare the traditional method with the new method proposed in the study, minimum errors for cone factors (N_{kt} = 20.97 and N_{ke} = 20.51) were determined using linear regression according to the "total" cone tip resistance method and effective cone resistance method, as shown in Figure 10a,b. In Figure 10c,d, cone factors N_{kt} and N_{ke} were calculated using Equations (9) and (11). By comparing Figure 10a–d, it is found that when considering the correlation between soil behavior type index and cone factors, the three predicted indicators (R^2, *RMSE*, *MAE*) show marked improvement. For instance, R^2 corresponding to N_{kt} (N_{ke}) increases from 0.09 (0.05) to 0.81 (0.78), respectively. *RMSE* corresponding to N_{kt} (N_{ke}) decreases from 0.124 (0.126) MPa to 0.056 (0.06) MPa, respectively. *MAE* corresponding to N_{kt} (N_{ke}) decreases from 0.0154 (0.016) MPa to 0.0032 (0.0036) MPa, respectively. The evaluated results show that considering the correlation between the soil behavior type index and cone factor can improve the prediction accuracy of undrained shear strength from CPTU.

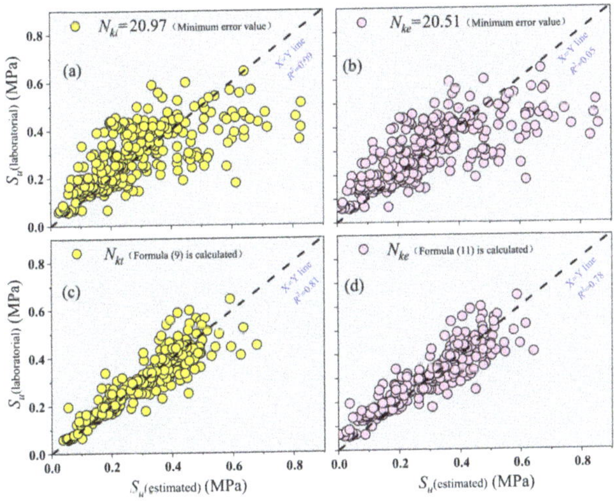

Figure 10. Predictions using traditional N_{kt} and N_{ke} methods in (**a**,**b**); predictions using the newly proposed N_{kt} and N_{ke} methods in (**c**,**d**).

An analysis of the CPTU test result plots in Figure 4 is presented to compare the difference between the traditional method and the new method, which incorporates the soil behavior type index in predicting the distribution of undrained shear strength along the depth direction. In the conventional method, the values of cone factors were kept the same as in the previous section, i.e., N_{kt} = 20.97 and N_{ke} = 20.51. Table 4 shows the soil cone factors obtained from the soil behavior type index. It is evident from Figure 11 that the undrained shear strength calculated using Equations (8) and (10) proposed in this study consistently has a smaller error than the conventional method. This indicates that the method incorporating the soil behavior type index has significant advantages in terms of prediction accuracy.

Table 4. Soil layer cone factor estimation.

Soil Number	Soil Depth (m)	N_{kt}	N_{ke}
1	27.5	13.23	15.17
2	30	46.05	50.47
3	40.5	15.2	17.34
4	41.7	30.67	34.11
5	44.5	22.36	25.16
6	48.3	26.28	29.39
7	57	11.93	13.73
8	68.2	15.75	17.94
9	69.7	26.69	29.83
10	70.9	17.27	19.61
11	78	24.96	27.97
12	81.2	27.1	30.27
13	86.2	9.4	10.91
14	91	23.97	26.9
15	93.52	13.48	15.45

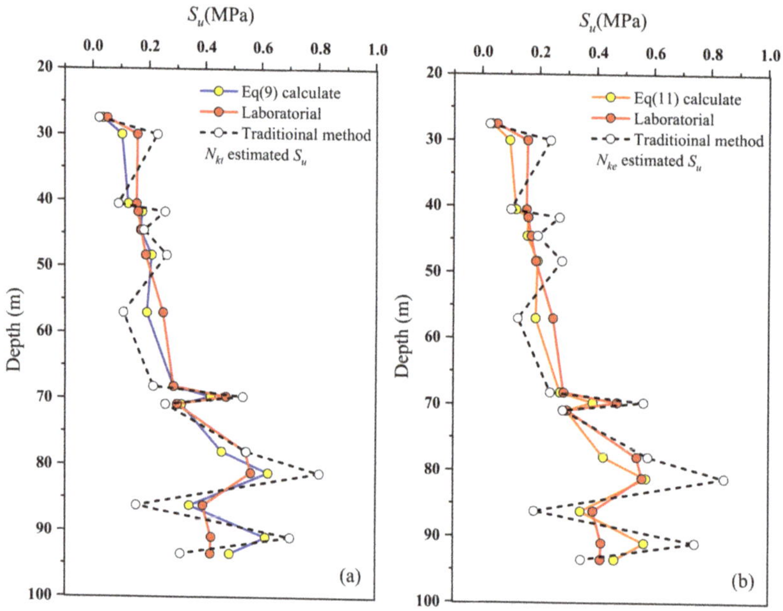

Figure 11. S_u forecasts along the depth direction ((**a**) with N_{kt}; (**b**) with N_{ke}).

To further validate the reliability of the proposed method, site B was predicted and analyzed. Soil layer data were divided into four groups with 19 data in each group. According to Equation (8), the undrained shear strength of 76 soil layers at site B within the range of the soil behavior index [2.20, 2.8] (see Figure 12) was predicted.

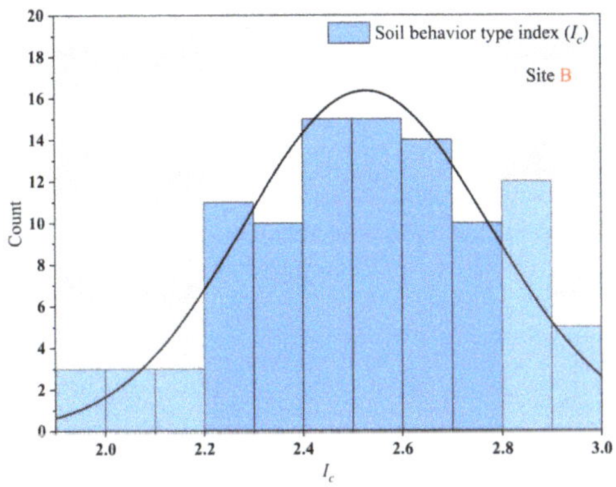

Figure 12. Histogram of I_c at site B.

Table 5 presents the evaluation results of the predicted values. It is evident that the precision of undrained shear strength prediction improves with the increase in I_c. Specifically, as I_c increases, the R^2 value rises while RMSE and MAE decrease accordingly. Figure 13 illustrates the predicted results for all soil layers, showing an R^2 value of 0.88, an RMSE of 0.0042 MPa, and an MAE of 0.0018 MPa. These results indicate that this method can effectively predict the undrained shear strength for the experimental site B. It should be noted that the data used in this study are from coastal areas near Zhejiang Province, China. Therefore, it is recommended to use Equations (8) and (10) under similar circumstances.

Table 5. Predicted accuracy of S_u corresponding to different intervals of I_c.

I_c	R^2	*RMSE* (MPa)	*MAE* (MPa)
2.20–2.37	0.79	0.0595	0.0035
2.38–2.51	0.84	0.0443	0.0019
2.51–2.62	0.92	0.0342	0.0011
2.63–2.80	0.94	0.0318	0.0010
2.20–2.80	0.88	0.0042	0.0018

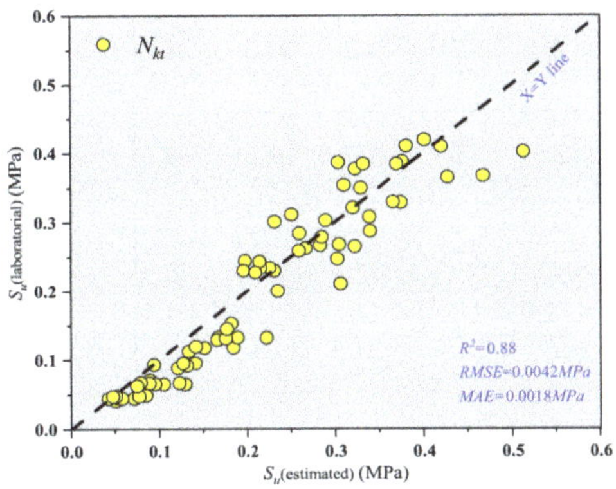

Figure 13. Analyzing the disparity between predicted and estimated values at site B.

4. Conclusions

This study aimed to improve the accuracy in predicting the undrained shear strength of fine-grained soil from CPTU. A new method for estimating the undrained shear strength of soil based on the soil behavior type index was proposed through in situ CPTU testing and an indoor consolidated quick direct shear test at one coastal area of Zhejiang Province in China. Then, a nearby site was selected to verify the accuracy and reliability of the developed method. The following conclusions were obtained:

1. The soil behavior type index I_c exhibits a negatively correlated exponential relation with the cone factors N_{kt}, N_{ke}.
2. Considering the correlation between cone factors N_{kt}, N_{ke} and the soil behavior type index I_c can enhance the prediction accuracy of undrained shear strength.
3. As I_c increases, the predicted accuracy of undrained shear strength from CPTU can be improved.
4. The functional relation between I_c and the N_{kt} can be applied to the adjacent regions of Zhejiang Province in China.

Author Contributions: Methodology, S.F.; software, S.F.; investigation, S.F.; data curation, S.F.; resources, S.F. and X.J.; formal analysis, Y.S.; writing—original draft, Y.S.; validation, Y.S. and X.J.; project administration, X.J.; conceptualization, Z.Z. and X.L.; supervision, Z.Z. and X.L.; Writing—review & editing, Z.Z. and X.L.; funding acquisition, X.L. All authors have read and agreed to the published version of the manuscript.

Funding: Xibin Li gratefully acknowledges the financial support provided by the Natural Science Foundation of Zhejiang Province under Grant No. LHZ21E090001.

Institutional Review Board Statement: Not applicable.

Informed Consent Statement: Not applicable.

Data Availability Statement: Data supporting the findings of this study, including tables, figures, and references, can be obtained from the corresponding author upon reasonable request.

Conflicts of Interest: Sai Fu and Xianlin Jia were employed by PowerChina Huadong Engineering Corporation Limited. The authors declare no conflicts of interest.

References

1. Bilgili, M.; Yasar, A.; Simsek, E. Offshore wind power development in Europe and its comparison with onshore counterpart. *Renew. Sust. Energy Rev.* **2011**, *15*, 905–915. [CrossRef]
2. Chen, J. Development of offshore wind power in China. *Renew. Sust. Energy Rev.* **2011**, *15*, 5013–5020. [CrossRef]
3. Qiao, H.; Liu, L.; He, H.; Liu, X.; Liu, X.; Peng, P. The practice and development of T-bar penetrometer tests in offshore engineering investigation: A comprehensive review. *J. Mar. Sci. Eng.* **2023**, *11*, 1160. [CrossRef]
4. Mbarak, W.K.; Cinicioglu, E.N.; Cinicioglu, O. SPT based determination of undrained shear strength: Regression models and machine learning. *Front. Struct. Civ. Eng.* **2020**, *14*, 185–198. [CrossRef]
5. Yang, Z.; Liu, X.; Guo, L.; Cui, Y.; Su, X.; Ling, X. Soil classification and site variability analysis based on CPT—A case study in the Yellow River subaquatic delta, China. *J. Mar. Sci. Eng.* **2021**, *9*, 431. [CrossRef]
6. Kang, X.; Sun, H.-M.; Luo, H.; Dai, T.; Chen, R.-P. A portable bender element-double cone penetration testing equipment for measuring stiffness and shear strength of in-situ soft soil deposits. *KSCE J. Civ. Eng.* **2020**, *24*, 3546–3560. [CrossRef]
7. Liu, X.; Shen, J.; Yang, M.; Cai, G.; Liu, S. Subsurface characterization of a construction site in Nanjing, China using ERT and CPTU methods. *Eng. Geol.* **2022**, *299*, 106563. [CrossRef]
8. Cai, G.; Liu, S.; Puppala, A.J. Evaluation of geotechnical parameters of a lagoonal clay deposit in Jiangsu Lixia River area of China by seismic piezocone tests. *KSCE J. Civ. Eng.* **2016**, *20*, 1769–1782. [CrossRef]
9. Duan, W.; Congress, S.S.C.; Cai, G.; Puppala, A.J.; Du, Y. Empirical Correlations of Soil Parameters based on Piezocone Penetration Tests (CPTU) for Hong Kong- Zhuhai-Macau Bridge (HZMB) Project. *Transp. Geotech.* **2021**, *10*, 100605. [CrossRef]
10. Drusa, M.; Gago, F.; Vlček, J. Contribution to estimating bearing capacity of pile in clayey soils. *Civ. Environ. Eng.* **2016**, *12*, 128–136. [CrossRef]
11. Salgado, R.; Lyamin, A.V.; Sloan, S.W.; Yu, H.S. Two and three dimensional bearing capacity of foundations in clay. *Géotechnique* **2004**, *54*, 297–306. [CrossRef]
12. Senneset, K.; Sandven, R.; Janbu, N. Evaluation of soil parameters from piezocone tests. *Transport. Res. Rec.* **1989**, *1235*, 24–37.
13. Teh, C.I.; Houlsby, G.T. An analytical study of the cone penetration test in clay. *Géotechnique* **1992**, *42*, 529–532. [CrossRef]
14. Cleary, M.P. Fundamental solutions for a fluid-saturated porous solid. *Int. J. Solids Struct.* **1977**, *12*, 785–806. [CrossRef]
15. Lunne, T.; Robertson, P.K.; Powell, J.J.M. *Cone-Penetration Testing in Geotechnical Practice*; Spon Press: London, UK, 2001.
16. Paniagua, P.; D'Ignazio, M.; L'Heureux, J.-S.; Lunne, T.; Karlsrud, K. CPTU correlations for Norwegian clays: An update. *AIMS Geosci.* **2019**, *5*, 82–103. [CrossRef]

17. Lunne, T.; Eide, O.; De, R.J. Correlations between cone resistance and vane shear strength in some Scandinavian soft to medium stiff clays. *Can. Geotech. J.* **1976**, *13*, 430–441. [CrossRef]
18. Aas, G.; Lacasse, S.M.; Lunne, T.A.; Høeg, K. Use of In Situ Tests for Foundation Design on Clay. In Proceedings of the ASCE Specialty Conference In Situ'86, Use of In Situ tests in Geotechnical Engineering, Blacksburg, VA, USA, 23–25 June 1986.
19. Shin, Y.J.; Kim, D. Assessment of undrained shear strength based on Cone Penetration Test (CPT) for clayey soils. *KSCE J. Civ. Eng.* **2011**, *15*, 1161–1166. [CrossRef]
20. Mayne, P.W.; Peuchen, J. CPTU bearing factor N_{kt} for undrained strength evaluation in clays. In Proceedings of the Cone Penetration Testing 2018: Proceedings of the 4th International Symposium on Cone Penetration Testing (CPT'18), Delft, The Netherlands, 21–22 June 2018.
21. *JTS/T 242-2020*; Technical Specification for Underwater Inspection Cone Penetration Engineering. China Communications Press Co., Ltd.: Beijing, China, 2020. (In Chinese)
22. Naeini, S.A.; Moayed, R.Z. Evaluation of undrained shear strength of loose silty sand soil using CPT results. *Int. J. Civ. Eng.* **2007**, *2*, 104–117.
23. Bol, E.; Önalp, A.; Özocak, A.; Sert, S. Estimation of the undrained shear strength of Adapazari fine grained soils by cone penetration test. *Eng. Geol.* **2019**, *261*, 105277. [CrossRef]
24. Bol, E. The influence of pore pressure gradients in soil classification during piezocone penetration test. *Eng. Geol.* **2013**, *157*, 69–78. [CrossRef]
25. Zhang, P.; Yin, Z.Y.; Jin, Y.F. Bayesian neural network-based uncertainty modelling: Application to soil compressibility and undrained shear strength prediction. *Can. Geotech. J.* **2022**, *59*, 546–557. [CrossRef]
26. Zhang, W.; Wu, C.; Zhong, H.; Li, Y.; Wang, L. Prediction of undrained shear strength using extreme gradient boosting and random forest based on Bayesian optimization. *Geosci. Front.* **2021**, *12*, 469–477. [CrossRef]
27. Baziar, M.H.; Saeedi Azizkandi, A.; Kashkooli, A. Prediction of pile settlement based on cone penetration test results: An ANN approach. *KSCE J. Civ. Eng.* **2015**, *19*, 98–106. [CrossRef]
28. Ma, H.; Chen, Z.; Yu, S. Correlations of soil shear strength with specific penetration resistance of CPT in Shanghai area. *Rock Soil Mech.* **2014**, *35*, 536–542. (In Chinese)
29. Li, Y.; Wang, Z.; Wang, H.; Xu, Z. Study on the undrained shear strength of Fujian Marine clay based on CPTU. *China Water Transp.* **2021**, *21*, 133–135. (In Chinese)
30. Robertson, P.K. Estimating in-situ soil permeability from CPT & CPTu. In Proceedings of the 2nd International Symposium on Cone Penetration Testing (CPT '10), Huntington Beach, CA, USA, 9–11 May 2010.
31. Fan, H.; Xu, G. Application of ROSON 200 kN Seabed Cone Penetration Test System. *Bull. Sci. Technol.* **2019**, *35*, 163–167. (In Chinese)
32. Robertson, P.K. Soil classification using the cone penetration test. *Can. Geotech. J.* **1990**, *27*, 151–158. [CrossRef]
33. Jefferies, M.G.; Davies, M.P. Use of CPTu to estimate equivalent SPT. *Geotech. Test. J.* **1993**, *16*, 458–468. [CrossRef]
34. Robertson, P.K. Interpretation of cone penetration tests a unified approach. *Can. Geotech. J* **2009**, *46*, 1337–1355. [CrossRef]
35. Du, G.; Gao, C.; Liu, S.; Guo, Q.; Luo, T. Evaluation method for the liquefaction potential using the standard penetration test value based on the CPTU soil behavior type index. *Adv. Civ. Eng.* **2019**, *2019*, 5612857. [CrossRef]
36. Wang, K.; Shen, K.; Wang, M.; Wang, H.; Guo, Z. Strength interpretation parameter of piezoncone penetration test for soft clay in offshore area of Hangzhou Bay. *Rock. Soil. Mech.* **2023**, *44* (Suppl. S1), 521–532. (In Chinese)

Journal of
Marine Science and Engineering

MDPI

Article

A Failed Rift in the Eastern Adventure Plateau (Sicilian Channel, Central Mediterranean)

Dario Civile [1,*], Giacomo Mangano [1,2], Aaron Micallef [3], Emanuele Lodolo [1] and Luca Baradello [1]

1 Istituto Nazionale di Oceanografia e di Geofisica Sperimentale-OGS, Borgo Grotta Gigante 42/C, 34010 Sgonico, Italy; giacomo.mangano@unical.it (G.M.); elodolo@ogs.it (E.L.); lbaradello@ogs.it (L.B.)
2 Department of Environmental Engineering, University of Calabria, 87036 Cosenza, Italy
3 Monterey Bay Aquarium Research Institute, Moss Landing, CA 95039, USA; amicallef@mbari.org
* Correspondence: dcivile@ogs.it

Abstract: Widespread volcanism has been known in the Sicilian Channel for a long time, even if some submarine volcanoes have only recently been discovered. Most of this volcanism formed along the NNE-trending transfer zone known as the Capo Granitola–Sciacca Fault Zone, while others, such as the islands of Pantelleria and Linosa, are associated with the continental rift zone that has developed since the early Pliocene in the central part of the Sicilian Channel through the formation of three deep tectonic troughs (Pantelleria, Linosa and Malta). However, the origin of a group of five volcanoes (here called "Tetide volcanic cluster") that form a NW-SE alignment on the eastern edge of the Adventure Plateau is not yet known. In this work, we hypothesize that this volcanic alignment may represent the remnants of a failed rift attempt that was unable to generate another tectonic trough in the Sicilian Channel. Based on seismic sections and gravimetric data, three phases in the formation of this volcanic alignment can be identified: (i) a major magmatic intrusion in the early Pliocene associated with a NW-SE normal fault that formed during the opening of the Pantelleria graben, leading to the uplift and deformation of the host sedimentary rocks; (ii) a late Pliocene-Quaternary tectono-magmatic quiescent phase; and (iii) a renewed magma intrusion through fissures or cracks that led to the formation of the volcanoes in the late Quaternary. This process was not able to cause significant extension and only limited volcanism, which is why the "Tetide volcanic cluster" is interpreted as the morphological expression of a failed rift.

Keywords: Sicilian Channel; volcanic cluster; failed rift; seismic sections; gravimetric data; tectono-magmatic evolution

1. Introduction

So-called failed rifts (also called aborted rifts) represent the initial stages of continental rifting that did not lead to the complete formation of an ocean basin. Their morphological and structural features reflect a history of attempted crustal extension and tectonic activity that ultimately failed to complete the rifting process, e.g., [1,2]. They typically exhibit elongated rift valleys containing significant sedimentary deposits, elevated blocks (horsts) and depressed blocks (grabens) as a result of normal faulting accommodating extensional strain, and generally display a topographic relief with steep fault scarps and large elevation differences between the rift floor and surrounding areas. Although not all failed rifts show magmatic activity, in many cases there is evidence of volcanic activity and in some cases large magmatic intrusions formed during the rifting process. In addition, they are often characterized by specific geophysical signatures such as gravity anomalies and seismic velocity changes that indicate differences in crustal density and composition as a result of the extensional processes, e.g., [3]. Among the most famous failed rifts are the Midcontinent Rift System in North America, the western branch of the East African Rift and the Rhine Rift in Europe.

The Sicilian Channel (Figure 1) hosts a continental rift that combines most of these morphological and tectonic features. The rifting process that led to the formation of the NW-trending tectonic grabens of Pantelleria, Linosa and Malta since the early Pliocene is considered an example of rifting generated as a passive response to a regional stress field [4,5]. It is characterized by diffuse volcanism, which led to the formation of the volcanic islands of Pantelleria and Linosa and to several submarine volcanic manifestations. Other volcanoes are located along the NNE-trending Capo Granitola–Sciacca lithospheric fault zone that crosses the central part of the rift (see inset in Figure 1). The grabens exhibit the morphological and structural characteristics of mature rifting, even if they have not developed into a complete oceanic spreading center.

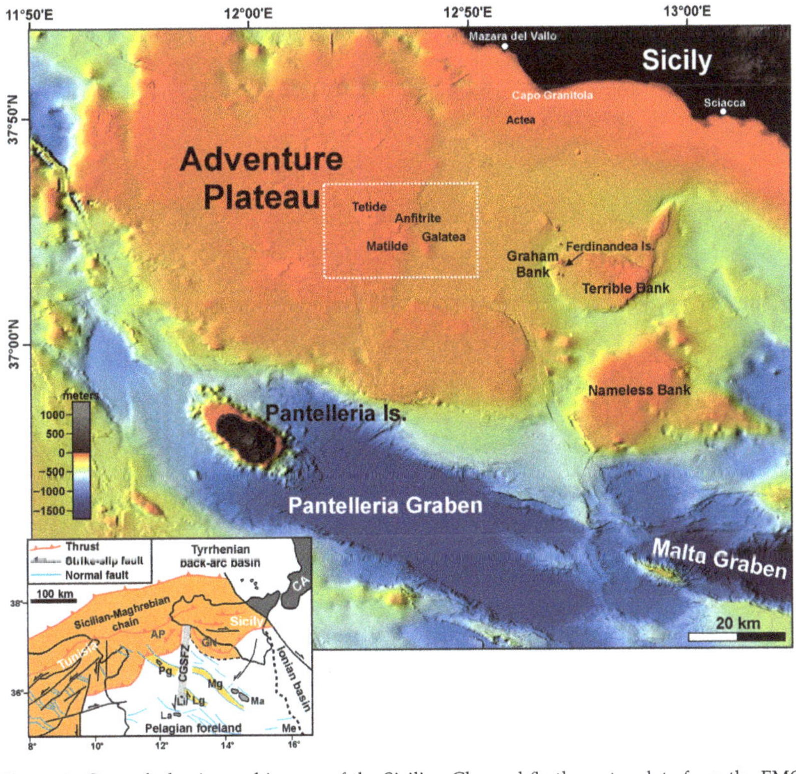

Figure 1. General physiographic map of the Sicilian Channel (bathymetry data from the EMOD-Net Bathymetry Portal, http://emodnet-bathymetry.eu/, accessed on 15 May 2024). The white rectangular box indicates the "Tetide volcanic cluster". In the lower-left corner, there is a tectonic sketch of the central Mediterranean Sea. AP = Adventure Plateau; CA = Calabrian Arc; CGSFZ = Capo Granitola–Sciacca Fault Zone; GN = Gela Nappe; Ma = Malta islands; Me = Malta escarpment; Pg = Pantelleria graben; Mg = Malta graben; Lg = Linosa graben; Li = Linosa Island; La = Lampedusa Island.

Here, we describe a group of small monogenic volcanoes (Tetide, Anfitrite, Matilde and Galatea) in the eastern sector of the Adventure Plateau [6] (Figure 1) that are aligned along the main direction of the rift grabens and whose origin is still unclear. This NNW–SSE-trending volcanic group, referred to in this work as the "Tetide volcanic cluster", is analyzed using a series of vintage multichannel seismic reflection profiles and gravimetric data, as well as recently acquired high-resolution seismic reflection profiles. It is interpreted as the possible surface expression of a failed rift zone where extension was unable to generate a tectonic depression.

2. Materials and Methods

The high-resolution seismic sections presented in this paper are both CHIRP and Boomer data (Figure 2). The CHIRP profiles were acquired from the R/V OGS Explora in November 2013 using a hull-mounted Chirp II Sub-Bottom Profiler, while the Boomer data were acquired with an AAE301 plate in October 2022 from a boat on which the system was installed.

The sweep length of the CHIRP data was 10 ms at frequencies between 2 and 7 kHz. The ping rate was 2 pings per second, the sampling rate 0.05 ms and the acquisition window 300 ms. During data acquisition, the ship speed was 4.5 knots, so the average measurement interval was ~1 m. The data were acquired in envelope mode and subsequently processed as follows: (1) derivative method to obtain analytical traces [7]; (2) spherical divergence and absorption compensation; (3) trace editing and de-spiking; (4) automatic gain control to homogenize the traces; and (5) Stolt time migration.

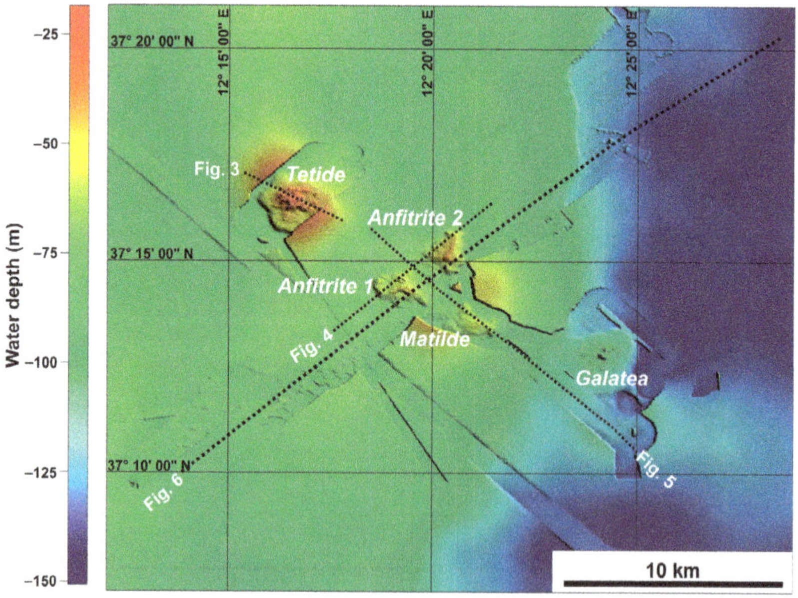

Figure 2. Bathymetric map of the area around the "Tetide volcanic cluster". Background data taken from the EMODNet Bathymetry Portal, http://emodnet-bathymetry.eu/ (accessed on 15 May 2024), integrated with multibeam high-resolution data from OGS [8] and multibeam high-resolution data acquired by the German vessel R/V Meteor [9].

The Boomer source consists of an electrodynamic transducer mounted on a catamaran frame and a pre-amplified fixed streamer equipped with 10 hydrophones connected in series. Each hydrophone can be switched on and off, extending the active range from 1 to 10 m. The distance between the source and the streamer (offset) was 20 m. The firing rate of the plate (the delivered energy unit is 300 J/pulse) was 2 shots per second, and the reflected signal was sampled at 0.05 ms with a time window of 400 ms. The seismic data were tracked using a navigation device connected to a differential GPS for accurate positioning. The data processing included the following: (1) DC (direct current) offset removal; (2) spherical divergence recovering and amplification; (3) time-varying bandpass filter; (4) predictive deconvolution to remove reverberation and increase the vertical resolution; and (5) Stolt time migration.

A network of 2D multichannel seismic reflection profiles, provided by the Italian Ministry of Economic Development in the framework of the project "Visibility of Petroleum Exploration Data in Italy "ViDEPI" (http://www.videpi.com, accessed on 15 May 2024), was used to build the contour map of the Messinian unconformity expressed in TWT (two-way traveltime) beneath the seafloor. To better constrain the position of this unconformity, information in the literature and the two offshore exploration wells closest to the study area, Onda 1 and Orione Est of the ViDEPI Project, were used [5,10,11].

During a recent geophysical–geological survey conducted between 16 July and 5 August 2023 by the German vessel R/V Meteor (M191), a significant number of multibeam and high-resolution seismic sections were acquired at several locations in the Sicilian Channel [9], particularly in the eastern sector of the Adventure Plateau (Tetide volcanic area), where a new small volcanic center (named Matilde) was discovered. The multibeam coverage realized during this cruise has complemented the map already produced in this area [8].

A complete map of the Bouguer gravity anomalies in the vicinity of the "Tetide volcanic cluster" was constructed from ship-borne gravity measurements made by the former Osservatorio Geofisico Sperimentale (OGS) during several pioneering surveys (from 1965 to 1972) in the Mediterranean Sea [12]. Details of the acquisition system can be found in Makris et al. [13], and the processing steps used to create the Bouguer anomaly map were described in detail in Lodolo et al. [14].

3. Geological Background of the Sicilian Channel and the Rift Zone

The Sicilian Channel is a broad shallow-water platform in the central Mediterranean between Sicily and Tunisia (see Figure 1). It hosts a foreland–foredeep chain system consisting of the Pelagian foreland, the Gela foredeep and the offshore part of the Neogene-Quaternary Maghrebian chain, e.g., [15,16], which was formed during the NNW–SSE-trending Neogene convergence between the African and European plates [17,18]. The central sector of the Pelagian foreland, part of the northern continental margin of the African plate, is characterized by the presence of three deep, NW–SE-trending tectonic grabens (Pantelleria, Malta and Linosa) that form the Sicilian Channel rift zone [4,19–25]. These grabens are bounded by kilometer-long NW–SE trending normal faults, which also affect the central-southern part of the Adventure Plateau [4,5,10].

The rift zone has developed since the early Pliocene and is divided into two independent sectors separated by the Capo Granitola–Sciacca Fault Zone [26], a NNE–SSW-trending lithospheric shear zone. The western sector comprises the Pantelleria graben, while the eastern one includes the Malta and Linosa grabens. The Capo Granitola–Sciacca Fault Zone extends for at least 200 km from the island of Linosa to the southern coast of Sicily, and is overall dominated by a transpressive tectonic regime [5,23,26–30].

The ~8000 km^2 Adventure Plateau is located along the north-western sector of the Sicilian Channel (Figure 1) and represents the shallowest part of the area, with depths ranging from 50 to 150 m [31]. It is characterized by an almost flat morphology that turns into a gentle slope towards the Pantelleria graben. The 6–7 km thick stratigraphic sequence of the Adventure Plateau consists of Triassic–Eocene, predominantly carbonate sequences and siliciclastic Oligocene–Quaternary sequences [10,11]. In particular, the Plio-Quaternary sequence consists of calcarenites and organogenic sands [6,31], which overlie an extensive erosional surface, the well-known Messinian unconformity [10,11]. The complex structural setting of the Adventure Plateau is the result of compressional tectonics associated with the following: (i) the late Miocene activity of ESE–WNW-trending thrusts and back-thrusts that form the outer sector of the Sicilian–Maghrebian chain, and (ii) the Pliocene to Quaternary evolution of the NW–SE-trending normal faults associated with the formation of the Pantelleria graben [4,10,27,28,32–35].

The Sicilian Channel is characterized by diffuse anorogenic volcanism that occurred mainly in the Plio-Pleistocene [6,36], although some eruptive events have been documented in historical times, such as the submarine eruption of Ferdinandea Island on Graham Bank (Figure 1) in 1831 [37,38] and the eruption ~5 km NW of Pantelleria Island in 1891 [39,40]. Holocene volcanic lavas have also been sampled at Actea Volcano (Figure 1) near the Sicilian coast of Capo Granitola [41,42]. The most voluminous volcanic edifices are found on the islands of Pantelleria and Linosa, most of which are below sea level [5,43–46]. Submarine volcanic centers have been recognized based on a combination of geophysical observations, direct sampling and interpretations of seafloor landforms in the Terrible, Nameless and Graham banks, off the coast of Capo Granitola–Sciacca, in the eastern sector of the Adventure Plateau and near the Pantelleria and Linosa grabens [5,6,36–38,41,47–49]. Most of the volcanism is probably related to the rifting processes in the central part of the Sicilian Channel and the magma upwelling along the Capo Granitola–Sciacca Fault Zone [5,26,41,48].

4. Tetide Volcanic Cluster

In this paragraph, the "Tetide volcanic cluster" is described based on information in the literature and new geophysical data. It consists of four submarine volcanoes (Tetide, Galtea, Matilde and Anfitrite, see Figure 1), which are quite close to each other and form a 16 km long NW–SE trending alignment roughly parallel to the axis of the Pantelleria graben.

Anfitrite is the only volcanic feature of the Tetide cluster that was mapped in detail [8,11] before the Meteor cruise (M191) that mapped Tetide, Galatea, and a newly discovered submarine feature named Matilde [9].

Tetide is the largest and northernmost volcano in the eastern part of the Adventure Plateau (Figure 3). It is ~9 km^2 in size and ~40 m high, and its summit is ~18 m below sea level [6]. Its morphology resembles a quasi-circular dome characterized by a flat top [6,11]. Sparker and Boomer profiles crossing the Tetide Volcano are reported in Calanchi et al. [6] and Civile et al. [11].

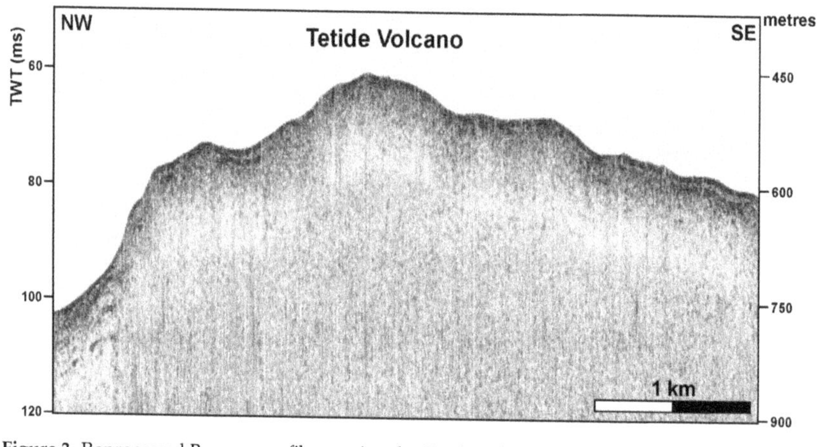

Figure 3. Reprocessed Boomer profile crossing the Tetide Volcano (modified from Civile et al. [11]). See location in Figure 2.

Anfitrite [32] consists of two different volcanic bodies, namely Anfitrite 1 and Anfitrite 2 [8] (Figure 4). Anfitrite 1 covers an area of ~1.8 km^2 and has a semicircular shape with a diameter of ~1.5 km. In its southern part, the volcanic edifice consists of a crater with a diameter of 500 m and a shallow north-eastern rim that lies 35 m below sea level. A semicircular ridge, which could be a remnant of a larger dismantled volcanic edifice, occupies the north-western part of Anfitrite 1. A recent volcanic neck has been identified along Anfitrite 1 [8]. Anfitrite 2 consists of a 1.5 km long and 1 km wide NNE–SSW trending ridge whose shallowest peak reaches a water depth of ~37 m. The deep structure of the Anfitrite volcanic complex was analyzed by Civile et al. [11] through the interpretation of multichannel seismic sections. These authors identified two magmatic outcrops generating Anfitrite 1 and 2, which are likely related to the presence of a large magmatic intrusion that fractures and deforms the sedimentary sequence into a dome-like geometry on a shallow crustal level.

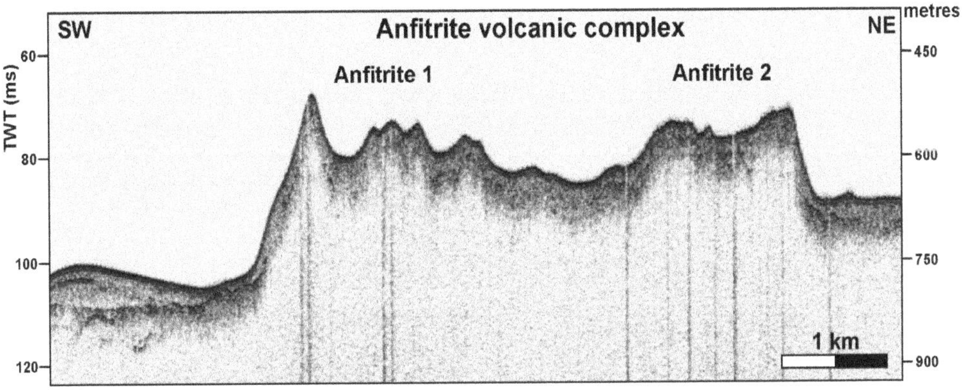

Figure 4. CHIRP profile crossing the two Anfitrite volcanoes. See location in Figure 2.

Between Anfitrite and Galatea is the newly documented Matilde volcanic edifice (Figure 5). It is 1.7 km wide and has a semicircular plan shape with an arcuate scar in its north-eastern sector (see the detailed description of this volcano in [9]). Galatea consists of an ellipsoidal, 3 km wide structure with margins at a water depth of 120 m. At the center of this structure is a 600 m long and 20 m high SW–NE-trending ridge, culminating in a well-preserved cone shape whose summit crater lies at a water depth of 74 m [6]. Sparker and sub-bottom profiles crossing the Galatea Volcano are reported in Calanchi et al. [6] and Miccallef et al. [9].

All the volcanic structures described above show a morphology indicative of long-lasting exposure conditions related to the dramatic drop in sea level during the Messinian period and especially since the Last Glacial Maximum [50].

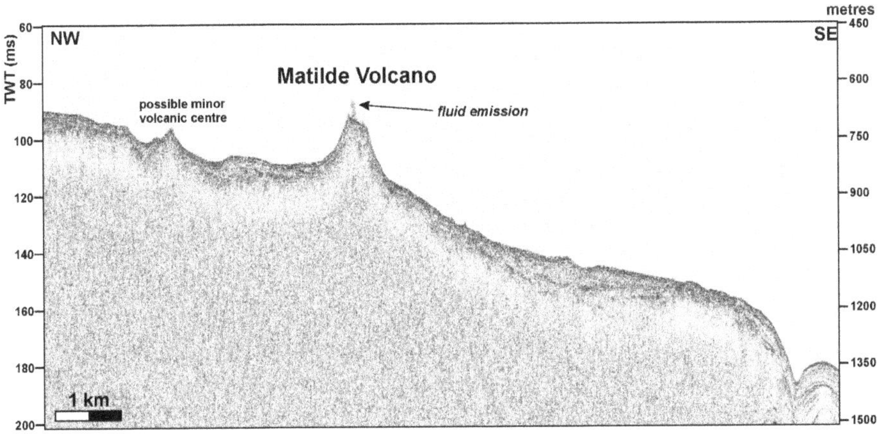

Figure 5. Boomer profile crossing the Matilde Volcano and another possible small volcanic center. A fluid emission escapes from the crater of the volcano. See location in Figure 2.

The only two volcanoes in the cluster that have been sampled to date are Tetide and Anfitrite. The magma types of these volcanic edifices include tholeiites (Tetide) and alkali basalts (Anfitrite) [6]. According to Calanchi et al. [6], these volcanoes were formed by the emplacement of mantle-derived magmas in a stress regime with an extension perpendicular to the orientation of the Pantelleria graben. Furthermore, the interpretation of the petrogenetic significance of the mineral chemistry and petrography of the volcanic rocks of the Tetide and Anfitrite volcanoes indicates the rapid ascent of the magmas from a medium-pressure reservoir [6].

5. Results and Discussion

The interpretation of nine multichannel seismic lines documents the gentle dome-like shape of the Messinian unconformity, which in turn is covered by sub-horizontal Plio-Pleistocene reflectors showing onlap terminations against the unconformity itself (Figure 6). This erosional surface is seismically expressed by an undulated, high-amplitude and continuous reflector that is generally well recognized on all seismic sections (e.g., Figure 6).

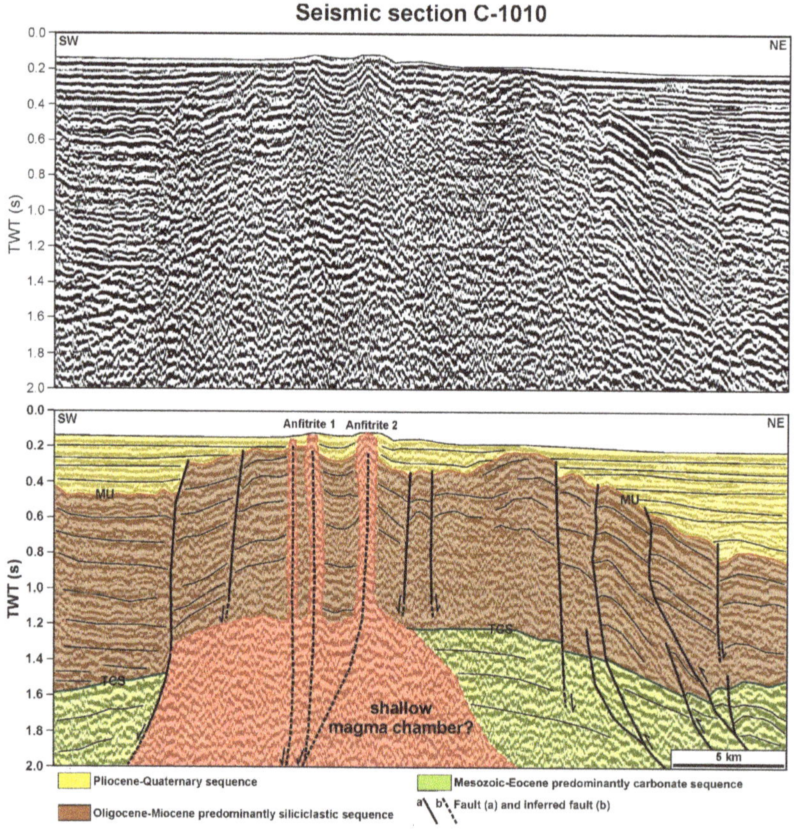

Figure 6. Multichannel seismic section C-1010 (available from the ViDEPI Portal, www.videpi.com, accessed on 15 May 2024) crossing Anfitrite 1 and 2 volcanic edifices and the corresponding line drawing (modified from Civile et al. [11]). MU = Messinian unconformity; TCS = Top of the predominantly carbonate sequence. See location in Figure 2.

We have chosen to show only seismic section C-1010 because it is the only one that crosses orthogonally the "Tetide volcanic cluster" and is of acceptable quality. It is therefore undoubtedly the most representative line for the objective of the work and has been shown both with and without interpretation.

The interpretation of the seismic section C-1010 (Figure 6) shows the presence of a doming structure interpreted as the result of a magma ascent along a NW–SE-trending normal fault characterized by some minor branches in the upper part and along which the "Tetide volcanic cluster" developed. The presence of a major normal fault with minor branches in the upper part, which represented a preferential path for the upward migration of magma, is not clearly recognizable from the seismic data.

Several normal faults with a NW–SE-trending parallel to the axis of the Pantelleria graben axis have been shown to affect part of the Adventure Plateau and are associated with the opening of this tectonic depression.

The Plio-Pleistocene sequence appears to be almost undeformed in the seismic section, indicating a period of probable tectonic inactivity during its deposition, with the exception of the area near the magma ascents where the reflectors are folded (Figure 6). The magma ascents seem to be very recent and may be associated with a reactivation of the upward propagation of the branches of the main NW–SE normal fault. The magma ascents forming the volcanic edifices of Anfitrite 1 and 2 are characterized by the pull-up effect of the reflectors, which makes it possible to identify the magmatic conduits that reach the seafloor.

The interpretation of seismic line C-1010 and in general the interpretation of seismic lines in this sector of the Sicilian Channel is based on the many years of experience gained by analyzing a large number of multichannel seismic reflection profiles acquired both by OGS and those available within the ViDEPI Project, as well as by studying the well data that allowed us to calibrate the seismic interpretation. This generally allowed us to identify two main marker horizons, the Messinian unconformity and the top of the Mesozoic-Cenozoic carbonate sequence. The Messinian unconformity is generally clearly recognizable in all seismic lines as a high-amplitude, undulating and continuous reflector. Instead, the top of the carbonate sequence is not always clearly recognizable due to the presence of highly tectonized zones.

The contour map of the Messinian unconformity (Figure 7) shows that this surface reaches values of about 0.2 s TWT (value in time from the sea floor to the top of the Messinian horizon) along the alignment of the volcanoes and then changes to values of about 0.34 s TWT to the NE and SW. The same occurs to the north and south of the volcanoes in a more pronounced form, indicating that the doming effect was caused by the magma ascent. Variations in the thickness of the Plio-Quaternary sequence from 220 m at the top of the doming structure to about 370 m at its edge can be hence observed (these values for thickness have been derived by applying an interval acoustic velocity of 2200 m/s for the whole sequence).

The Bouguer anomaly map (Figure 8) shows the presence of a WNW–ESE-trending positive anomaly (+62 mGals) located a few kilometers east of the "Tetide volcanic cluster". This is the most significant anomaly in the vicinity of the volcanic area. The maximum value of this anomaly is found in the south-eastern sector of the volcanic group, in a rather decentralized position with respect to its center. This could indicate that the magma chamber that feeds the volcanic group is limited and restricted by the tectonic structures in this sector of the edge of the Adventure Plateau. A few kilometers to the east, in the western part of the Terrible Bank, there is another positive anomaly known to host several volcanic features [48].

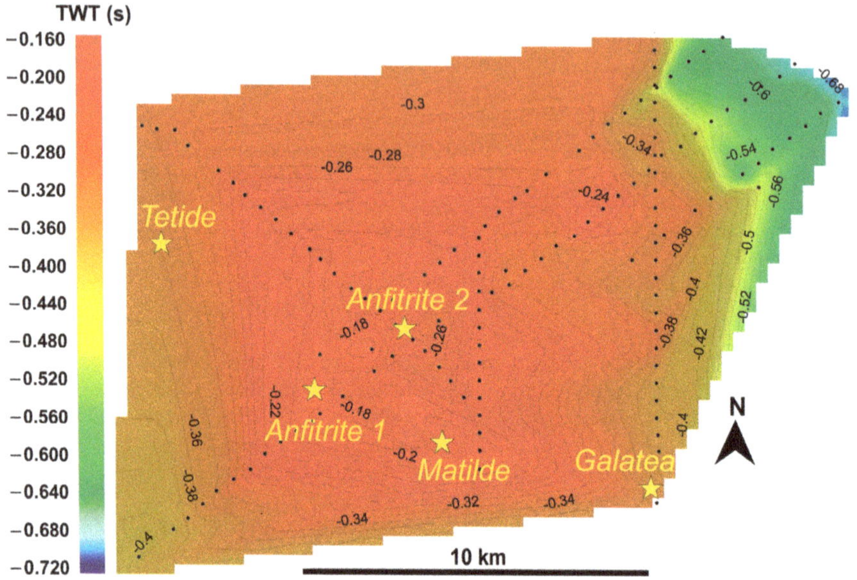

Figure 7. Contour map of the top of the Messinian unconformity derived from the seismic data interpretation. Interpreted points indicated by the black small circles. Values in TWT (two-way travel time) below the seafloor. Yellow stars indicate the positions of the individual volcanoes.

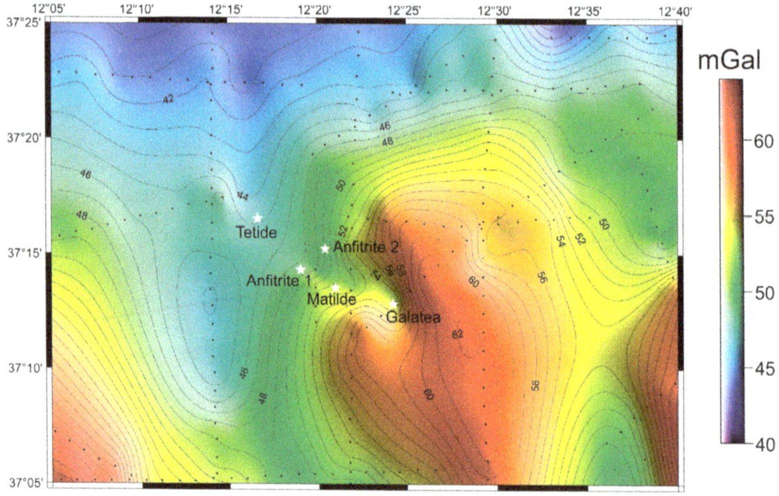

Figure 8. Complete Bouguer anomaly map of the area surrounding the "Tetide volcanic cluster". White stars indicate the positions of the individual volcanoes.

Based on the above results, a possible evolution of the "Tetide volcanic cluster" is proposed (Figure 9):

Phase 1: Swelling in the early Pliocene associated with a magma ascent that probably occurred along a presumed NW–SE-trending normal fault.
Phase 2: A phase of volcanic and tectonic stasis during which the Plio-Pleistocene sequence was deposited.

Phase 3: A late Quaternary new upward migration of magma along zones of weakness developed almost up to the surface above the doming zone as the result of a possible reactivation of the main normal fault. This made possible the formation of monogenic volcanic edifices on the seafloor.

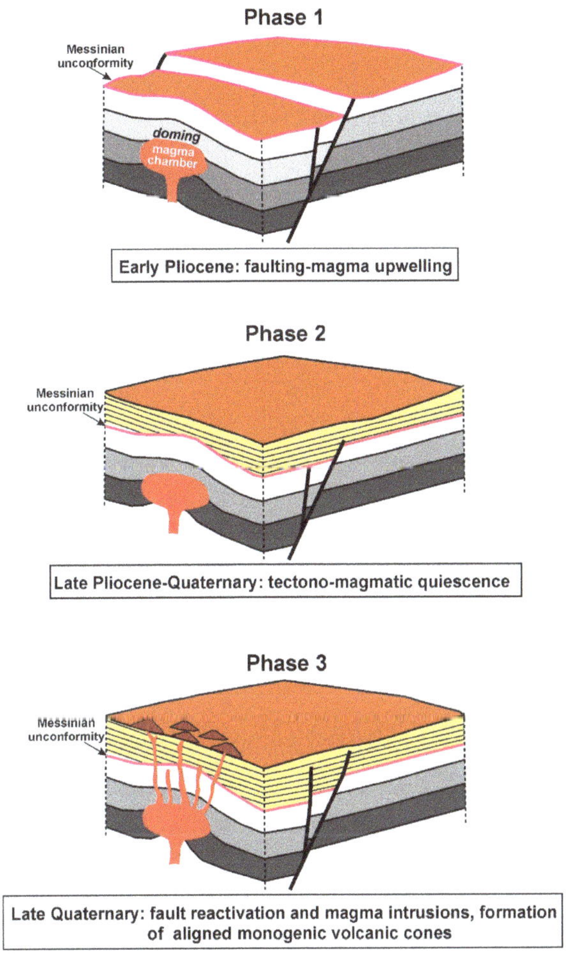

Figure 9. Simplified sketch of the evolution of the "Tetide volcanic cluster" drawn according to the data presented in this work.

On the basis of the presented data, the following basic aspects can be highlighted: (1) the alignment of the volcanic edifices of the "Tetide volcanic cluster" along a NW–SE trend similar to that of the Pantelleria graben; (2) the presence of a gravimetric anomaly, albeit weak, together with the documented dome-like shape of the Messinian unconformity and the doming structure visible in the seismic lines C-1010 allowed us to infer the possible presence of a shallow magma chamber beneath the "Tetide volcanic cluster", which is tentatively associated with the presence of a major NW–SE normal fault. These faults have been documented in this sector of the Adventure Plateau; (3) the formation of the volcanic bodies on the seafloor during the late Quaternary by the ascent of magma possibly driven by the reactivation of the NW–SE-trending normal fault. Although the data presented in this work do not allow a definitive conclusion, we have hypothesized that the "Tetide volcanic cluster" could be the product of a failed rift process. It was not able to cause significant extension in this area along with the formation of a deep trough, and only produced limited volcanism. The causes for the premature abandonment of this structure are not known from a tectonic perspective and may need to be sought in the broader context of the region's deformation style, which instead led to the development of a well-defined tectonic depression to the south (i.e., the Pantelleria graben), but which never reached the final stage of spreading with the formation of oceanic crust. For the hypothesis that the "Tetide volcanic cluster" could be considered a failed rift, we have no evidence regarding why it did not develop like the other deep tectonic depressions of the Sicilian Channel, but a hypothesis can be put forward. The eastern part of the Sicilian Channel rift zone hosts two sites of crustal extension, the Linosa and Malta grabens, while in the western part, there is only the Pantelleria graben, which is the largest tectonic depression of the rift zone. The extension was probably concentrated in this site, while elsewhere, as in the case of the "Tetide volcanic cluster", it was not sufficient to generate a trough.

An example of an incipient and prematurely abandoned rift that bears remarkable morphological and structural similarities to the "Tetide volcanic cluster" is illustrated by a linear chain of submarine volcanic edifices that parallels the present-day western sector of the Scotia–Antarctic plate boundary. They were deposited in a rift-like environment related to the activity of the former western segment of the Scotia–Antarctica plate boundary [51]. The former segment was abandoned when the Phoenix–Antarctic ridge segments NW of the South Shetland Islands ceased spreading by about 3.3 Ma, and it jumped to the present-day location of the Bransfield Strait, an actively expanding marginal basin at the transition from mature rifting to incipient, punctuated spreading.

6. Conclusions

The presence of a series of volcanic edifices in the eastern sector of the Adventure Plateau, aligned parallel with the trend of the Pantelleria graben, suggests that they may represent the morphological expression of a failed rift, which was therefore unable to generate significant extension as well as a well-developed tectonic depression. The analysis of a series of multichannel seismic sections has shown a significant bulging of the Messinian unconformity and the underlying sequences, related to the formation of a shallow magma chamber, and then a series of magmatic intrusions that reached the seafloor forming the present-day volcanoes. The presence of a shallow magma chamber can also be inferred from a significant positive gravimetric anomaly in the complete Bouguer map. The data allowed us to reconstruct the evolutionary processes of this group of volcanoes, here referred to as the "Tetide volcanic cluster" (including Tetide, Anfitrite 1 and 2, Galatea and the recently discovered Matilde Volcano). The phase of inflation, which occurred in the early Pliocene, was probably associated with the presence of a main NW–SE-trending normal fault developed during the opening of the Pantelleria graben. This was followed by a quiescent phase during which the Plio-Quaternary sedimentary sequence was deposited. Then, following a new magmatic input, the lavas pushed upward through sub-vertical structural discontinuities related to a possible reactivation of the main NW–SE normal fault, generating the volcanic edifices of the cluster on the seafloor.

There is no evidence as to why this sector did not develop like the other tectonic depressions of the Sicilian Channel. One possible hypothesis is that the extension in the western part of the Sicilian Channel was concentrated on the largest tectonic depression of the rift zone, the Pantelleria graben, while elsewhere, as in the case of the "Tetide volcanic cluster", it was not sufficient to generate a trough, and only generated localized volcanism. Instead, two sites of crustal extension, the Linosa and Malta grabens, developed in the eastern part of the Sicilian Channel rift zone.

A quite similar example of an early-stage failed rift is the western sector of the Scotia–Antarctica plate boundary, where a series of submarine volcanic edifices that are consistent with the trend of the Bransfield Basin further north, where rifting has evolved into a punctiform spreading, have been identified.

Author Contributions: Conceptualization, D.C. and E.L.; Methodology, E.L., D.C. and G.M.; Software, L.B. and G.M.; Validation, E.L., D.C. and A.M.; Formal analysis, D.C, G.M. and A.M.; Investigation, E.L., A.M. and L.B.; Writing—original draft preparation, D.C. and E.L.; Writing—review and editing, D.C., E.L., G.M., A.M. and L.B.; Visualization, D.C, G.M. and L.B.; Data curation, D.C., E.L. and L.B. All authors have read and agreed to the published version of the manuscript.

Funding: This research received no external funding.

Institutional Review Board Statement: Not applicable.

Informed Consent Statement: Not applicable.

Data Availability Statement: The multichannel seismic sections used in this study are available from the portal http://www.videpi.com (accessed on 15 May 2024). The H-R seismic sections presented here are available from the first author. Bathymetric data may be downloaded from the portal https://emodnet.ec.europa.eu/en/bathymetry.

Conflicts of Interest: The authors declare no conflicts of interest.

References

1. Burke, K. Intracontinental rifts and aulacogens. In *Continental Tectonics*; National Academy of Sciences: Washington, DC, USA, 1980; pp. 42–49.
2. Elling, R.; Stein, S.; Stein, C.A.; Gefeke, K. Three major failed rifts in Central North America: Similarities and differences. *GSA Today* **2022**, *32*, 4–11. [CrossRef]
3. Stein, S.; Stein, C.A.; Elling, R.; Kley, J.; Keller, G.R.; Wysession, M.; Moucha, R. Insights from North America's failed Midcontinent Rift into the evolution of continental rifts and passive continental margins. *Tectonophysics* **2018**, *744*, 403–421. [CrossRef]
4. Civile, D.; Lodolo, E.; Accettella, D.; Geletti, R.; Ben-Avraham, Z.; Deponte, M.; Facchin, L.; Ramella, R.; Romeo, R. The Pantelleria graben (Sicily Channel, Central Mediterranean): An example of intraplate 'passive' rift. *Tectonophysics* **2010**, *490*, 173–183. [CrossRef]
5. Civile, D.; Brancolini, G.; Lodolo, E.; Forlin, E.; Accaino, F.; Zecchin, M.; Brancatelli, G. Morphostructural setting and tectonic evolution of the central part of the Sicilian Channel (central Mediterranean). *Lithosphere* **2021**, *2021*, 7866771. [CrossRef]
6. Calanchi, N.; Colantoni, P.; Rossi, P.L.; Saitta, M.; Serri, G. The Strait of Sicily continental rift systems: Physiography and petrochemistry of the submarine volcanic centres. *Mar. Geol.* **1989**, *87*, 55–83. [CrossRef]
7. Baradello, L.; Battaglia, F.; Vesnaver, A. Fast method to transform chirp envelope data into pseudo-seismic data. *Mar. Geophys. Res.* **2021**, *42*, 14. [CrossRef]
8. Civile, D.; Lodolo, E.; Zecchin, M.; Ben-Avraham, Z.; Baradello, L.; Accettella, D.; Cova, A.; Caffau, M. The lost Adventure Archipelago (Sicilian Channel, Mediterranean Sea): Morpho-bathymetry and Late Quaternary palaeogeographic evolution. *Glob. Planet. Chang.* **2015**, *125*, 36–47. [CrossRef]
9. Micallef, A.; Geldmacher, J.; Watt, S.F.L.; Ferrante, G.M.; Ford, J.; Lodolo, E.; Civile, D.; Hodgetts, A.G.E.; Felgendreher, M.; Licari, J.G.; et al. Submarine volcanism in the Sicilian Channel revisited. *Mar. Geol.* **2024**, *474*, 107342. [CrossRef]
10. Civile, D.; Lodolo, E.; Alp, H.; Ben-Avraham, Z.; Cova, A.; Baradello, L.; Accettella, D.; Burca, M.; Centonze, J. Seismic stratigraphy and structural setting of the Adventure Plateau (Sicily Channel). *Mar. Geophys. Res.* **2014**, *35*, 37–53. [CrossRef]
11. Civile, D.; Lodolo, E.; Caffau, M.; Baradello, L.; Ben-Avraham, Z. Anatomy of a submerged archipelago in the Sicilian Channel (central Mediterranean Sea). *Geol. Mag.* **2016**, *153*, 160–178. [CrossRef]
12. Morelli, C.; Gantar, C.; Pisani, M. Bathymetry, gravity and magnetism in the Strait of Sicily and in the Ionian Sea. *Boll. Geofis. Teor. Appl.* **1975**, *17*, 39–58.
13. Makris, J.; Morelli, C.; Zanolla, C. The Bouguer gravity map of the Mediterranean Sea (IGCM-G). *Boll. Geofis. Teor. Appl.* **1998**, *39*, 79–98.
14. Lodolo, E.; Zampa, L.; Civile, D. The Graham and Terrible volcanic province (NW Sicilian Channel): Gravimetric constraints for the magmatic manifestations. *Bull. Volcanol.* **2019**, *81*, 17. [CrossRef]
15. Catalano, R.; Di Stefano, P.; Sulli, A.; Vitale, F.P. Paleogeography and structure of the central Mediterranean: Sicily and its offshore area. *Tectonophysics* **1996**, *260*, 291–323. [CrossRef]
16. Catalano, R.; Merlini, S.; Sulli, A. The structure of western Sicily, Central Mediterranean. *Pet. Geosci.* **2002**, *8*, 7–18. [CrossRef]
17. Gueguen, E.; Doglioni, C.; Fernandez, M. On the post-25 Ma geodynamic evolution of the western Mediterranean. *Tectonophysics* **1998**, *298*, 259–269. [CrossRef]
18. Carminati, E.; Lustrino, M.; Doglioni, C. Geodynamic evolution of the central and western Mediterranean: Tectonics vs. igneous petrology constraints. *Tectonophysics* **2012**, *579*, 173–192. [CrossRef]
19. Finetti, I.R. Geophysical study of the Sicily Channel Rift Zone. *Boll. Geof. Teor. Appl.* **1984**, *26*, 3–28.
20. Reuther, C.D.; Eisbacher, G.H. Pantelleria rift-crustal extension in a convergent intraplate setting. *Geol. Rundsch.* **1985**, *74*, 585–597. [CrossRef]
21. Boccaletti, M.; Cello, G.; Tortorici, L. Transtensional tectonics in the Sicily Channel. *J. Struct. Geol.* **1987**, *9*, 869–876. [CrossRef]

22. Cello, G. Structure and deformation processes in the Strait of Sicily "rift zone". *Tectonophysics* **1987**, *141*, 237–247. [CrossRef]
23. Argnani, A. The Strait of Sicily rift zone: Foreland deformation related to the evolution of a back-arc basin. *J. Geodyn.* **1990**, *12*, 311–331. [CrossRef]
24. Dart, C.J.; Bosence, W.J.; McClay, K.R. Stratigraphy and structure of the Maltese graben system. *J. Geol. Soc.* **1993**, *150*, 1153–1166. [CrossRef]
25. Maiorana, M.; Artoni, A.; Le Breton, E.; Sulli, A.; Chizzini, N.; Torelli, L. Is the Sicily Channel a simple rifting zone? New evidence from seismic analysis with geodynamic implications. *Tectonophysics* **2023**, *864*, 230019. [CrossRef]
26. Civile, D.; Lodolo, E.; Accaino, F.; Geletti, R.; Schiattarella, M.; Giustiniani, M.; Fedorik, J.; Zecchin, M.; Zampa, L. Capo Granitola-Sciacca Fault Zone (Sicilian Channel, Central Mediterranean): Structure vs magmatism. *Mar. Pet. Geol.* **2018**, *96*, 627–644. [CrossRef]
27. Argnani, A.; Cornini, S.; Torelli, L.; Zitellini, N. Neogene-Quaternary foredeep system in the Strait of Sicily. *Mem. Soc. Geol. Ital.* **1986**, *36*, 123–130.
28. Antonelli, M.; Franciosi, R.; Pezzi, G.; Querci, A.; Ronco, G.P.; Vezzani, F. Paleogeographic evolution and structural setting of the northern side of the Sicily Channel. *Mem. Soc. Geol. Ital.* **1988**, *41*, 141–157.
29. Calò, M.; Parisi, L. Evidences of a lithospheric fault zone in the Sicily Channel continental rift (southern Italy) from instrumental seismicity data. *Geophys. J. Int.* **2014**, *199*, 219–225. [CrossRef]
30. Ferranti, L.; Pepe, F.; Barreca, G.; Meccariello, M.; Monaco, C. Multi-temporal tectonic evolution of Capo Granitola and Sciacca foreland transcurrent faults (Sicily Channel). *Tectonophysics* **2019**, *765*, 187–204. [CrossRef]
31. Colantoni, P.; Cremona, G.; Ligi, M.; Borsetti, A.M.; Cati, F. The Adventure Bank (off south-western Sicily): A present-day example of carbonate shelf sedimentation. *Giorn. Geol.* **1985**, *47*, 165–180.
32. Argnani, A. Neogene tectonics of the Strait of Sicily. In *Geological Development of the Sicilian-Tunisian Platform, Proceedings of International Scientific Meeting*; Unesco Report in Marine Science; Max, M.D., Colantoni, P., Eds.; University of Urbino: Urbino, Italy, 1993; Volume 58, pp. 55–60.
33. Argnani, A. Neogene basins in the Strait of Sicily (Central Mediterranean): Tectonic settings and geodynamic implications. In *Recent Evolution and Seismicity of the Mediterranean Region*; Boschi, E., Mantovani, E., Morelli, A., Eds.; Kluwer Academic Publication: Dordrecht, The Netherlands, 1993; pp. 173–187.
34. Lentini, F.; Carbone, S.; Catalano, S. Main structural domains of the central Mediterranean region and their tectonic evolution. *Boll. Geofis. Teor. Appl.* **1994**, *36*, 103–125.
35. Grasso, M. The Apenninic-Maghrebian orogen in southern Italy, Sicily and adjacent areas. In *Anatomy of an Orogen: The Apennines and Adjacent Mediterranean Basins*; Vai, G.B., Martini, I.P., Eds.; Kluwer Academic Publishers: Norwell, MA, USA, 2001; pp. 255–286.
36. Rotolo, S.G.; Castorina, F.; Cellula, D.; Pompilio, M. Petrology and geochemistry of submarine volcanism in the Sicily Channel. *J. Geol.* **2006**, *114*, 355–365. [CrossRef]
37. Colantoni, P.; Del Monte, M.; Gallignani, P.; Zarudzky, E.F.K. Il Banco Graham: Un vulcano recente nel Canale di Sicilia. *Giorn. Geol.* **1975**, *40*, 141–162.
38. Carapezza, M.; Ferla, P.; Nuccio, P.M.; Valenza, M. Caratteri metrologici e geochimica delle vulcaniti dell'Isola Ferdinandea. *Rend. Ital. Soc. Mineral. Pet.* **1979**, *35*, 377–388.
39. Washington, H.S. The submarine eruption of 1831 and 1891 near Pantelleria. *Am. J. Sci.* **1909**, *27*, 131–150. [CrossRef]
40. Conte, A.M.; Martorelli, E.; Calarco, M.; Sposato, A.; Perinelli, C.; Coltelli, M.; Chiocci, F.L. The 1891 submarine eruption offshore Pantelleria Island (Sicily Channel, Italy): Identification of the vent and characterization of products and eruptive style. *Geochem. Geophys. Geosyst.* **2014**, *15*, 2555–2574. [CrossRef]
41. Lodolo, E.; Civile, D.; Zecchin, M.; Zampa, L.S.; Accaino, F. A series of volcanic edifices discovered a few kilometers off the coast of SW Sicily. *Mar. Geol.* **2019**, *416*, 105999. [CrossRef]
42. Lodolo, E.; Renzulli, A.; Cerrano, C.; Calcinai, B.; Civile, D.; Quarta, G.; Calcagnile, L. Unraveling past submarine eruptions by dating lapilli tuff-encrusting coralligenous (Actea Volcano, NW Sicilian Channel). *Front. Earth Sci.* **2021**, *9*, 664591. [CrossRef]
43. Rossi, P.I.; Tranne, C.A.; Calanchi, N.; Lanti, E. Geology, stratigraphy and vulcanological evolution of the island of Linosa (Sicily Channel). *Acta Vulcan.* **1996**, *8*, 73–90.
44. Civetta, L.; D'Antonio, M.; Orsi, G.; Tilton, G.R. The Geochemistry of Volcanic Rocks from Pantelleria Island, Sicily Channel: Petrogenesis and Characteristics of the Mantle Source Region. *J. Petrol.* **1998**, *39*, 1453–1491. [CrossRef]
45. Romagnoli, C.; Belvisi, V.; Innangi, S.; Di Martino, G.; Tonielli, R. New insights on the evolution of the Linosa Volcano (Sicily Channel) from the study of its submarine portions. *Mar. Geol.* **2020**, *419*, 106060. [CrossRef]
46. White, J.C.; Neave, D.A.; Rotolo, S.G.; Parker, D.F. Geochemical constraints on basalt petrogenesis in the Strait of Sicily Rift Zone (Italy): Insights into the importance of short length scale mantle heterogeneity. *Chem. Geol.* **2020**, *545*, 119650. [CrossRef]
47. Beccaluva, L.; Colantoni, P.; Di Girolamo, P.; Savelli, C. Upper-Miocene submarine volcanism in the Strait of Sicily (Banco senza Nome). *Bull. Volcan.* **1981**, *44*, 573–581. [CrossRef]
48. Coltelli, M.; Cavallaro, D.; D'Anna, G.; D'Alessandro, A.; Grassa, F.; Mangano, G.; Patane, D.; Gresta, S. Exploring the submarine Graham Bank in the Sicily Channel. *Ann. Geophys.* **2016**, *59*, S0208. [CrossRef]
49. Cavallaro, D.; Coltelli, M. The Graham volcanic field offshore southwestern Sicily (Italy) revealed by high-resolution seafloor mapping and ROV images. *Front. Earth Sci.* **2019**, *7*, 311. [CrossRef]

50. Lodolo, E.; Loreto, M.F.; Melini, D.; Spada, G.; Civile, D. Palaeo-shoreline configuration of the Adventure Plateau (Sicilian Channel) at the Last Glacial Maximum. *Geosciences* **2022**, *12*, 125. [CrossRef]
51. Lodolo, E.; Pérez, L.F. An abandoned rift in the southwestern part of the South Scotia Ridge (Antarctica): Implications for the genesis of the Bransfield Strait. *Tectonics* **2015**, *34*, 2451–2464. [CrossRef]

Journal of
Marine Science and Engineering

MDPI

Article

Regional Difference in Distribution Pattern and Morphological Characteristics of Embayed Sandy Beaches in Zhejiang Province, Eastern China

Junli Guo [1,2], Lianqiang Shi [1,2,*], Min Zhang [1], Zhaohui Gong [1], Wei Chen [3] and Xiaoming Xia [1,2,*]

1 Second Institute of Oceanography, Ministry of Natural Resources of China, Hangzhou 310012, China; guojl@sio.org.cn (J.G.); cheungminn@163.com (M.Z.); 13433279480@163.com (Z.G.)
2 Key Laboratory of Ocean Space Resource Management Technology, Ministry of Natural Resources of China, Hangzhou 310012, China
3 Institute of Coastal System-Analysis and Modeling, Helmholtz-Zentrum Hereon, 21502 Geesthacht, Germany; wei.chen@hereon.de
* Correspondence: lqshi@sio.org.cn (L.S.); xiaxm@sio.org.cn (X.X.)

Abstract: The distribution pattern and the morphology of sandy beaches have been extensively studied, while those in turbid coastal environments near large river estuaries are still unclear. This study analyzes the distribution pattern, morphological characteristics, and influencing factors of Zhejiang sandy beaches using statistical analysis, based on field data and historical records. Results show that the mean grain size distribution of Zhejiang sandy beaches ranges from fine sand to very coarse sand, and the beach slope and sediment grain size correspond well with the wave heights in the three regions of Zhejiang. The extent of beach headlands in central Zhejiang appeared the largest, suggesting an increased susceptibility to wave erosion due to the less sheltered headlands. Most sandy beaches in Zhejiang formed on the islands and the areas far from the estuaries, showing quantity difference in beach distribution. The comparison of the regional difference in Zhejiang sandy beaches shows that embayment is the main factor affecting the beach distribution pattern and morphological characteristics. The different embayment characteristics provide the space for beach formation and the interaction with the coastal process, the sediment supply, the nearshore hydrodynamic environment, and human intervention also have influence on the morphological characteristics of Zhejiang beaches.

Keywords: embayed sandy beaches; beach distribution; morphological characteristics; regional difference; influencing factors

1. Introduction

Sandy coasts constitute over one-third of the world's non-frozen coasts [1] and possess significant socio-economic and ecological value. These coasts provide habitats for coastal flora and fauna, and sandy beaches are popular tourist destinations, generating substantial benefits for coastal areas [2]. Furthermore, sandy coasts serve as crucial buffers during extreme events, such as storms and coastal flooding, providing a natural protective barrier for coastal residents. However, sandy beaches have experienced erosion in recent years, with 15% of them retreating landward at a rate of 1 m/a or even faster over the past decades [1]. To address this issue, numerous beach ecological protection and restoration projects have been implemented worldwide since the early 20th century [3–5], making it crucial to understand the formation conditions and morphological characteristics of sandy beaches.

The distribution pattern and morphological characteristics of sandy beaches have been extensively studied using coastal morphodynamic methods and statistical analysis techniques [6–9]. Wave action is the primary coastal dynamic factor [10,11], with additional influences from tides, currents, and winds [12–14]. Apart from the aforementioned coastal forces, headland (embayment) characteristics, sediment supply, and sea level changes can also play a role in beach formation and evolution [6,15,16]. Among the influencing factors, the geological setting exerts significant control over beach sediments and morphology [17–20], shaping the distribution pattern of beach landforms. With the increasing demand for beach tourism, human activities such as the construction of seawalls, artificial headlands, artificial sandbars, and breakwaters, and beach nourishment have been undertaken to prevent storm erosion and maintain recreational functions [21,22]. There interventions can also change the nearshore hydrodynamics, and even cause long-term disturbance [23,24], thus influencing beach evolution.

Embayed beaches constitute more than 50% of sandy beaches [25] and exhibit distinct behavior compared to beaches without headland control [7,18,19,26]. This distinction is crucial for beach management and conservation. Assessing and categorizing headland morphology has been proven effective in organizing embayment. The indentation of embayment has impacts on sediment transport and influences morphological patterns [27–29]. Previous studies have focused on the classification of embayed beaches [28,30–32], while the distribution pattern and morphological characteristics of embayed sandy beaches in different regions require further understanding.

Most of the sandy beaches in China are embayed beaches, and they are mainly distributed in the south and in the Shandong Peninsula, while they are rare in eastern China [15,33]. Zhejiang, with a total coastline of 6910 km—the longest among China's coastal provincial administrative regions—only has 122 km of gravel–sandy coastline and limited embayed sandy beach resources [34]. Tides, waves, and coastal currents promote the southward transport and diffusion of Yangtze River sediments [35–37], resulting in the turbid coastal environment of Zhejiang province with sufficient fine-grained sediment supply [38,39]. Current studies on this sandy beach-poor area are focused on the morphodynamic processes on individual beaches [40,41], while the basic characteristics and regional differences in sandy beaches formed on this coast with sufficient fine sediment supply remain unclear.

Therefore, this study focuses on the Zhejiang sandy beaches in the turbid coastal environment near the Yangtze River estuary, and the aims of this study are (1) to systematically analyze the distribution pattern and morphological characteristics, (2) to find out the regional difference in the sandy beaches in Zhejiang, and (3) to further explore the influencing factors of the distribution pattern and morphological characteristics of sandy beaches in a similar area.

2. Study Area

2.1. Geographical Location and Geological Setting

Zhejiang province is situated on the eastern coast of China (Figure 1a), in a subtropical monsoon climate zone, characterized by seasonal changes in prevailing wind direction and precipitation, and frequently affected by typhoon events (approximately six typhoons per year [42]).

The total length of the Zhejiang coastline is 6910 km, including 2410 km of mainland coastline and 4500 km of island coastline [34]. The Zhejiang coastal zone is in the South China Uplifted Belt. A series of NE-SW and NW-SE trending fault structures formed in the Cenozoic era intersect with the South China coastline obliquely, and mountains or hills rise to the sea, thus forming a series of meandering coasts, which is the basis for the formation of the embayed coast. The northern coast of Zhejiang province is located in the southern margin of the Yangtze River delta plain, while the other coast is located in the volcanic hills of eastern and southern Zhejiang province [33]. Its coastal regions can be broadly divided into northern Zhejiang (NZ), central Zhejiang (CZ), and southern Zhejiang (SZ) from north to south. The NZ coastal region is located at the southern end of the Yangtze River delta plain, including Zhoushan city, which has a wide range of islands and intersecting waterways. The CZ coastal region mainly covers Ningbo city and Taizhou city, with a small number of scattered islands. The SZ coastal region is mostly under the jurisdiction of Wenzhou city, which has a large number of relatively concentrated islands [43]. The southward transport of fine sediments from the Yangtze River has a significant impact on the extensive development of tidal flats in Zhejiang, while the sandy–gravel coastline only accounts for 4% [34]. This study focuses on 75 embayed sandy beaches along the Zhejiang coast, including 30 mainland beaches and 45 island beaches (Figure 1b and Table S1).

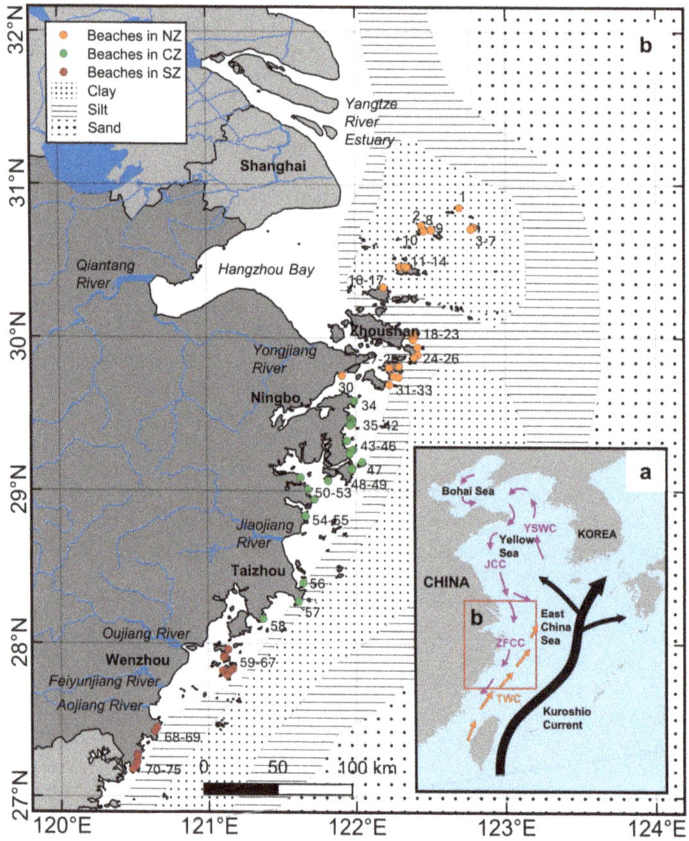

Figure 1. Geographical location of study area and the coastal current system (**a**) and the locations of the 75 accessible embayed sandy beaches in Zhejiang province (**b**). YSWC, JCC, ZFCC, and TWC in (**a**) represent the Yellow Sea Warm Current, Jiangsu Coastal Current, Zhejiang–Fujian Coastal current, and Taiwan Warm Current, respectively. Numbers in this figure show the ID of the studied beaches (refer to Table S1; the ID numbers of the beaches are arranged from the north according to latitude). The coastal current system and the surficial sediment type distribution were modified after [36].

Apart from the influence of the southward transport of Yangtze River sediment, Zhejiang rivers also have impacts on the coast. The main rivers of Zhejiang, including Qiantangjiang, Yongjiang, Jiaojiang, Oujiang, Feiyunjiang, and Aojiang, flow into the East China Sea and are all mountainous rivers with short streams, forming the basis of the headland coast of Zhejiang and providing part of the sediment source for the coast (Figure 1b) [33,44]. Among the Zhejiang rivers, Qiantangjiang and Oujiang are the largest and the second large river, with a multiyear averaged sediment discharge of 668.7×10^4 t and 225.6×10^4 t, respectively [33].

The surficial sediments in the Zhejiang sea area are mainly silt and clay (Figure 1b), resulting from the southward transport of Yangtze River sediment and the local small–moderate rivers [36]. The total sediment volume of the clinoform in the Zhejiang–Fujian mud wedge is estimated to be about 4.5×10^{11} m^3, which represents about 32% the total Yangtze-derived mud to the sea. The suspended sediment concentration (SSC) in the sea area of NZ is between 0.02 kg/m^3 and 0.57 kg/m^3 during winter. The SSC distribution in summer is generally lower than that in winter, but the distribution pattern is basically similar, characterized by a low concentration in the eastern islands far from the mainland. The SSC in the CZ sea area is generally low, and the SSC in winter is much higher than that in summer. The SSC in CZ during winter is mostly between <0.1 kg/m^3 and 0.69 kg/m^3, while the SSC is low during summer, with values mostly below 0.05 kg/m^3. The SSC in the sea area of SZ is between <0.1 kg/m^3 and 1.66 kg/m^3, with the most turbid value occurring in the Oujiang estuary. In summer, the SSC is obviously centered on the Oujiang estuary, and the SSC shows a downward trend from south to east. Compared with winter, the range of low-value areas is greatly expanded, and the range of high-value areas is correspondingly reduced. The SSC in the estuary area is greater than that in the offshore area in SZ [43].

2.2. Coastal Hydrodynamic Environment

The wave direction in the Zhejiang coastal area is generally consistent with the seasonal wind direction. During summer, the waves mainly originate from the south in the coastal regions of Zhejiang, while northeasterly waves are predominant during spring and autumn. The multi-annual average wave heights along the Zhejiang coast range from 0.3 m to 1.2 m, with the average wave periods being between 1.4 s and 5.6 s [43,45–47]. The annual averaged wave height of NZ, CZ, and SZ are 0.79 m, 0.52 m, and 0.69 m [43,45–47], with the main wave directions from S-SE during summer and N-NE during winter, respectively.

Tides also show significant regional difference in the Zhejiang coastal area. The NZ region is situated in a mixed sea area with both regular and irregular semi-diurnal tides, while CZ and SZ belong to the regular semi-diurnal tidal sea area [48]. The distribution of tidal range along the Zhejiang coast exhibits distinct regional characteristics, with the average tidal range ranging from 1.91 m to 4.48 m. The annual averaged tidal range of NZ, CZ, and SZ is 2.48 m, 3.29 m, and 4.23 m, respectively [43]. The tidal range in the nearshore and island areas gradually increases from north to south and from east to west [43,49].

The coastal current in Zhejiang is mainly the Zhejiang–Fujian Coastal Current (ZFCC) flowing southward (Figure 1a), and this current intensifies in winter, carrying the Yangtze's brackish water and sediment discharge southward along the inner shelf [36]. Offshore, there is a northward flow of warm and saline middle-deep water flow, the Taiwan Warm Current (TWC, Figure 1). In summer, under the prevailing southeast monsoon, the northward TWC intensifies and, correspondingly, the southward ZFCC weakens.

3. Materials and Methods

3.1. Collection and Processing of Beach Slope and Sediment

In this study, we conducted field measurements on 75 accessible sandy beaches in Zhejiang with lengths greater than 100 m, including 30 mainland beaches and 45 island beaches (Figure 1 and Table S1). To be specific, we established three (in the north, central, and south parts of beaches), two (in the north and south parts of beaches), and one (in the central of beaches) profiles on beaches with lengths longer than 500 m, 300 m to 500 m, and shorter than 300 m, respectively (Table S1). The profiles were measured at low tide on the measurement day from July to August 2020 using RTK GPS connected to Continuously Operating Reference Stations (CORS), with a vertical and horizontal accuracy of ±15 mm and ±8 mm, respectively. The profiles were perpendicular to the coastline, extending from the base (seawall or bedrock) to the daily low tide waterline, and a total of 147 profiles were collected (Table S1). The elevation data were then corrected to the 1985 Yellow Sea Elevation

Datum. The average slope (i) of the beach was obtained by calculating the arithmetic mean of slope using the elevation data from all profiles on each beach (Figure S1).

During the measurement of beach slope, we also collected surficial sediment along the profiles (Figure S1), and a total of 441 sediment samples were obtained (Table S1). For beaches with berms, samples were collected along the profile at the backshore, beach berm, and the beach face, respectively. For beaches without berms, the sampling stations were located at the higher, middle, and lower intertidal beach along the profile, respectively. The samples were pre-treated according to laboratory standards [50] and then analyzed by SFY-D sonic vibratory automatic sieve size analyzer (Nanjing Zhonghu Ltd., Nanjing, China). Grain size parameters such as the mean grain size (M_z), sorting coefficient (σ), skewness (S_k), and kurtosis (K_u) of the beach sediments were calculated according to the graphical method [51]. M_z can represent the change in beach sediment characteristics in previous studies on the Zhejiang coast [40] and in the statistical analysis [9], and so we only used this in the subsequent statistical analysis.

3.2. Collection and Processing of Beach Length and Dry Beach Width

In this study, we utilized the beach length (l) and averaged dry beach width (w) (Figure S1) to characterize the beach morphology. The beach length (l) was defined as the coastline length of the sandy beach in this study and determined using ArcGIS 10.8, while the dry beach width (w) was calculated as the arithmetic average of the width between the base (the landward rocky cliff foot or seawall foot) and the waterline at low tide, measured from each beach profile (Figure S1).

3.3. Collection and Processing of Embayment Characteristic Parameters

The beaches in this study are all embayed beaches, characterized by the upper and lower control headlands, a sheltered section, and a tangential section (sometimes divided into transitional sections) [52]. The embayment parameters used in this study were derived from previous research by [6,7,9,52]. These parameters include the headland length (S), the bay mouth orientation angle (γ), the spiral tangent angle (β), the maximum indentation (a), the bay mouth chord length (b), the ratio of the maximum indentation to the bay mouth chord length (a/b), and the tangential section length (L) (Figure 2).

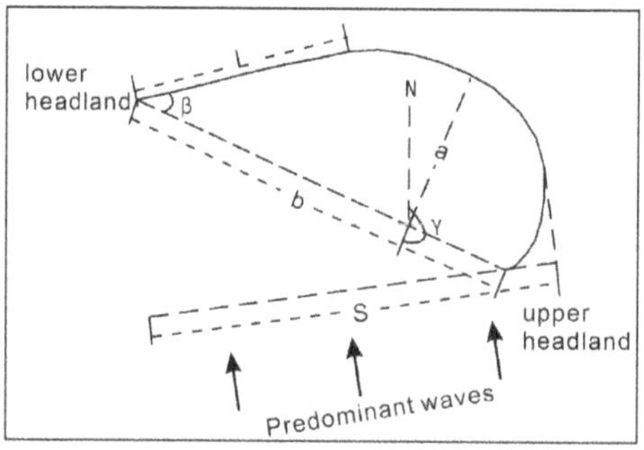

Figure 2. Embayment parameters used in this study according to [52].

The perennial dominant wave is taken as the vertical line, which is tangent to the arc curve of the shielding section, and the length between the tangent line and the upper headland is S. S can reflect the shadow size of the headland. γ is calculated from the N direction, measured clockwise, reflecting the tectonic setting of the coast. Spiral tangent angle β is the angle between the straight section of the bay and the chord of the bay. It is not only related to the degree of bay erosion retreat, but also related to the bay chord length. a is defined as the longest segment length of the intersection of the vertical line connecting the upper and lower headlands and the curved shoreline. It represents the degree of erosion and the retreat of the coastline under the action of waves. b was defined as the distance between the upper and lower headlands, reflecting the spatial scale of the bay and is related to geological structure. a/b is a ratio which can represent the relative strength of the wave force and the headland sheltering ability. L is the length of the straight shore segment of the curved bay. It indicates the beach segment that is not obscured by the headland, and the incident wave reaches the shoreline directly without diffraction, which is related to the chord length of the bay mouth [6,9,52].

Statistical analyses were performed on those embayment parameters, along with the average wave height and average tidal range obtained from [43,45,46,49].

3.4. Correlation Analysis and Principal Component Analysis

We defined the beach slope (i), mean grain size (M_z), and embayment parameters as the morphological factors and took them into the Pearson correlation analysis. To explore the factors affecting the distribution patten and morphological characteristics of embayed sandy beaches in Zhejiang, we further conducted principal component analysis (PCA) by region. Apart from the morphological factors, we also included the mean tidal range (TR) and mean wave height ($\overline{H}$) obtained from [34,43,45–47,49] in PCA, and all of the factors used in PCA are defined as morphodynamic factors in this study. The eigenvalues, variance contribution, and cumulative variance contribution of each principal component were obtained through PCA. A principal component was considered to reflect the main information represented by the morphodynamic factors when the cumulative variance contribution was greater than or equal to 70–85% [53].

4. Results

4.1. Beach Slope and Sediment Grain Size

The average slope (i) of sandy beaches in Zhejiang varied from 0.02 to 0.13 (Figure 3), with a mean value of 0.06. Among these, the inclination of the slope was found to be gentlest at Yangshashan beach and steepest at Miaogan beach. The averaged beach slope was steepest in NZ, while it was gentlest in CZ.

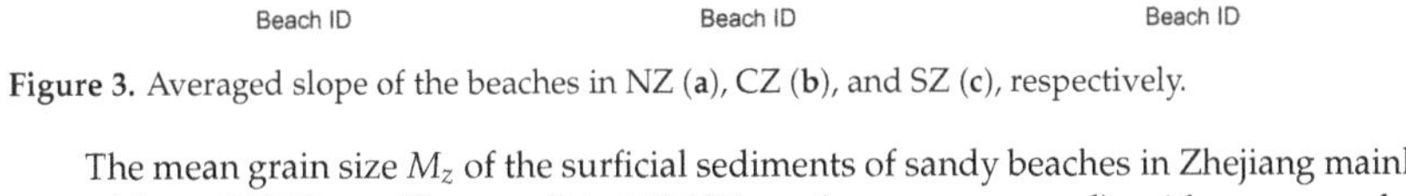

Figure 3. Averaged slope of the beaches in NZ (**a**), CZ (**b**), and SZ (**c**), respectively.

The mean grain size M_z of the surficial sediments of sandy beaches in Zhejiang mainly ranged from 128.93 μm (fine sand) to 1486.74 μm (very coarse sand), with a mean value of 309.07 μm (medium sand). The average value of mean grain size of sandy beaches in NZ was the largest, followed by beaches in SZ, and the finest sediments were observed in CZ beaches (Figure 4a–c). Specifically, in NZ, the M_z distribution ranged from 128.93 μm (fine sand) to 732.91 μm (coarse sand), with a mean value of 335.70 μm (medium sand), while in CZ, it varied between 134.69 μm (fine sand) and 609.30 μm (coarse sand), with a mean value of 240.09 μm (fine sand). In SZ, M_z ranged between 186.45 μm (fine sand) and 1486.74 μm (very coarse sand), with a mean value of 358.80 μm (medium sand).

The sediment sorting coefficients (σ) of beach surficial sediments ranged from 0.29 to 1.89, with an average of 0.81. This indicates that the sediments ranged from poorly sorted to well sorted. The best and worst sorting were observed on Qiansha beach and Huangsha beach (Figure 4d–f), respectively. The mean σ values of surficial sediments on sandy beaches in NZ, CZ, and SZ were 0.81, 0.84, and 0.78, respectively, suggesting that the sediments of SZ beaches had the best sorting.

The skewness coefficient (S_k) of the sediments ranged from −0.48 to 0.20 (Figure 4g–i), with a mean value of −0.20. Predominantly, negative skewness was observed, with extremely negative skewness, negative skewness, and near-symmetry being the main types. The averaged S_k values of surficial sediments on sandy beaches in NZ, CZ, and SZ were −0.16, −0.28, and −0.16, respectively, indicating that negative skewness is the predominant performance.

The kurtosis (K_u) of surficial sediments of studied beaches ranged from 0.78 to 2.92, with a mean value of 1.32, generally showing moderate and narrow peaks (Figure 4j–l). The averaged K_u values of surficial sediments on sandy beaches in NZ, CZ, and SZ were 1.19, 1.56, and 1.24, respectively.

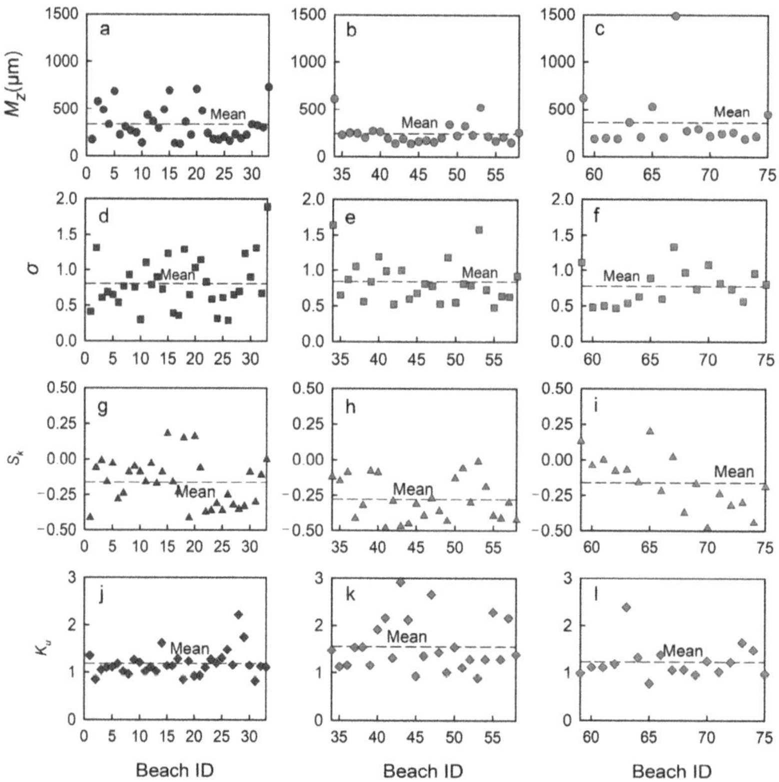

Figure 4. Grain size characteristics of sandy beaches in Zhejiang from north to south: 33 beaches in NZ (**a**,**d**,**g**,**j**), 25 beaches in CZ (**b**,**e**,**h**,**k**), and 17 beaches in SZ (**c**,**f**,**i**,**l**).

4.2. Beach Length and Dry Beach Width

It is evident from Figure 5 that the lengths of sandy beaches varied significantly, ranging from 112.61 m (Si'ao beach) to 2381.46 m (Huangcheng beach). The sandy beaches in CZ were the longest, with most of them being in the range of 200 m to 1000 m. In contrast, NZ predominantly featured beaches shorter than 1000 m, while those in SZ were mostly shorter than 500 m. The widest and narrowest dry beach widths among the studied beaches were Houcao beach (281.57 m) and Huangsha beach (21.69 m) in NZ, respectively. The average dry beach width (w) of sandy beaches in NZ ranged from 21.69 m to 334.60 m, with the majority measuring smaller than 180 m. The averaged dry beach widths of sandy beaches in CZ were relatively wider, ranging from 37.63 to 268.20 m, and were concentrated in the range from 50 m to 200 m. Meanwhile, w ranged from 32.06 m to 281.57 m in the south, with w smaller than 200 m prevailing. There was significant correlation between the beach width and the beach length, as shown in Figure 6, indicating a trend where longer beaches tended to have wider widths.

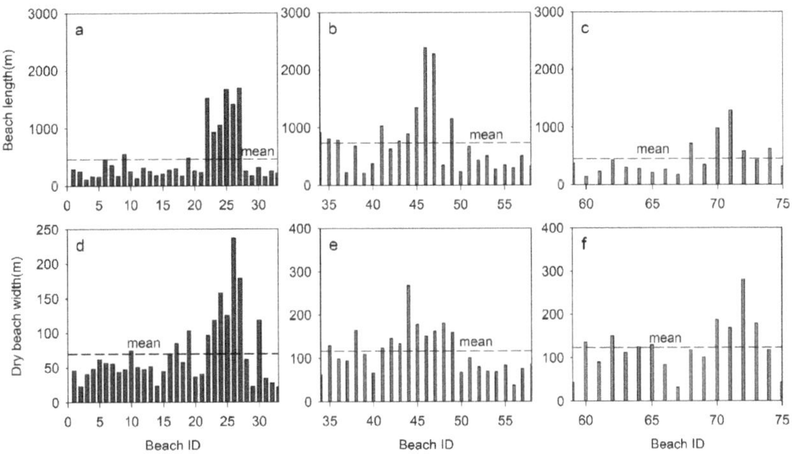

Figure 5. Beach length and averaged dry beach width of sandy beaches in Zhejiang: 33 beaches in NZ (**a**,**d**), 25 beaches in NZ (**b**,**e**), and 17 beaches in SZ (**c**,**f**).

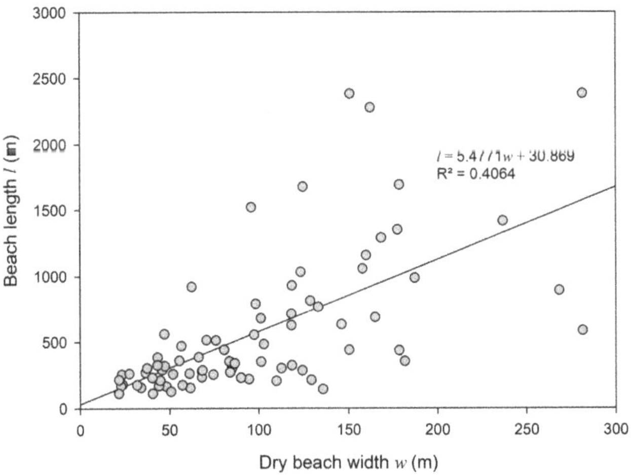

Figure 6. Correlation between dry beach width and beach length.

4.3. Embayment Characteristics

The headland length (S), bay mouth orientation (γ), and spiral tangential angle (β) are shown in Figure 7. The beach headlands in CZ were the largest among the three regions in Zhejiang, suggesting the higher susceptibility to wave erosion due to the less sheltered headlands. In CZ, the headland length (S) was the largest and exhibited the widest variation, ranging from 7.82 m to 494.65 m, indicating a broader sheltering range. Meanwhile, in NZ, the averaged S value was the second largest, ranging from 4.48 m to 611.18 m, with most beaches having an S shorter than 160 m. Conversely, the averaged S value in SZ was the smallest, with an average shorter than 40 m, ranging between 6.25 m and 117.47 m. The bay mouth orientation (γ) of sandy beaches varied across regions. In NZ, the orientations varied significantly due to the presence of numerous islands, with most facing northeast and northwest. CZ beaches also had an orientation facing northwest and northeast, with γ values mainly ranging from 0° to 90° and 270° to 360°, while in SZ, the γ values mainly ranged from 180° to 360°, with the majority facing southwest and northwest. The spiral tangential angles of sandy beaches in Zhejiang ranged from 2.92° to 168.93°, showing no significant regional difference. The distribution range of β of beaches in CZ was relatively concentrated, while those in NZ and SZ were more dispersed.

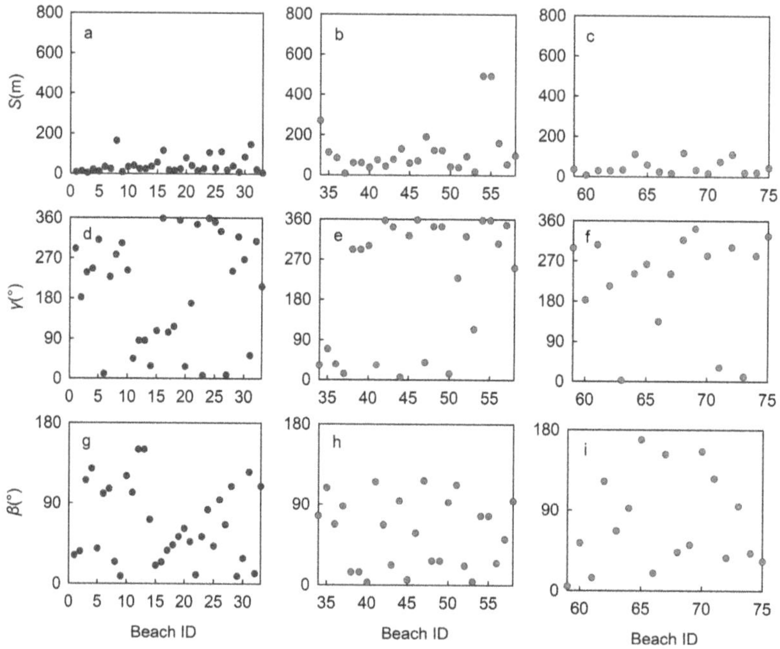

Figure 7. Headland lengths (S), bay mouth direction angles (γ), and spiral tangent angles (β) of sandy beaches in Zhejiang: 33 beaches in NZ (**a**,**d**,**g**), 25 beaches in CZ (**b**,**e**,**h**), and 17 beaches in SZ (**c**,**f**,**i**).

The maximum indentation (a) of sandy beaches in NZ ranged from 67.12 to 2743.84 m, with a mean value of 502.55 m, and most of the values were smaller than 900 m (Figure 8a). In CZ, sandy beaches had the largest mean maximum indentation (815.02 m), with values ranging from 171.65 to 1828.58 m, and the distribution of a values was more dispersed (Figure 8b). In SZ, a varied between 42.18 m and 1278.92 m, with a mean value of 440.67 m (Figure 8c), showing the smallest indentation in the three parts of Zhejiang coastal regions.

The bay mouth chord length (b) of sandy beaches in NZ varied from 123.12 m to 2839.65 m, with the most between 123.12 m and 900 m (Figure 8d). In CZ, the headland opening degree was larger, and the shading effect was weaker. Here, b was mainly concentrated between 250 and 1500 m, with a variation range of 290.85 to 3096.37 m (Figure 8e). In SZ, sandy beaches varied from 176.98 to 2175.09 m, with most b values less than 600 m (Figure 8f), indicating a relatively small headland opening degree.

The ratio of the maximum indentation to the bay mouth chord length (a/b) was mostly distributed between 0.25 and 1.00, with the range of variation from 0.22 to 1.60 in NZ, 0.49 to 1.75 in CZ, and 0.18 to 1.59 in SZ (Figure 8g–i). In CZ, the ratios were mostly greater than 0.50, reflecting larger headland indentation, stronger wave action, and a smaller shadowing effect in CZ.

The tangential section length L of sandy beaches in NZ varied from 48.18 m to 1077.91 m, with most of the L values shorter than 400 m (Figure 8j). In CZ, the tangential section was the longest, with the L values scattered between 71.76 m and 448.97 m (Figure 8k). In SZ, L varied from 66.55 m to 834.53 m, with most values distributed within 300 m (Figure 8l).

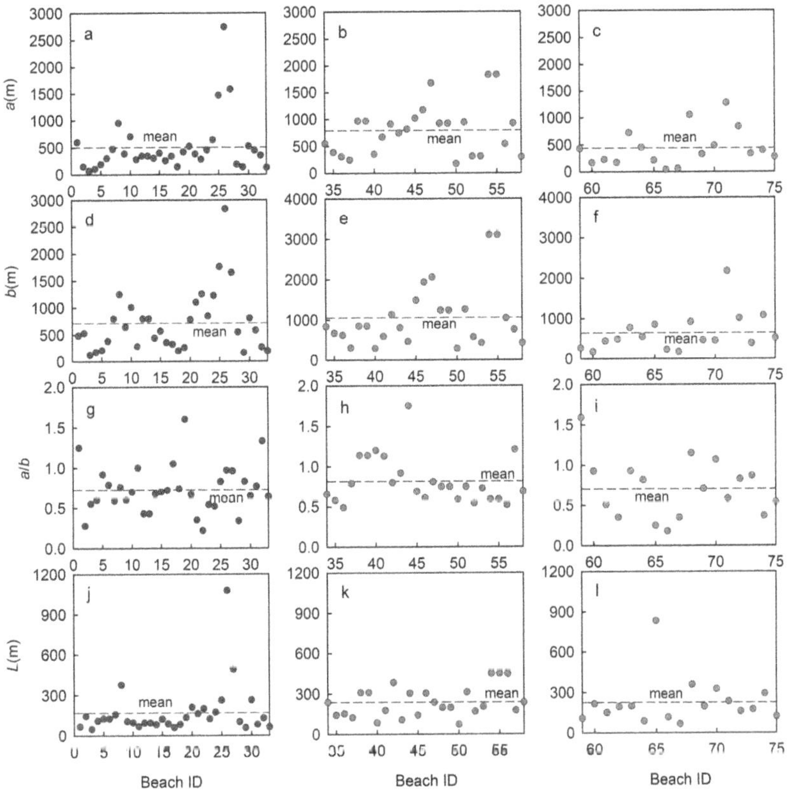

Figure 8. Maximum indentation (a), bay mouth arc length (b), ratio of a and b, and tangential section length (L) of sandy beaches in Zhejiang: 33 beaches in NZ (**a**,**d**,**g**,**j**), 25 beaches in CZ (**b**,**e**,**h**,**k**), and 17 beaches in SZ (**c**,**f**,**i**,**l**). The dash lines in each panel show the mean value.

5. Discussion

5.1. Difference in the Distribution and Morphological Characteristics of Beaches in Zhejiang

All beaches in NZ are distributed on the islands, and most of them are located on the east coast of the islands facing the predominant waves. No sandy beaches have formed along the continental coastline shaded by islands. The beaches in NZ have the coarsest sediment grain size and the steepest beach slope, which corresponds to the maximum nearshore wave height among the three regions in Zhejiang. Meanwhile, the beaches in NZ are affected by the sediments from the Yangtze River and Qiantang River, and fine sediments are abundant. The beaches in NZ can be summarized as sandy beaches affected by rivers and only developed on islands.

There are relatively few islands in the CZ, and a few beaches are distributed on the mainland coast without island shelter in the seaward. The beach headland in CZ is the longest, and the average values of beach lengths and widths in the region are also the largest among the three regions along the Zhejiang coast. Sediment grain size is the finest and beach slope is the gentlest among the three regions, corresponding to the minimum wave height in the CZ sea area.

The beaches in SZ are the least distributed, mainly formed on the islands far away from the Oujiang estuary and the southernmost end of the Zhejiang mainland coast (Cangnan). Affected by the sediment of the Oujiang River, the turbidity in the estuary area of SZ is very high, and there is no sandy beach near the estuary.

To summarize, the distribution pattern and the morphological characteristics of sandy beaches in NZ, CZ, and SZ are different (Figures 3–8). All accessible sandy beaches are embayed beaches, with most of the sandy beaches formed on the islands (especially in NZ) and the regions far from the estuaries, showing quantity difference in beach distribution (Figure 9).

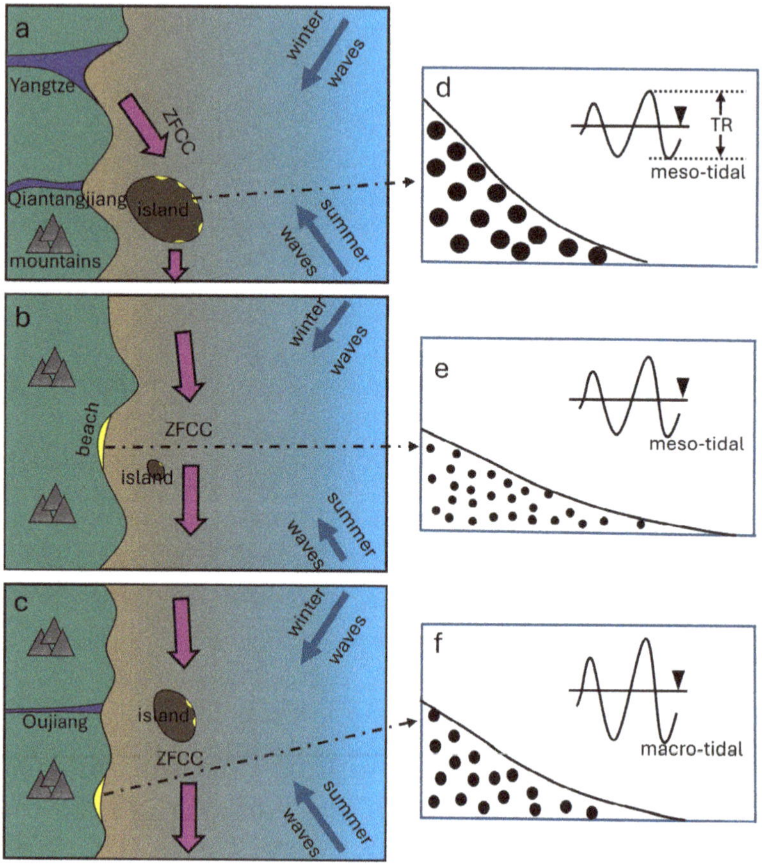

Figure 9. Conceptual model of difference in the distribution (**a**–**c**) and morphological characteristics (**d**–**f**) of NZ beaches (**a**,**d**), CZ beaches (**b**,**e**), and SZ beaches (**c**,**f**). The black dots in (**d**–**f**) show the conceptual sediment grain size.

5.2. Controlling Factors of the Different Distribution and Morphology among Zhejiang Beaches

To further explore the controlling factors on the difference in Section 5.1, a correlation analysis and a PCA were conducted. The correlation results (Table 1) show that beach slope (i) and the mean grain size (M_z) exhibited a significant positive correlation, indicating that coarser beach sediment particles are associated with steeper beach slopes in this study. The maximum indentation (a) is positively correlated with the length of the tangential section (L), the chord length of the bay mouth (b), and the length of the headland (S), suggesting that a is influenced by multiple parameters. This also implies that the intensity of wave erosion is related to the distance between headlands, resulting from wave refraction and diffraction effects on headland bay geometries [16,19,20,26]. Specifically, as the distance between headlands increases, the erosion ability of waves increases, leading to stronger wave bypassing and refraction with spatial gyration, and ultimately resulting in greater beach indentation.

The chord length of the bay mouth (b), the tangential section length (L), and the headland length (S) were significantly and positively correlated with each other, indicating that the headland distance of the beach is related to the geological structure, which in turn affects the tangential section length and headland length of the beach.

Table 1. Pearson correlation coefficients of the morphodynamic parameters of the studied sandy beaches in Zhejiang.

Parameter	i	M_z	a	L	b	S	β	γ	a/b
i	1.000								
M_z	0.523	1.000							
a	−0.230	0.109	1.000						
L	−0.036	0.098	0.554	1.000					
b	−0.197	0.015	0.756	0.575	1.000				
S	−0.068	0.003	0.587	0.533	0.595	1.000			
β	0.035	−0.039	−0.226	−0.098	−0.224	−0.223	1.000		
γ	0.080	0.150	0.020	−0.012	−0.079	0.096	−0.274	1.000	
a/b	−0.058	0.168	0.343	0.036	−0.182	0.048	−0.132	0.084	1.000

Given the correlations among the above morphodynamic parameters, a principal component analysis was conducted by region to further explore the factors influencing the distribution pattern and morphological characteristics of sandy beaches in Zhejiang, following the methods of previous studies [6,9,54].

5.2.1. NZ Beaches

Four principal components were extracted after the analysis of the morphodynamic factors of beaches in NZ, with the cumulative contribution of the total variance reaching 72.97% (Table S2). These four principal components were then used to analyze the main morphodynamic features. The load values of each factor in the first four principal components were also extracted (Table 2). In the first principal component, the load values of maximum indentation a, tangential section length L, and bay mouth chord length b are relatively large, 0.25, 0.23, and 0.23, respectively. Headland morphology is the primary geological control on the hydrodynamic and morphological processes of embayed beaches [19,55–57]. The three embayment characteristic factors (a, b, and L) reflect the geological settings of the headland coast. Several mountainous rivers with short streams enter the East China Sea, forming the basis of the headland coast of Zhejiang. A series of NE-SW and NW-SE trending fault structures formed in the Cenozoic are oblique to the coastline of South China, and mountains or hills rise to the sea, thus forming a series of meandering coasts, which is the basis for the formation of these embayed coasts [33]. The first principal component well reflects this feature, which is consistent with previous research results [9]. The beach slope i and the mean sediment grain size M_z correlate well with the second principal component, both with a load value of 0.32. The second principal component mainly reflects the influence of sediment sources in the formation of the beach. The main source of the sandy beach in Zhejiang is the local erosion of the headland and the sandy sediment from the river discharge [58], the large quantities of islands in NZ, and the Qiantang River can provide the sediment source. Among the third and fourth principal components, the factor loads of the ratio a/b and the average wave height are the largest, with values of 0.61 and 0.68, respectively. The ratio of a/b reflects the relative strength of wave force and the shadowing ability of the headland [6], indicating that these two parameters illustrate the important role of wave action in the formation of sandy beaches.

Table 2. The load of morphodynamic factors in principal components of sandy beaches in NZ, CZ, and SZ, respectively.

PC	i	M_z	a	L	b	S	γ	β	a/b	TR	$\overline{H}$
NZ1	−0.173	−0.130	0.246	0.230	0.232	0.095	0.028	0.191	0.015	−0.065	−0.065
NZ2	0.323	0.321	0.063	0.173	0.096	0.021	−0.283	0.250	−0.164	0.312	−0.016
NZ3	0.066	0.127	0.113	−0.014	−0.200	0.047	0.275	0.130	0.613	0.295	−0.353
NZ4	−0.015	−0.329	0.178	−0.057	−0.114	−0.223	−0.237	0.179	0.351	0.260	0.681
CZ1	−0.050	−0.027	0.255	0.229	0.268	0.241	0.089	0.095	−0.044	0.120	0.024
CZ2	0.502	0.449	−0.017	0.005	0.019	0.016	0.208	−0.074	−0.141	0.096	0.099
CZ3	0.082	0.174	−0.017	0.035	−0.115	0.073	−0.442	0.476	0.175	0.120	0.220
CZ4	0.064	0.258	0.227	0.037	0.022	0.008	0.009	0.062	0.439	−0.333	−0.491
SZ1	−0.188	−0.081	0.265	0.180	0.290	0.157	−0.123	0.137	0.003	−0.125	0.194
SZ2	−0.142	−0.271	0.167	−0.106	0.088	0.149	0.180	−0.293	0.268	0.186	−0.197
SZ3	0.433	0.431	0.211	−0.053	0.090	0.237	0.060	0.014	0.328	0.193	0.184
SZ4	−0.121	−0.066	−0.171	0.219	−0.123	0.102	−0.391	0.408	0.238	0.505	−0.227

Note: NZ1, CZ1, and SZ1 represent the first principal component in NZ beach PCA, respectively.

5.2.2. CZ Beaches

The principal component analysis results for sandy beaches in CZ are shown in Tables S3 and 2. The first four principal components were used to analyze the main morphodynamic features, as their accumulated contribution reached 78.103%. In the first principal component, the loads of maximum indentation a, tangential section length L, bay mouth chord length b, and headland length S are the largest, with values of 0.26, 0.23, 0.27, and 0.24, respectively. This is consistent with the result of NZ, reflecting the important influence of headland characteristics on beach development. In the second principal component, the load of beach slope i and the mean sediment grain size M_z accounted for a relatively large proportion, 0.50 and 0.45, respectively, closely related to the sediment source. In the third and fourth principal components, the factor loads of the spiral tangent angle β, the ratio of a/b, and the average wave height $\overline{H}$ are the largest, with values of 0.48, 0.44, and −0.49, respectively. The spiral tangent angle reflects the sheltering ability of the headland on the beach, while a/b reflects the relative strength of wave power and the sheltering ability of the headland. Therefore, these three factors are closely related to wave forcing.

5.2.3. SZ Beaches

The principal component analysis results of sandy beaches in SZ are shown in Tables S4 and 2. The first four principal components, as their accumulated contribution rate reached 72.98%, were used to analyze the main morphodynamic features. In the first principal component, the loads of maximum indentation and bay mouth chord length are the largest, with values of 0.27 and 0.29, respectively, which are similar to those in NZ and CZ. In the second principal component, the load values of the maximum indentation and the ratio of a/b are the largest, with values of −0.29 and 0.27, respectively. In the third principal component, the load value of beach slope i and sediment mean grain size M_z are relatively large (with the same value of 0.43), further indicating that the sediment source contributes to the formation of the sandy beach. In the fourth principal component, the mean tidal range TR has the largest load, with a value of 0.51, reflecting the influence of tides on the beach. Although sandy beaches are mainly controlled by waves, this component may indicate that the influence of tides cannot be ignored in a macrotidal coast like SZ. The flood and ebb tides affect the wave action on the beach, affecting the distribution of net sediment transport in space and time, and ultimately causing differences in the beach morphological characteristics.

5.3. Influence of Hydrodynamic Environment on Beach Morphological Characteristics

The morphological characteristics of sandy beaches are primarily driven by wave action [10,11]. The deformation and breaking of waves as they propagate from deep water to shallow water dominate the hydrodynamic processes near the beach [11,59]. Wave breaking is a key driver of various coastal processes, including sediment transport and morphodynamic changes [26,55,60]. Wave action in the coastal regions of islands is typically stronger than that in the sheltered mainland coastal regions along Zhejiang [47], resulting in a significantly higher number of sandy beaches on islands compared to the mainland coast (Table S1). The annual averaged wave height of NZ, CZ, and SZ is 0.79, 0.52, and 0.69, respectively [43,45–47], showing that NZ has the strongest wave force. This may be the reason why NZ has the majority of the beaches among the three regions with similar sediment supply, although the NZ coast can obtain more sediment from the Yangtze River [36]. The average values of the beach slope (Figure 3) and the mean grain size (Figure 4) in NZ, CZ, and SZ correspond well with the wave height. The larger the wave height, the coarser the mean grain size, and the steeper the beach slope.

Tides are also an important factor in controlling the position of wave action and the resulting changes in beach morphology [12,14,61]. The magnitude of the tidal range can indicate the strength of the coastal tidal dynamic environment. Based on the magnitude of the tidal range [62], coasts can be classified as micro-tidal ($0 < TR < 2$ m), meso-tidal ($2 \text{ m} < TR < 4$ m), or macro-tidal ($TR > 4$ m). The mean tidal ranges in the sea areas of NZ, CZ, and SZ are 2.48 m (meso-tidal), 3.29 m (meso-tidal), 4.23 m (macro-tidal), respectively, showing a gradual increase from north to south and from east to west in the mainland nearshore and island areas [49]. The distribution of wave height and tidal range along the Zhejiang coast exhibits significant regional differences, and the combination of wave and tide forcings can produce diverse beach morphological characteristics.

Waves, tides, and coastal currents promote the southward transport and diffusion of Yangtze River sediments [35–37], creating a turbid coastal environment (Figure 10) with sufficient fine sediment supply supplemented by the Qiantangjiang and Oujiang rivers [38,39]. This may also explain the quantity difference between island beaches and mainland beaches and the quantity difference between the estuary region and other areas (Figure 1b). The sandy sediments are mainly from the local small–moderate rivers, as well as the erosion of rocks from the headlands [58]. The construction of seawalls (Table S1) and the abundance of fine sediments have led to insufficient sand supply, causing some beaches to have the potential to become muddy [63,64]. In this study, 5 mainland sandy beaches among the 75 studied beaches exhibit mud on the subaerial beach at low tide (Table 3 and Figure 1). Similar phenomena were also found in sandy beaches along the Fujian coast of China [65,66], Cayenne beach in French Guiana [67], Casino beach in Brazil [68], Dassari beach in South Korea [69], and a mesotidal beach near the Mekong River Delta in Vietnam [70]. These beaches are all in a hydrodynamic environment with abundant fine sediments, large tidal range, and small wave height, highlighting the need to consider the risk of becoming muddy in the future development and protection of these beaches. The difference is that those beaches with mud occurrence in the above studies are usually open beaches near the river mouth, while the five mudding beaches in this study are embayed beaches, which may be related to the effect of abundant sediment supply from the Yangtze River and the Zhejiang rivers.

Table 3. Information of five mainland beaches with mud on the subaerial beach.

ID	Name	w (m)	l (m)	M_z (μm)	S (m)	β (°)	γ (°)	a/b	L (m)	TR (m)	$\overline{H}$
30	Yangshashan	118.52	322.9	335.26	82.89	28.39	267.08	0.65	262.88	2.06	0.40
34	Changshayucun	62.43	919.89	609.3	270.1	77.07	32.8	0.66	237.46	2.45	0.40
50	Hutoushanzui	68.13	234.58	218.78	42.25	92.38	14.01	0.59	71.76	4.17	0.50
52	Mushao	80.75	440.55	225.92	92.92	21.39	321.48	0.54	168.28	4.17	0.50
54	Longwan	68.53	286.98	212.25	494.65	77.2	358.16	0.59	448.97	4.17	0.50

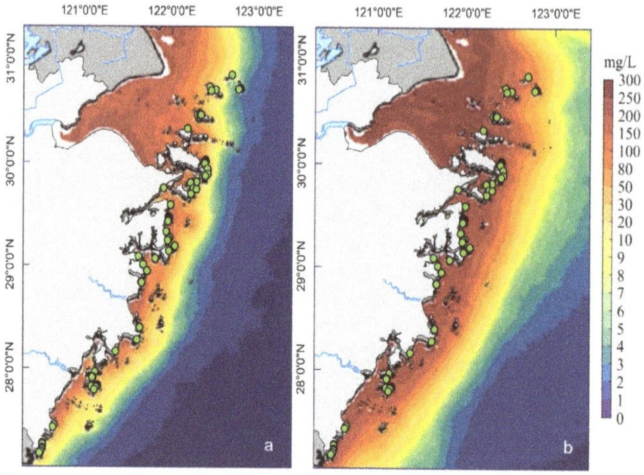

Figure 10. Multiyear averaged surficial suspended sediment concentration (mg/L) distribution in the coastal regions of Zhejiang and the adjacent sea areas during summer (**a**) and winter (**b**) (modified from [71]), in which the green circles show the sandy beaches in this study.

5.4. Influence of Human Activities on Beach Evolution

The coastal zone is a fragile and complex dynamic system threatened by the combined pressure of anthropogenic constraints and climate change. Sandy beaches, a significant type of coast, are increasingly at risk of erosion. Typhoons represent the most severe meteorological disaster in Zhejiang province, with an average of six typhoons per year [42], significantly impacting beach morphology and threatening the tourism industry. To mitigate typhoon-induced erosion and meet the increasing tourism demands, Zhejiang has seen an increasing number of beach protection or restoration projects [72] (Table S1), including both hard engineering and soft engineering. The most common approach adopted in Zhejiang to protect beaches is hard engineering, primarily due to the frequent influence of typhoons (e.g., the construction of seawalls and breakwaters; Table S1). Although seawalls provide protection against coastline retreat and the associated loss of coastal property, they may have negative impacts on the adjacent beach and potentially aggravate beach erosion. Additionally, accommodation space is important for beaches facing sea level rise due to climate change [73]. Over half of the studied beaches in Zhejiang have seawalls (Table S1), resulting in insufficient accommodation space and increasing the risk of beach extinction (Figure 11a). In addition to hard engineering, 10 beaches with nourishment projects have been documented (Table S1 and Figure 11b). Beach nourishment could significantly broaden beach width and maintain recreation functions [74,75]. However, nourishment projects may change the nearshore hydrodynamics [76] and further affect the long-term evolution of sandy beaches [23]. Therefore, human intervention is also an important factor influencing sandy beach development in Zhejiang.

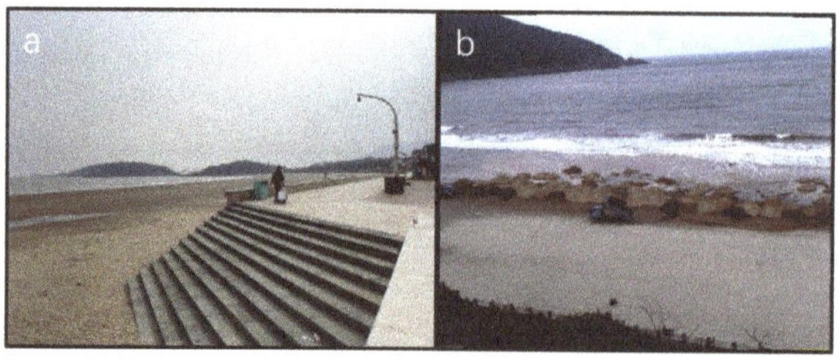

Figure 11. Example of human activities on the Zhejiang beaches: seawall construction (**a**) and beach nourishment (**b**).

6. Conclusions

Based on the analyses of beach sediment, slope, and planform characteristics, combined with the annual averaged tidal range and wave height, this study revealed regional differences in beach distribution, morphological characteristics, and influencing factors of 75 accessible beaches in Zhejiang, eastern China, and the following main conclusions were obtained.

The averaged slope of sandy beaches in Zhejiang varies from 0.02 to 0.13, with the beach slope in NZ being the steepest and that in CZ being the gentlest. The studied beaches are all embayed beaches, with most located on the islands, especially in NZ. The mean grain size (M_z) distribution of sandy beach sediments along the Zhejiang coast ranges from fine sand (128.93 μm) to very coarse sand (1486.74 μm), with the coarsest sediment occurring in SZ and the finest occurring in CZ. Beach sediment grain size is closely related to beach slope, showing that steeper beach slopes correspond to coarser sediment grain size.

The length of sandy beaches in Zhejiang ranges from 112.61 m to 2381.46 m, with the average length of sandy beaches in CZ being the longest. The dry beach width of the studied beaches ranges from 21.69 m to 281.57 m, with the widest beach occurring in CZ. The beach scale in the CZ coast is larger than those in NZ and SZ. In general, the longer the beach length, the wider the beach width.

The beach headlands in CZ are the largest among the three regions in Zhejiang, suggesting the higher susceptibility to wave erosion in the latter regions due to the less sheltered headlands. The distribution pattern and the morphological characteristics of sandy beaches in NZ, CZ, and SZ are different. All accessible sandy beaches are embayed beaches, and most of them formed on the islands (especially in NZ) and the regions far from the estuaries, showing the quantity difference in beach distribution.

Through a comparison of the regional difference in the distribution pattern and morphological characteristics of embayed sandy beaches in NZ, CZ, and SZ, we found that there is a gradual shift from more wave influence in the north to more tidal influence in the south. Although waves are the primary factor controlling sandy beach evolution, the embayment characteristic is the main influencing factor affecting the beach distribution pattern and the morphological characteristics in Zhejiang beaches. The different characteristics of embayment provide the space for beach formation and the interaction with the coastal process, the sediment supply, the nearshore hydrodynamic environment, and human intervention also have influence on the morphological characteristics of Zhejiang beaches.

Supplementary Materials: The following supporting information can be downloaded at: https://www.mdpi.com/article/10.3390/jmse12071223/s1, The Supplementary Materials include Tables S1–S4 and Figure S1. Table S1 shows the location, measurement information, and human intervention information of the studied beaches. Tables S2–S4 show the principal component eigenvalues, variance contribution rates, and cumulative contribution rates of sandy beaches in NZ, CZ, and SZ, respectively. Figure S1 shows the profile and sampling site setting examples for beaches with different length and the calculation of beach slope and dry beach width. More details can be found in the Supplementary Materials.

Author Contributions: Conceptualization, J.G. and L.S.; methodology, J.G. and M.Z.; software, J.G. and M.Z.; validation, J.G., L.S. and M.Z.; formal analysis, J.G. and M.Z.; investigation, J.G., M.Z. and Z.G.; resources, L.S.; data curation, J.G., M.Z. and Z.G.; writing—original draft preparation, J.G. and L.S.; writing—review and editing, J.G., L.S., M.Z., Z.G., W.C. and X.X.; visualization, J.G.; supervision, L.S.; project administration, J.G. and L.S.; funding acquisition, J.G. and L.S. All authors have read and agreed to the published version of the manuscript.

Funding: This research was funded by the National Key R&D Program of China, grant number 2022YFC3106200, the Scientific Research Funds of the Second Institute of Oceanography, MNR (JG2315&XRJH2309), and the Zhejiang Provincial Natural Science Foundation of China: LHZ22D060001.

Institutional Review Board Statement: Not applicable.

Informed Consent Statement: Not applicable.

Data Availability Statement: Data can be obtained by contacting the corresponding authors.

Acknowledgments: We thank Daheng Zhang for the help in field investigation and Yang Chang for the help in data processing. We also thank the reviewers and the editors for their constructive suggestions on improving our manuscript.

Conflicts of Interest: The authors declare no conflicts of interest. The funders had no role in the design of the study; in the collection, analyses, or interpretation of data; in the writing of the manuscript; or in the decision to publish the results.

References

1. Luijendijk, A.; Hagenaars, G.; Ranasinghe, R.; Baart, F.; Donchyts, G.; Aarninkhof, S. The State of the World's Beaches. *Sci. Rep.* **2018**, *8*, 6641. [CrossRef] [PubMed]
2. Park, K.; Reisinger, Y.; Kang, H. Visitors' Motivation for Attending the South Beach Wine and Food Festival, Miami Beach, Florida. *J. Travel Tour. Mark.* **2008**, *25*, 161–181. [CrossRef]
3. Castelle, B.; Turner, I.L.; Bertin, X.; Tomlinson, R. Beach Nourishments at Coolangatta Bay over the Period 1987–2005: Impacts and Lessons. *Coast. Eng.* **2009**, *56*, 940–950. [CrossRef]
4. Hamm, L.; Capobianco, M.; Dette, H.H.; Lechuga, A.; Spanhoff, R.; Stive, M.J.F. A Summary of European Experience with Shore Nourishment. *Coast. Eng.* **2002**, *47*, 237–264. [CrossRef]
5. Hanson, H.; Brampton, A.; Capobianco, M.; Dette, H.H.; Hamm, L.; Laustrup, C.; Lechuga, A.; Spanhoff, R. Beach Nourishment Projects, Practices, and Objectives—A European Overview. *Coast. Eng.* **2002**, *47*, 81–111. [CrossRef]
6. Dai, Z.; Li, C. Coastline Configuration and Geomorphologic Development Mode of Arc-Shaped Coast in South China. *Acta Geogr. Sin.* **2004**, *59*, 92–95.
7. Fellowes, T.E.; Vila-Concejo, A.; Gallop, S.L. Morphometric Classification of Swell-Dominated Embayed Beaches. *Mar. Geol.* **2019**, *411*, 78–87. [CrossRef]
8. Gallop, S.L.; Kennedy, D.M.; Loureiro, C.; Naylor, L.A.; Fellowes, T.E. Geologically Controlled Sandy Beaches: Their Geomorphology, Morphodynamics and Classification. *Sci. Total Environ.* **2020**, *731*, 139123. [CrossRef] [PubMed]
9. Li, Z.; Li, W.; Chen, Z.; Zhu, Y. Influencing Factors and Classifications of Arc-Shaped Coasts in South China. *Acta Geogr. Sin.* **2014**, *69*, 595–606.
10. Ding, Y.; Yu, J.; Cheng, H. Long-Term Wave Climate Characteristics and Potential Impacts on Embayed Beaches along the West Guangdong Coastline. *Reg. Stud. Mar. Sci.* **2019**, *30*, 100741. [CrossRef]
11. Komar, P.D. *Beach Processes and Sedimentation*, 2nd ed.; Prentice-Hall: New Jersey, NJ, USA, 1976; pp. 270–327.
12. Butt, T.; Russell, P. Hydrodynamics and Cross-Shore Sediment Transport in the Swash-Zone of Natural Beaches: A Review. *J. Coast. Res.* **2000**, *16*, 255–268.
13. Iglesias, G.; López, I.; Carballo, R.; Castro, A. Headland-Bay Beach Planform and Tidal Range: A Neural Network Model. *Geomorphology* **2009**, *112*, 135–143. [CrossRef]
14. Wright, L.D.; Short, A.D. Morphodynamic Variability of Surf Zones and Beaches: A Synthesis. *Mar. Geol.* **1984**, *56*, 93–118. [CrossRef]
15. Li, Z.; Lei, H. Distribution and Existing Problems of Sandy Coast in China. *Mar. Geol. Front.* **2006**, *22*, 1–4.
16. da Silva, G.M.; Mousavi, S.M.S.; Jose, F. Wave-Driven Sediment Transport and Beach-Dune Dynamics in a Headland Bay Beach. *Mar. Geol.* **2012**, *323–325*, 29–46. [CrossRef]
17. Backstrom, J.T.; Jackson, D.W.T.; Cooper, J.A.G. Contemporary Morphodynamics of a High-Energy Headland-Embayment Shoreface. *Cont. Shelf Res.* **2009**, *29*, 1361–1372. [CrossRef]
18. Bluck, B.J. Structure of Gravel Beaches and Their Relationship to Tidal Range. *Sedimentology* **2011**, *58*, 994–1006. [CrossRef]
19. Jackson, D.W.T.; Cooper, J.A.G. Geological Control on Beach Form: Accommodation Space and Contemporary Dynamics. *J. Coast. Res.* **2009**, *I*, 69–72.
20. Pranzini, E.; Rosas, V.; Jackson, N.L.; Nordstrom, K.F. Beach Changes from Sediment Delivered by Streams to Pocket Beaches during a Major Flood. *Geomorphology* **2013**, *199*, 36–47. [CrossRef]
21. Cai, F.; Dean, R.G.; Liu, J. Beach Nourishment in China: Status and Prospects. *Coast. Eng. Proc.* **2011**, *1*, 31. [CrossRef]
22. Luo, S.; Liu, Y.; Jin, R.; Zhang, J.; Wei, W. A Guide to Coastal Management: Benefits and Lessons Learned of Beach Nourishment Practices in China over the Past Two Decades. *Ocean Coast. Manag.* **2016**, *134*, 207–215. [CrossRef]
23. Paris, P.; Leach, A.; Corbett, D.R. Potential Long-Term Disturbance Associated with Beach Nourishment—Insights and Observations from Pea Island National Wildlife Refuge, Outer Banks, North Carolina. *Heliyon* **2023**, *9*, e12816. [CrossRef] [PubMed]
24. Ramesh, M.; Swathy Krishna, P.S.; Amrutha Raj, V.; Sheela Nair, L. Coupled Coastal Monitoring Framework for the Analysis of Beach Stability and Nearshore Hydrodynamics of a Structure Influenced Medium Energy Coast in India. *Ocean Coast. Manag.* **2023**, *239*, 106619. [CrossRef]
25. Short, A.D.; Masselink, G.G. Embayed and Structurally Controlled Beaches. In *Handbook of Beach and Shoreface Morphodynamics*; Short, A.D., Ed.; Wiley: Chichester, UK, 1999; pp. 230–250. ISBN 978-0-471-96570-1.

26. Castelle, B.; Robinet, A.; Idier, D.; D'Anna, M. Modelling of Embayed Beach Equilibrium Planform and Rotation Signal. *Geomorphology* **2020**, *369*, 107367. [CrossRef]
27. George, D.A.; Largier, J.L.; Pasternack, G.B.; Barnard, P.L.; Storlazzi, C.D.; Erikson, L.H. Modeling Sediment Bypassing around Idealized Rocky Headlands. *J. Mar. Sci. Eng.* **2019**, *7*, 40. [CrossRef]
28. Horta, J.; Oliveira, S.; Moura, D.; Ferreira, Ó. Nearshore Hydrodynamics at Pocket Beaches with Contrasting Wave Exposure in Southern Portugal. *Estuar. Coast. Shelf Sci.* **2018**, *204*, 40–55. [CrossRef]
29. Valiente, N.G.; Masselink, G.; Scott, T.; Conley, D.; McCarroll, R.J. Role of Waves and Tides on Depth of Closure and Potential for Headland Bypassing. *Mar. Geol.* **2019**, *407*, 60–75. [CrossRef]
30. Bowman, D.; Guillén, J.; López, L.; Pellegrino, V. Planview Geometry and Morphological Characteristics of Pocket Beaches on the Catalan Coast (Spain). *Geomorphology* **2009**, *108*, 191–199. [CrossRef]
31. da Fontoura Klein, A.H.; de Menezes, J.T. Beach Morphodynamics and Profile Sequence for a Headland Bay Coast. *J. Coast. Res.* **2001**, *17*, 812–835.
32. Prodger, S.; Russell, P.; Davidson, M.; Miles, J.; Scott, T. Understanding and Predicting the Temporal Variability of Sediment Grain Size Characteristics on High-Energy Beaches. *Mar. Geol.* **2016**, *376*, 109–117. [CrossRef]
33. Cai, F. *Overview of Beach Resources in China*; China Ocean Press: Beijing, China, 2019.
34. Xia, X. *Thematic Investigation and Research Report on Shoreline Geomorphology and Silting Dynamics*; Second Institute of Oceanography, State Oceanic Administration: Hangzhou, China, 2011.
35. Liu, J.P.; Li, A.C.; Xu, K.H.; Velozzi, D.M.; Yang, Z.S.; Milliman, J.D.; DeMaster, D.J. Sedimentary Features of the Yangtze River-Derived along-Shelf Clinoform Deposit in the East China Sea. *Cont. Shelf Res.* **2006**, *26*, 2141–2156. [CrossRef]
36. Liu, J.P.; Xu, K.H.; Li, A.C.; Milliman, J.D.; Velozzi, D.M.; Xiao, S.B.; Yang, Z.S. Flux and Fate of Yangtze River Sediment Delivered to the East China Sea. *Geomorphology* **2007**, *85*, 208–224. [CrossRef]
37. Wang, L.; Zhou, Y.; Shen, F. Suspended Sediment Diffusion Mechanisms in the Yangtze Estuary Influenced by Wind Fields. *Estuar. Coast. Shelf Sci.* **2018**, *200*, 428–436. [CrossRef]
38. Hu, R.; Wu, J.; Li, G.; Zhu, L.; Ma, F. Characteristics of Sediment Transport in the Zhoushan Archipelago Sea Area. *Acta Oceanol. Sin.* **2009**, *28*, 116–127.
39. Li, G.; Gao, S.; Wang, Y.; Li, C. Sediment Flux from the Zhoushan Archipelago, Eastern China. *J. Geog. Sci.* **2018**, *28*, 387–399. [CrossRef]
40. Guo, J.; Shi, L.; Chen, S.; Castelle, B.; Chang, Y.; Cheng, W. Sand-Mud Transition Dynamics at Embayed Beaches during a Typhoon Season in Eastern China. *Mar. Geol* **2021**, *441*, 106633. [CrossRef]
41. Jiang, Z.; Shi, L.; Guo, J.; Zhang, M. Storm Response and Recovery Process after Nourishment of Dongsha Beach, Zhujiajian Island, Zhejiang Province. *J. Coast. Res.* **2020**, *95*, 231–236. [CrossRef]
42. Lu, J. *Research Report on the Regional Climate of 908 Special Island Survey in Zhejiang Province*; Zhejiang Ocean Monitoring and Forecast Center: Hangzhou, China, 2010; pp. 54–67.
43. Wang, K. *Comprehensive Survey of Island Resources in Zhejiang Province*; Second Institute of Oceanography, State Oceanic Administration: Hangzhou, China, 1994.
44. Wang, J.; Li, J. Analysis of Geomorphic Environment on the Formation of Coastal Cultural Resources in Zhejiang Province. *J. Jingchu Univ. Technol.* **2010**, *25*, 63–66.
45. CCRBC (Compilation Committee of Records of Bays in China). *Records of Bays in China, Volume 5, Bays in Shanghai and North Zhejiang*; China Ocean Press: Beijing, China, 1992.
46. CCRBC (Compilation Committee of Records of Bays in China). *Records of Bays in China, Volume 6, Bays in South Zhejiang*; China Ocean Press: Beijing, China, 1993.
47. Yu, S. Study on the Influence of Coastal Storm Surge Level on Typhoon Wave in Zhejiang Province. Master's Thesis, Zhejiang University, Hangzhou, China, 2012.
48. Zhang, H. *Basic Status of Marine Environmental Resources in Zhejiang Province*; China Ocean Press: Beijing, China, 2013.
49. Chen, Q.; Huang, D.; Zhang, B.; Wang, M. The Research of the Tidal Features in the Coastal Zone of Zhejiang Province. *Donghai Mar. Sci.* **2003**, *21*, 1–12.
50. Carver, R.E. *Procedures in Sedimentary Petrology*; Wiley-Interscience: New York, NY, USA, 1971.
51. Folk, R.L.; Ward, W.C. Brazos River Bar [Texas]; A Study in the Significance of Grain Size Parameters. *J. Sediment. Res.* **1957**, *27*, 3–26. [CrossRef]
52. Hsu, J.R.C.; Evans, C. Parabolic Bay Shapes and Applications. *ICE Proc.* **1989**, *87*, 557–570. [CrossRef]
53. Chen, D. *Econometric Geography: Methods and Applications*; Nanjing University Press: Nanjing, China, 2011.
54. Hsu, J.R.C.; Silvester, R.; Xia, Y.M. Application of Headland Control. *J. Waterw. Ports Coast. Ocean Eng.* **1989**, *115*, 299–310. [CrossRef]
55. Castelle, B.; Coco, G. The Morphodynamics of Rip Channels on Embayed Beaches. *Cont. Shelf Res.* **2012**, *43*, 10–23. [CrossRef]
56. Loureiro, C.; Ferreira, Ó.; Cooper, J.A.G. Geologically Constrained Morphological Variability and Boundary Effects on Embayed Beaches. *Mar. Geol.* **2012**, *329–331*, 1–15. [CrossRef]
57. Masselink, G.; Scott, T.; Poate, T.; Russell, P.; Davidson, M.; Conley, D. The Extreme 2013/2014 Winter Storms: Hydrodynamic Forcing and Coastal Response along the Southwest Coast of England. *Earth Surf. Proc. Land.* **2016**, *41*, 378–391. [CrossRef]
58. Cai, F. *Brief Introduction of Chinese Beach Resources*; Ocean Press: Beijing, China, 2019.

59. Gomes, E.R.; Mulligan, R.P.; Brodie, K.L.; McNinch, J.E. Bathymetric Control on the Spatial Distribution of Wave Breaking in the Surf Zone of a Natural Beach. *Coast. Eng.* **2016**, *116*, 180–194. [CrossRef]
60. Aagaard, T.; Greenwood, B.; Hughes, M. Sediment Transport on Dissipative, Intermediate and Reflective Beaches. *Earth-Sci. Rev.* **2013**, *124*, 32–50. [CrossRef]
61. Sreenivasulu, G.; Jayaraju, N.; Reddy, B.C.S.R.; Prasad, T.L.; Lakshmanna, B.; Nagalakshmi, K. Coastal Morphodynamics of Tupilipalem Coast, Andhra Pradesh, Southeast Coast of India. *Curr. Sci. India.* **2017**, *112*, 823–829. [CrossRef]
62. Davies, J.L.; Gideon, J.D.; Davies, M.C. A Morphogenic Approach to World Shorelines. *Z. Geomorphol.* **1964**, *8*, 127–142. [CrossRef]
63. Huang, S.; Yao, W.; Liu, X.; Wu, C. Profile Characteristics of the Beaches Adjacent to Muddy Seabed in the Headland Bays. *Chin. Coast. Eng.* **2016**, *35*, 1–9.
64. Li, M.; Sun, Z.; Gong, Y.; Hu, Y. Numerical Simulation of Sand-Mud Coexisting Beach Based on Delft-3D. *Bull. Sci. Technol.* **2019**, *35*, 109–113.
65. Li, Y.; Zhang, C.; Zhao, S.; Qi, H.; Cai, F.; Zheng, J. Equilibrium Configurations of Sandy-Muddy Transitional Beaches on South China Coasts: Role of Waves in Formation of Sand-Mud Transition Boundary. *Coast. Eng.* **2024**, *187*, 104401. [CrossRef]
66. Zhao, S.; Cai, F.; Qi, H.; Liu, J.; Cao, C.; Liu, G.; Lei, G. Sediment Source-to-Sink Process Variations of Sandy-Muddy Transitional Beaches and Their Morphological Indications. *Acta Oceanol. Sin.* **2023**, *42*, 10–24. [CrossRef]
67. Anthony, E.J.; Dolique, F. The Influence of Amazon-Derived Mud Banks on the Morphology of Sandy Headland-Bound Beaches in Cayenne, French Guiana: A Short- to Long-Term Perspective. *Mar. Geol.* **2004**, *208*, 249–264. [CrossRef]
68. Calliari, L.J.; Machado, A.A.; Marroig, P.; Vinzon, S.; Gianuca, N. Mud Deposits at Cassino Beach: Role of Dredging. *Geo-Mar. Lett.* **2020**, *40*, 1031–1043. [CrossRef]
69. Chang, T.S.; Hong, S.H.; Chun, S.S.; Choi, J.-H. Age and Morphodynamics of a Sandy Beach Fronted by a Macrotidal Mud Flat along the West Coast of Korea: A Lateral Headland Bypass Model for Beach-Dune Formation. *Geo-Mar. Lett.* **2017**, *37*, 361–371. [CrossRef]
70. Tamura, T.; Horaguchi, K.; Saito, Y.; Nguyen, V.L.; Tateishi, M.; Ta, T.K.O.; Nanayama, F.; Watanabe, K. Monsoon-Influenced Variations in Morphology and Sediment of a Mesotidal Beach on the Mekong River Delta Coast. *Geomorphology* **2010**, *116*, 11–23. [CrossRef]
71. Qiao, L.; Liu, S.; Xue, W.; Liu, P.; Hu, R.; Sun, H.; Zhong, Y. Spatiotemporal Variations in Suspended Sediments over the Inner Shelf of the East China Sea with the Effect of Oceanic Fronts. *Estuar. Coast. Shelf Sci.* **2020**, *234*, 106600. [CrossRef]
72. Cai, F.; Liu, G. Beach Nourishment Development and Technological Innovations in China: An Overview. *J. Appl. Oceanogr.* **2019**, *38*, 452–463.
73. Cooper, J.A.G.; Masselink, G.; Coco, G.; Short, A.D.; Castelle, B.; Rogers, K.; Anthony, E.; Green, A.N.; Kelley, J.T.; Pilkey, O.H.; et al. Sandy Beaches Can Survive Sea-Level Rise. *Nat. Clim. Chang.* **2020**, *10*, 993–995. [CrossRef]
74. de Schipper, M.A.; Ludka, B.C.; Raubenheimer, B.; Luijendijk, A.P.; Schlacher, T.A. Beach Nourishment Has Complex Implications for the Future of Sandy Shores. *Nat. Rev. Earth Environ.* **2021**, *2*, 70–84. [CrossRef]
75. Guo, J.; Shi, L.; Pan, S.; Ye, Q.; Cheng, W.; Chang, Y.; Chen, S. Monitoring and Evaluation of Sand Nourishments on an Embayed Beach Exposed to Frequent Storms in Eastern China. *Ocean Coast. Manag.* **2020**, *195*, 105284. [CrossRef]
76. Kuang, C.; Mao, X.; Gu, J.; Niu, H.; Ma, Y.; Yang, Y.; Qiu, R.; Zhang, J. Morphological Processes of Two Artificial Submerged Shore-Parallel Sandbars for Beach Nourishment in a Nearshore Zone. *Ocean Coast. Manag.* **2019**, *179*, 104870. [CrossRef]

Journal of
Marine Science and Engineering

MDPI

Article

Study of Physical and Mechanical Relationships during the Natural Dewatering of River Sediments and a Kaolin

Dalel Azaiez [1,2], Beatriz Boullosa Allariz [2] and Daniel Levacher [2,*]

[1] Laboratoire de Recherche Ingénierie Géotechnique, LRI4ES03, Ecole Nationale d'Ingénieurs de Tunis, Université de Tunis El Manar, BP37, Le Belvédère, Tunis 1002, Tunisia; dalelazaiez@gmail.com
[2] M2C, UMR 6143 CNRS, University of Caen Normandy, 14000 Caen, France; bea.boullosa.allariz@gmail.com
* Correspondence: daniel.levacher@unicaen.fr

Abstract: This paper investigated the relationships of some physical and mechanical parameters of sediments and a typical clay during a natural dewatering process. Four sediments from different French river dams sampled by the Électricité De France group (EDF group) and a commercial kaolin clay used for comparative purposes were the focus of this study. Continuous dewatering was monitored in a laboratory by quantifying the percentage of water remaining in sediments or clay, drained and evaporated. Undrained shear strength was also assessed during the sediment or clay dewatering process, using the laboratory vane shear test. The samples were controlled along different dimensions during the dewatering process throughout the whole experiment. The results showed a certain interdependence between the physical parameters and the water content (ω), which was normalized by the liquidity limit (ω/L_L) over time. This led to sigmoidal and exponential correlations when considering the percentage of water drained. The percentage of water remaining in the sediments or clay was characterized using the normalized water content, leading to exponential and power correlations. Both exponential and linear correlations were perfect for describing the evolution of the percentage of water evaporated. Other correlations were established for variations in void index, dry unit weight/solid unit weight ratio and undrained shear strength during the dewatering process.

Keywords: sediments; undrained shear strength; dewatering; water content; vane shear testing; correlations

1. Introduction

Sediment dredging is an essential and inevitable operation for ensuring the safety and navigability of ships in shipping lanes and their access to ports. It involves maintenance and investment dredging for many ports. Throughout the world, a significant quantity of sediment is dredged annually. Each year, it is around 50×10^6 m^3 in France, 300×10^6 m^3 in Europe and more than double worldwide [1–4]. In Tunisia, it averages just 8.5×10^6 m^3 [5]. The dumping of dredged sediments at sea is regulated by national and international conventions, such as the London Convention [6]. It is an easy way and an economic solution for sediment management even if it implies the need for a thorough study of its impact on the aquatic environment. In France, around 90–95% of marine and estuarine dredged sediments are immersed in the sea [7]. However, regulations on dumping of sediments in the sea are becoming stricter in every country, and onshore sediment management needs to be implemented. Moreover, sediments can be a renewable resource and a source of materials for long-term recycling. Many beneficial uses of dredged sediments have been investigated worldwide such as materials in construction and for filling and materials in road engineering [8–20].

Onshore sediment management involves storing a large volume of dredged sediment with a very highwater content, which makes it difficult to transport, for instance, to storage areas for recycling. This necessarily means that the sediment has to be dewatered.

Mechanical dewatering processes have been recently developed for dredged sediments but the rate of volume of sediments to be dewatered is limited. Moreover, the placement of the machinery must be near the dredging operations that pose a problem [21–23]. Natural dewatering is considered the most economical and environmentally friendly way to eliminate water from dredged sediments even if this technique requires significant onshore deposit areas. This is a time-consuming method, but it is possible to accelerate the dewatering by returning the sediment. Such an operation requires earthmoving machinery on the site and a certain bearing capacity of deposited sediments to ensure safe and smooth traffic for the machinery. Loss of water induces reduced volume and weight for the final transport towards the recycling area and could be considered as an economic and environmental benefit.

If the natural dewatering technique is a sustainable, eco-friendly and economical solution, it is important to understand how the sediment is dewatered and what the physical and mechanical parameters which govern dewatering are. Laboratory dewatering tests were performed on four river sediments and a typical clay to study their behavior throughout the dewatering process by assessing the quantity of water drained, evaporated and remaining in the sediments and clay. The choice of river sediments was made because their dumping at sea is not economically feasible due to the distance from the coastline, and mechanical dewatering could not be performed due to difficult access near the rivers. During these tests, other physical parameters were measured such as the dimensions of tested samples, geotechnical parameters and the undrained shear strength Su [24–26]. Relationships between physical and mechanical parameters were proposed in such a way that it represents the dewatering process of river sediments and reduces the duration of laboratory tests which can take more than one month.

2. Materials and Methods

2.1. Samples

The four river sediments were proposed by Electricité de France (EDF Group) because they were located near river dams (IS, RA and SA) and near a nuclear power station (RH) where dredging operations have been planned in the short term. The IS sediment came from samples extracted in the Sautet dam, located in the region of Auvergne-Rhone Alps. These alpine sediments were dredged in 2012 from 50 m under water. The SA sediment from St-Aignan in Brittany region (West of France) were sampled from the downstream part of the Guerlédan dam from a 2 m depth. For these sediments, the geotechnical campaign was performed in March 2015 [23]. Estuarine sediment was also included; the RA sediment was from the Rance dam located at the mouth of the coastal river of Rance between the municipalities of Richardais and Saint-Malo. These sediments were extracted during November 2014 and March 2015 to be desalted. They were delivered in 2017 after being completely dried [22]. RH sediment from the Great East region (Rhin river), exactly from the Marckolsheim river dam near a nuclear power station, were also investigated in this study. These RH sediments were sampled in 2012 from a depth that ranges between 1.3 m and 9 m under the water [27]. In addition, a kaolin (KA) sample completed the four sediment samples. It was chosen as a typical laboratory clay for comparing the different behaviors during the dewatering process. These four river sediments and kaolin will be referenced as "all samples" in the following sections.

2.2. Testing Methods

2.2.1. Natural Dewatering Test Procedure

The experimental procedure involves dewatering of the sediments or clay in the open air, which consists of eliminating water contained in samples through evaporation and drainage. It decreases the sample water content. The dewatering process is accompanied by volume variation which was quantified based on the evolution of void ratio. These variations depend on the physiochemical and mineralogical characteristics of the soils or sediments tested.

Different samples were prepared and tested to study their behavior during the dewatering process at a small scale. Saturated samples were put into sieves of sizes as indicated in Table 1, allowing water drainage and evaporation; see the rationale in Figure 1a. The investigated sediments and clay were first oven-dried and then weighed (Figure 1b) to provide the quantity to be hydrated at twice the liquid limit of the sediment or clay.

Table 1. All samples' tested sizes.

Samples References	IS	RA	RH	SA	KA
Sample diameter (mm)	210	185	185	165	420
Sample height (mm)	70	70	70	70	75

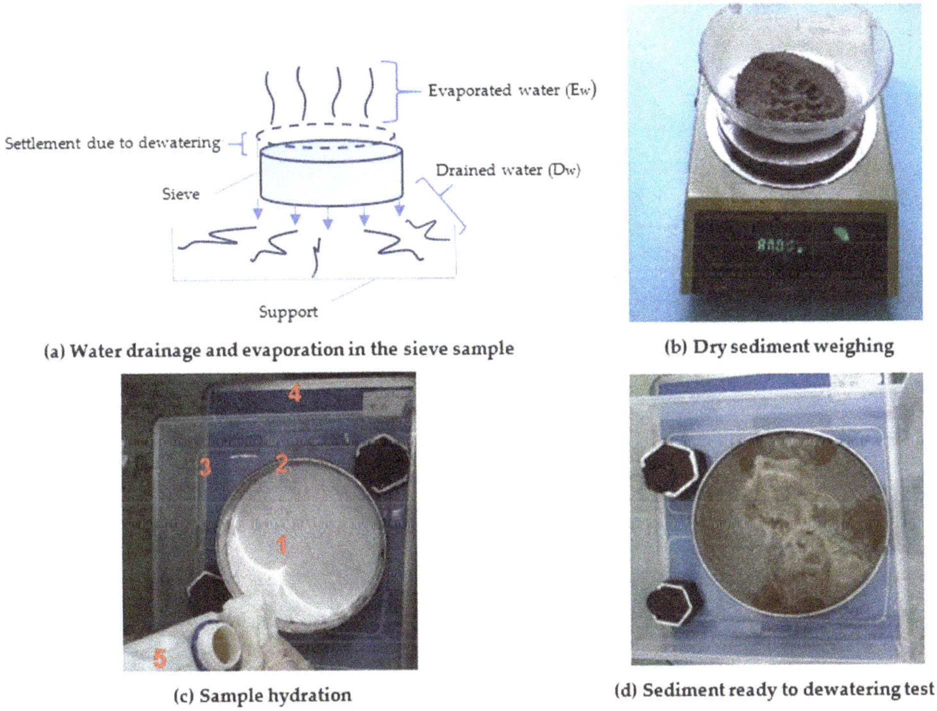

Figure 1. Natural dewatering process: principle, procedure and monitoring.

As illustrated in Figure 1c, the sediment or clay for a water content of $\omega_0 = 2\ L_L$ is placed in a sieve (2) with a non-woven polypropylene filtration geotextile sheet of 160 g/m^2 at its bottom (1) to ensure sediment drainage without any particle loss and to prevent any particles from blocking the 10 mm sieve mesh. The sieve mesh and geotextile guarantee fluid water drainage from the dewatered samples.

The drained water is to be recovered into the plastic support (3). The lubricating oil (5) is applied to facilitate the sample extraction from the sieve when being completely dried. All the equipment used are weighed before starting the dewatering; once all settled as shown in Figure 1d, they are put all together on the balance weight (4) and have their weight reset to 0 kg for finding out the weight of the necessary sample to fill the sieve with. The sieve filled with the sample will be kept permanently in the plastic support at an ambient room temperature of 20 °C ± 2 °C. It is removed once a day to be weighed which allows for measuring the water remaining in the material, E_1. The support is also weighed with the drained water E_2 to determine and deduce the evaporated water E_3.

The consistency of samples initially saturated with water content superior to the liquid limit evolves from a liquid state to a dry state when they are completely dewatered [24]. During the dewatering process, the sample goes through three steps: (a), (b) and (c). When the sediment is at water content equal to its liquid limit, during the first step (a), the water for the experiment is completely contained in the sediment (E_{1a}, mass of water contained in the sediment during the first stage of dewatering (a)); meanwhile, the masses of the drained and evaporated water E_{2a} and E_{3a} are null (E_{2a}, E_{3a}: mass of water drained and mass of water evaporated from the sediment during the first stage of dewatering (a)). During the following steps (b) and (c), the sample has a plastic consistency. The second step (b) corresponds to the drained water quantity E_{2b} (mass of water drained from the sediment during the second stage of dewatering (b)) when it predominates until reaching a maximum value over the evaporated water E_{3b} (mass of water evaporated from the sediment during the second stage of dewatering (b)) that is negligible. Then, in the third step (c), the quantities of evaporated water E_{3c} (mass of water evaporated from the sediment during the third stage of dewatering (c)) increase remarkably, while the water likely to be drained E_{2c} (mass of water drained from the sediment during the third stage of dewatering (c)) becomes very insignificant. Finally, the samples reach the shrinkage limit, and crack openings are observed in the sample.

During the dewatering process, as the sample dimensions vary, the sample settlement and diameter are measured. A vernier caliper that provides an accuracy of 0.01 mm is used to make these measurements. The diameter is measured within three equidistant positions on the sieve perimeter. As the used sieves were completely filled, it has allowed us to consider the sieve's upper surface as a reference for the settlement measurement. The void ratio and state parameters are then determined. The unit weight of solid particles Y_s is considered equal to 26.5 kN/m^3 for data analysis.

2.2.2. Undrained Shear Strength Monitoring

All samples' shear strength Su was measured using the laboratory vane shear test. It is well known that undrained shear strength governs the short-term stability mechanisms and failure of landfills, foundations and dams. Its measurement on site could be useful at different steps of the sediment dewatering to manage their storage time, their handling, their removal, etc. During the sample dewatering test, Su was quantified using the vane shear test (VST) as the sediment dried. Considering the evolution of sample resistance measured with the vane shear test while dewatering, different blades and springs were used. The undrained shear strength was determined according to the ASTM standard [28] as given in Equation (1).

$$Su = (2\ Mmax)/(\pi D^2\ (H + D/3)) \quad (1)$$

where Su is the undrained shear strength; H is the height of the vane blade, D is the vane diameter, and Mmax is the maximum recorded torque at failure.

3. Results and Discussion

3.1. Sediment Characterization

Using the laser apparatus (Beckman LS320 type, Beckman Coulter, Brea, CA, USA), the grain size distribution is shown in Figure 2. The uniformity coefficient c_u and gradation coefficient c_c were determined with a grain size analysis. For all tested samples, it was found that $Cu > 5$ and $1 < Cc < 3$—see Table 2—and they were well graded. According to the Soil Survey Manual and soil classification system of the United States Department of Agriculture [29] (Figure 3), it is confirmed that the RH, IS and RA sediments have a silt loam structure; the sandy loam structure is attributed to the SA sediment, whereas the kaolin has a silt clay loam structure.

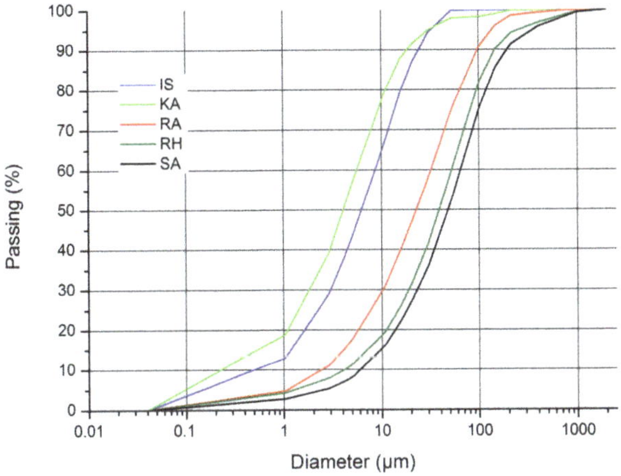

Figure 2. Grain size distribution of sediments.

Table 2. Sediment physiochemical characteristics.

Samples	Gradation Coefficients		OM (%)	$CaCO_3$ (%)	ω_0 (%)	Atterberg Limits		
	Cu	Cc				L_L (%)	P_L (%)	P_I (%)
IS	19.07	2.36	3.71	18.19	92.0	46	30	16
RA	13.48	1.38	9.18	17.91	108.0	54	50	4
RH	12.38	1.56	6.76	20.26	142.0	71	46	25
SA	10.63	1.33	6.61	0.70	101.8 *	65	54	11
KA	26.65	2.97	0.12	0.45	110.0	55	30	20

Note: Cu: uniformity coefficient; Cc: curvature coefficient; OM: organic matter; ω_0: initial water content; L_L: liquidity limit; P_L: plasticity limit; P_I: plasticity index; * case study where ω_0 = 1.57 L_L.

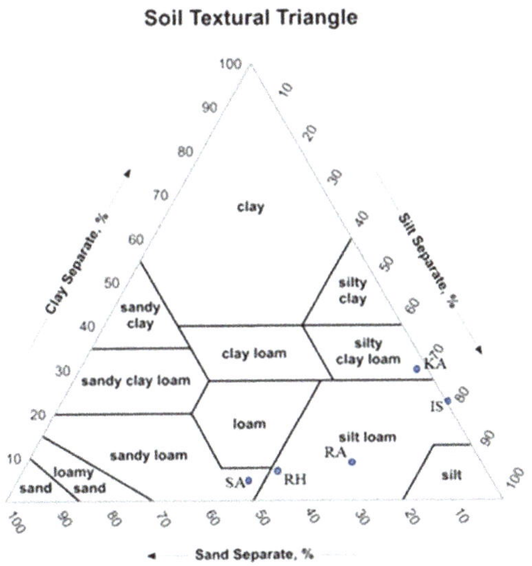

Figure 3. Sediments classification based on their texture, adapted from USDA [29].

Organic matter content was determined using the ignition temperature of 550 °C for dredged sediments or sludge, according to [30]. The $CaCO_3$ content was obtained from a calcimetry test. Atterberg limits were determined using Casagrande apparatus, and all results are summarized in Table 2.

3.2. *Dewatering Characterization*

3.2.1. Characterization of Changes in Water Content

Changes in water content were characterized by monitoring the remaining water content percentage ω_1, drained water percentage ω_2 and evaporated water percentage ω_3. The remaining water percentage ω_1 is the percentage of the initial weight of water added initially to all the samples. The drained water percentage ω_2 is the percentage of the water released from all the samples and gathered into the support, from the beginning of the dewatering process until the time of measurement. Likewise, the evaporated water percentage ω_3 is the water evaporated from all samples in the same conditions. So, the different water percentages give Equation (2).

$$\omega_1 + \omega_2 + \omega_3 = 1 \quad (2)$$

The evolution of ω_1 is directly dependent on water content as one can deduce from Equation (3). The sediment water content ω at any given day during the dewatering process is then characterized through a normalized expression.

$$\omega/L_L = (w_w\ \omega_1/w_s)/L_L \quad (3)$$

where ω/L_L is the water content to liquidity limit; w_w is the initial weight of water added to all samples and w_s is the corresponding dry weight. The change in drained water is plotted in Figure 4. Quick drainage of the IS, RA and RH sediments is observed compared to the others; see Figure 4a. In particular, these sediments have high drainage capacity where ω_2 has reached 32%, 28% and 46%, respectively, compared to 4% and 8.5% for the SA sediment and the kaolin (KA), respectively; see Figure 4b,c.

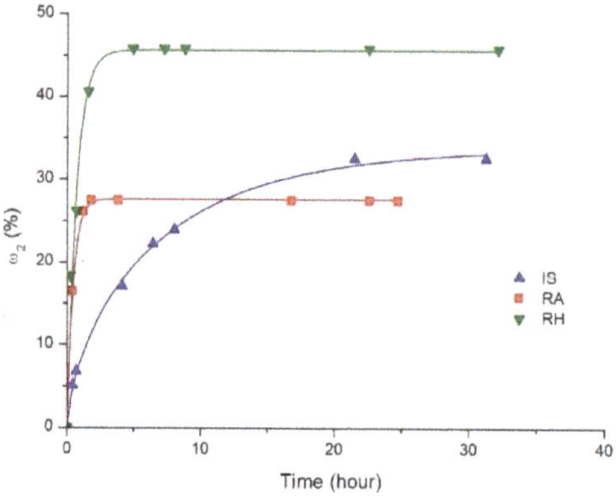

(a)

Figure 4. *Cont.*

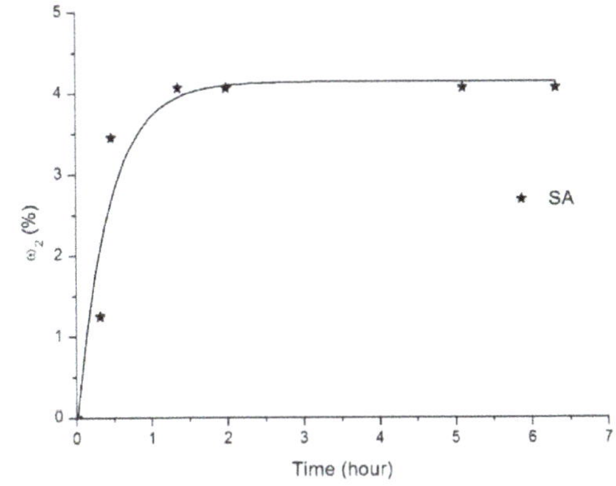

(**b**)

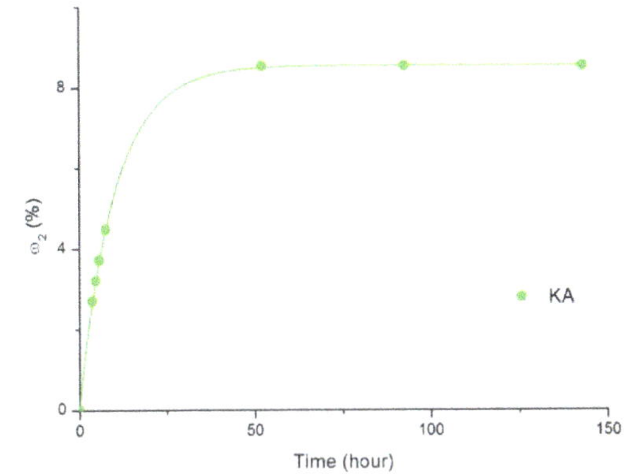

(**c**)

Figure 4. Cumulative percentage of drained water ω_2 measured in the support for all samples: (**a**) IS, RA and RH sediments, (**b**) SA sediment and (**c**) kaolin clay (KA).

A maximum amount of drained water is reached (Figure 4a) for sediments with coarser soil texture, which is the case for the RA and RH sediments when compared to the IS sediment. Having the highest liquid limit—and therefore having the highest initial water content and consequently the highest void ratio—also explains the important quantity of drained water for the RH sediment. On the same figure, we observe, for the IS sediment, a finer soil texture than that of the RA sediment; the water drained was lesser for the RA sediment because of greater organic matter content; see Table 2. Sediment SA, with a significant amount of organic matter content (6.61%), contributed also to shortening the drainage time, and a low quantity of water (4%) was drained despite its coarse soil texture and minimum $CaCO_3$ content (Table 2). Once the plateau is observed on the graphs, the drainage phenomenon tends to stop. We also note that the drainage lasts from a few hours for sediments to several days for kaolin clay (between 2 and 3 days; see Figure 4c) for a low water volume of around 8% of the initial water content. This is due to the fine colloidal texture of the kaolin that contributes to water retention, even with low percentages of $CaCO_3$ and organic matter content. Drained water graphs fit well with an exponential function over time that corresponds to Equations (4)–(6); see Table 3.

Table 3. Possible relationships of drained water vs. time for all samples.

From Figure 4a ω_2 [%] = f(t)[hour], for RA and RH sediments Equation (4): $\omega_2 = a(1 - b^t)$, type: Exponential, BoxLucas 1 Mod				
Sediment	a	b	r^2	
RA	27.61	0.11	0.99	
RH	45.64	0.23	0.99	
From Figure 4a,c, for IS sediment and for kaolin clay (KA) Equation (5): $\omega_2 = a(1 - e^{-bt})^c$, type: Exponential, Chapman				
Sediment	a	b	c	r^2
IS	33.81	0.10	0.64	0.99
Kaolin clay	8.55	0.09	0.94	0.99
From Figure 4b, ω_2[%] = f(t)[hour], for SA sediment Equation (6): $\omega_2 = a - bc^t$, type: Exponential, Asymptotic				
Sediment	a	b	c	r^2
SA	4.14	4.28	0.10	0.92

Figure 5 shows how the drained water ω_2 evolves versus the normalized water content ω/L_L. Drainage ends when the curve becomes constant. A similar trend is observed from the beginning of drainage till the end when the plateau appears. It is confirmed by all samples with normalized water content $\omega/L_L = 2$. The level of the plateau depends on the factors previously mentioned and govern the drainage behavior. Since the dewatering of SA sediment began from normalized water content lesser than the other tested samples ($\omega/L_L = 1.56$), it seems that the initial slope is not so different from the other samples as illustrated in Figure 5. It is evident that the drainage depends on the initial water content of the sample as well as its consistency. The evolution of ω_2 versus normalized water content $\omega/(L_L)$ is well characterized by a sigmoidal-type function; see Equation (7) in Table 4.

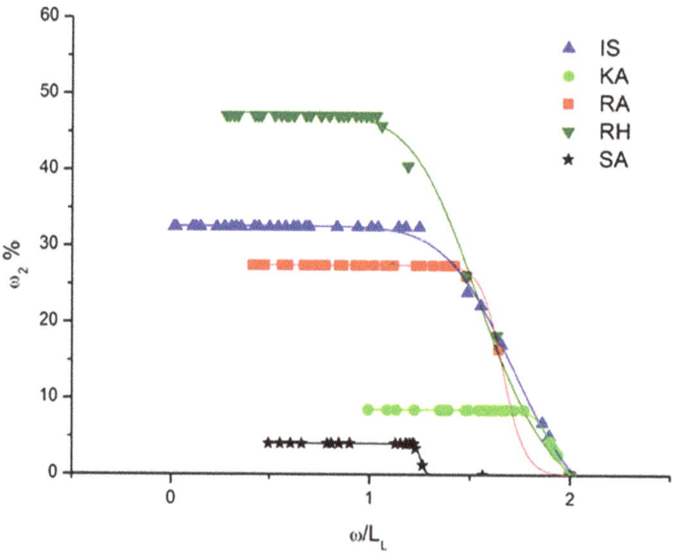

Figure 5. Cumulative percentage of drained water ω_2 versus normalized water content to liquid limit for all samples.

Table 4. Possible relationships of drained water vs. normalized water content for all samples.

From Figure 5, ω_2 = [%] = f(ω/L_L) Equation (7): $\omega_2 = A_2 + (A_1 - A_2)/(1 + ((\omega/L_L)/((\omega/L_{L0.5})^{p})))$ type: Sigmoidal, Logistic					
Sediment	A_1	A_2	$\omega/L_{L0.5}$	p	r^2
IS	32.65	−8.32	1.75	9.36	0.99
KA	8.55	−1.25	1.92	43.64	0.99
RA	27.52	−0.10	1.66	30.47	0.99
RH	46.17	−4.98	1.58	9.00	0.99
SA	4.08	0.00773	1.26	108.07	0.99

Figure 6 illustrates how the normalized water content evolves for all samples over time. When reaching a specific value in less than a week, indicated by pink diamonds, the normalized water content shows an approximately uniform rate of evolution. These specific values correspond to the end of drainage, which is observed in Figure 4. Hence, in addition to the remaining water—see Equation (3)—the normalized-water-content-to-liquid-limit ratio ω/L_L depends also on the ω_2 percentage rather than the samples' physical states.

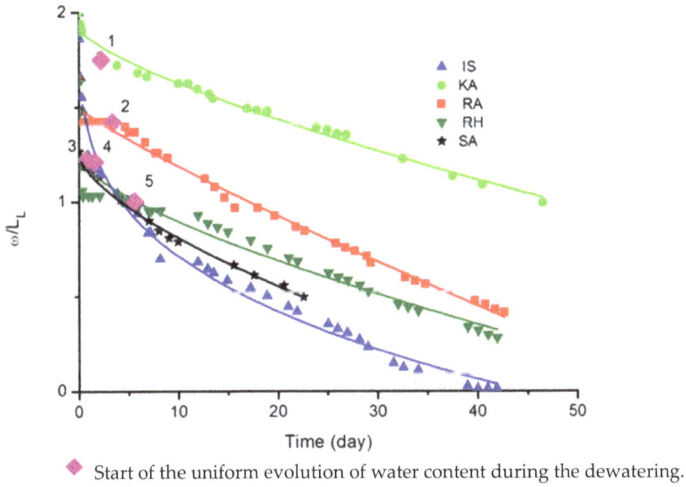

Figure 6. Water-content-to-liquid-limit ratio versus time for all samples.

The normalized-water-content-to-liquid-limit ratio ω/L_L seems to fit with a power type evolution with time for all the samples, as detailed in Equation (8); see Table 5.

Table 5. Possible relationships of water content to liquid limit vs. time for all samples.

From Figure 6, ω/L_L = [%] = f(t) Equation (8): $\omega/L_L = a + bt^c$, type: Power, Allometric				
Sediment	a	b	c	r^2
IS	2.00	−0.70	0.28	0.99
RH	1.23	−0.07	0.69	0.90
KA	1.92	−0.05	0.73	0.99
RA	1.50	−0.04	0.87	0.99
SA	1.25	−0.09	0.66	0.99

Considering the evaporated water in Figure 7, a linear increase of ω_3 over time is observed fitted with Equation (9) in Table 6.

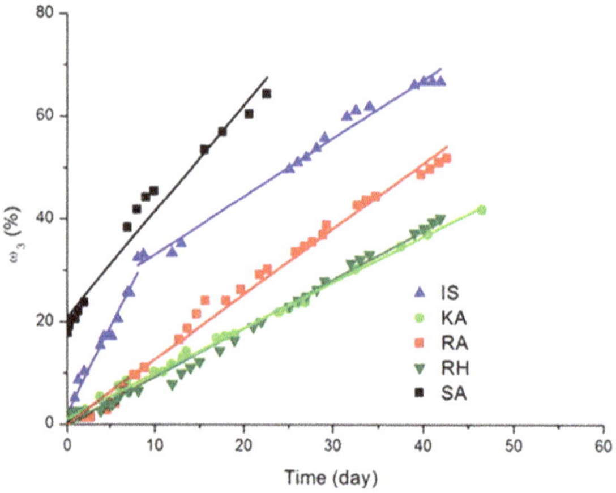

Figure 7. Percentage of evaporated water ω_3 versus time for all samples.

Table 6. Possible relationships of evaporated water vs. time for all samples.

From Figure 7, ω_3 [%] = f(t)[day] Equation (9): ω_3 = at + b type: Linear, Linear			
Sediment	a	b	r^2
IS (inf part)	2.05	3.36	0.97
IS (sup part)	1.15	21.34	0.99
KA	0.89	0.10	0.99
RH	0.95	−0.27	0.99
RA	1.27	−0.09	0.99
SA	2.07	20.72	0.97

Beyond the end of drainage, if ω_3 is plotted as a function of ω/L_L, we also observe a linear variation. For the RH sediment, the linear increase starts when the sediment reaches the liquid limit; see Figure 8.

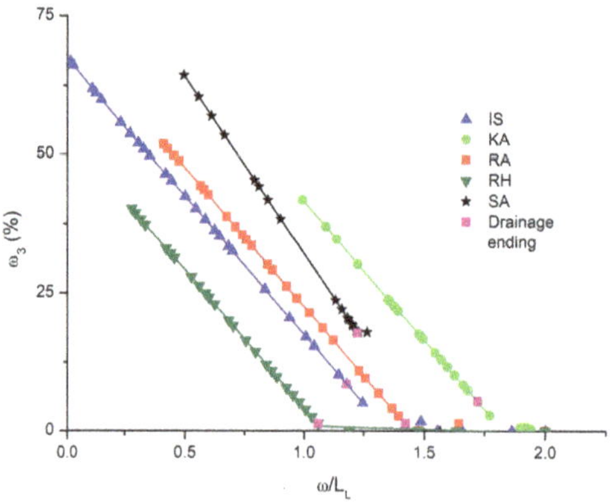

Figure 8. Percentage of evaporated water ω_3 versus normalized water content for all samples.

If we observe all the quantities of water, ω_1, ω_2 and ω_3, during the dewatering process in Figure 9, it is confirmed that (1) the evaporated water increases linearly when drainage stops, and (2) the approximate slope change in water content in sediments is roughly noted. This is more perceptible for the sediments in Figure 9a–d than it is for kaolin, as illustrated in Figure 9e.

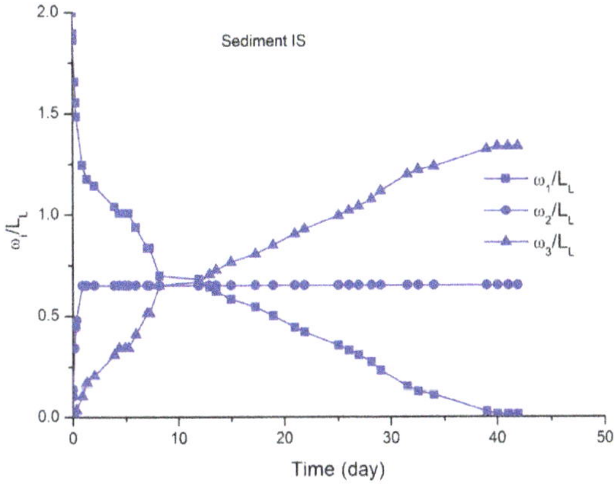

(**a**) IS sediment

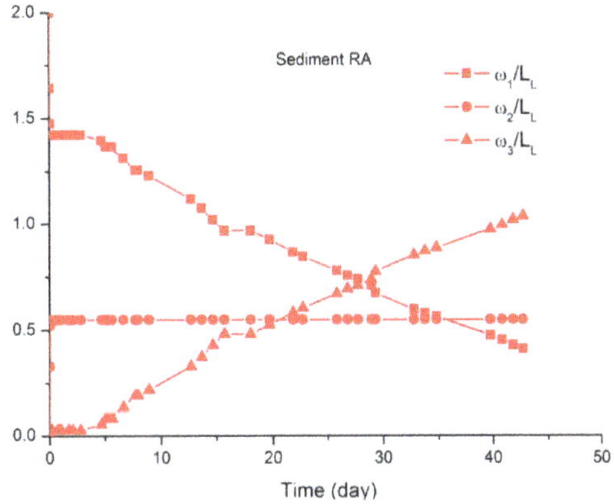

(**b**) RA sediment

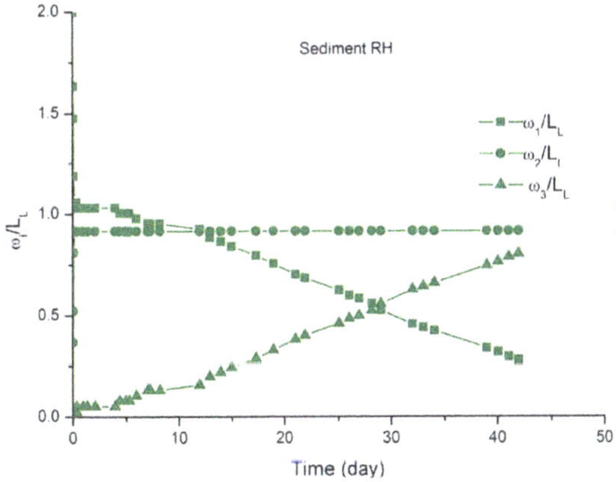

(**c**) RH sediment

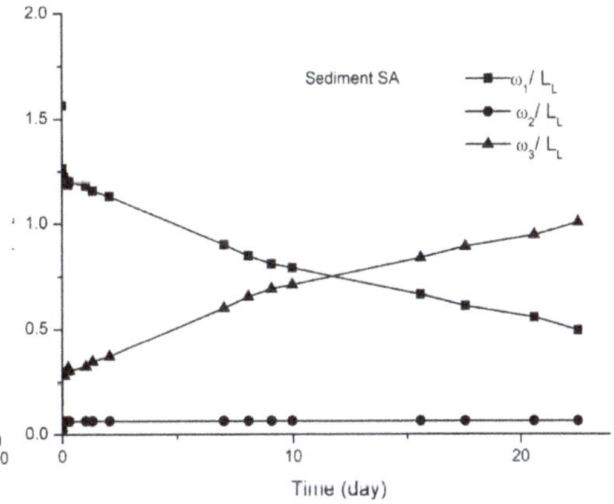

(**d**) SA sediment

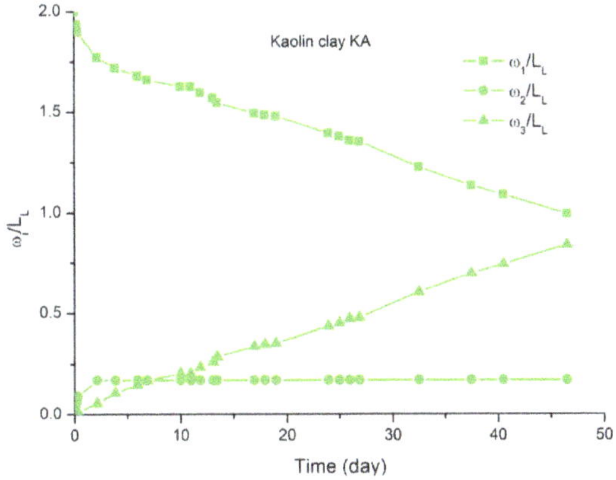

(**e**) KA kaolin clay

Figure 9. Ratio of the content of water to liquid limit versus time.

3.2.2. Void Ratio Evolution

The void ratio e is usually calculated from Equation (10):

$$e = \gamma_s/\gamma_d - 1 = \gamma_s/(gM_s/V_T) - 1 \quad (10)$$

where $\gamma_s = 26.5\ kN/m^3$; $g = 9.81\ m/s^2$; γ_d is the dry unit weight; Ms is the mass of dry solid particles; and V_T is the total volume. The dewatering phenomenon resulted in sediment shrinkage. Shrinkage characterization requires the determination of the void ratio by finding out the initial weight of the sample and its initial volume as given in Table 1. The variation in the diameter and height of the sample versus time is assessed as explained in Section 2.2.1; it allows one to determine the void ratio. Considering Figure 10, the RH sediment underwent significant shrinkage over 5 hours; its void ratio decreased from 3.7 to 2.5. This is explained by the high sand fraction of about 41.7% which enhances water drainage and then subsequent grain rearrangement. The same shrinkage behavior is also observed for the RA sediment, which contain 26.3% of sand fraction. The decrease in void ratio was from 2.7 to 2.1 over two days, as illustrated in Figure 10. Such an observation could be explained by the initial water content ω_0, considering that the higher the water content, the higher the void ratio is. For the RH sediment, initial water content equals to 142%, which makes it 1.3 times that of the RA sediment's ω_0. The IS sample, with the lowest organic matter, shows a slow decrease in void ratio versus time. Note its drainage needed more time than the other sediments as seen in Figure 4a. Sediment SA, with its $\omega_0 = 101.8\%$ and $\omega_0 = 1.57\ L_L$, sustained little volume change within an hour; its void ratio varied from 2.5 to 2.3 before stabilizing.

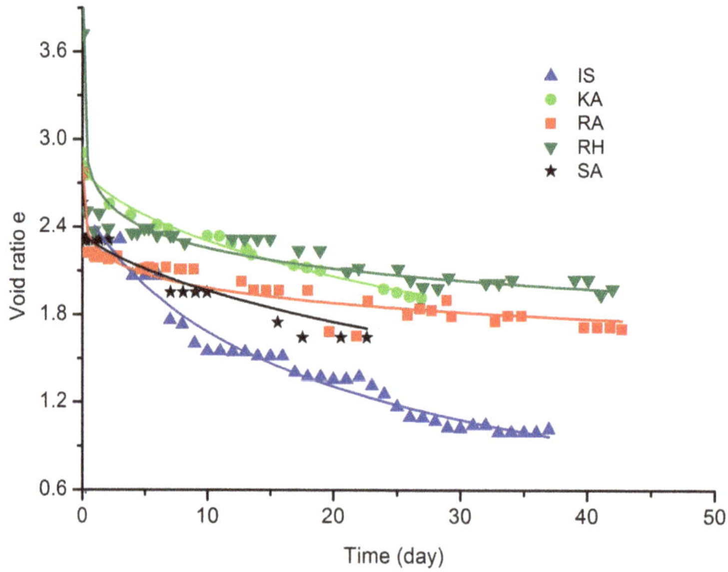

Figure 10. Void ratio versus time during the first three days for all samples.

Figure 10 shows the instantaneous rearrangement of particles of sandy soil highlighted with a rapid decrease in void ratios for sediments RA and RH. In turn, the colloidal behavior of the clayey materials in the IS sediment and kaolin clay (KA) resulted in a slow decrease. The variation in void ratio versus time illustrated in Figure 10 is correlated by a power-type evolution given by Equation (11) given in Table 7.

Table 7. Possible relationships of void ration vs. time for all samples.

From Figure 10, e = f(t)[day], for all sediments and kaolin clay Equation (11): $e = (a + bt^c)^{-1}$ type power, Harris				
Sediment	a	b	c	r^2
IS	0.39	0.02	0.88	0.97
KA	0.36	0.015	0.71	0.99
SA	0.46	0.03	0.95	0.97
RA	0.35	0.09	0.24	0.84
RH	0.24	0.13	0.19	0.79

Based on the variation in void ratio for all samples with comparable initial water content, it is concluded that soil texture mostly governs soil particle rearrangement during drainage.

3.3. Geotechnical Characterization

3.3.1. Variation in Dry Unit Weight

Unit weight has an important role in sediment transportation as a parameter interfering with the bulk factor of soil and hence its volume. It also affects soil consolidation and, consequently, its undrained shear strength [31]. Touiti et al. [32] proposed for Tunis soft clay characterization a linear correlation (slope variation). All samples of Tunis soft clays were tested at liquid limit L_L with varying clay content ranging from 20% to over 70%. As shown in Figure 11, a linear variation is more or less observed for a range of water content beyond the liquidity limit. Their evolution is perfectly characterized by a linear function; see Equation (12) given in Table 8. Fitting is based on data corresponding to Figure 11 with a minimal regression coefficient $r^2 = 0.83$. The clay content in the sediments could explain the difference of the coefficients (a) relative to the slope; see Table 8. The slope is more pronounced with Tunis soft clays [32]. It is observed that the RA, RH and SA sediments have a similar slope due to their comparable soil texture. The IS sediment, whose texture is much finer compared to other tested sediments, has a different evolution; see Figure 11 and Table 8.

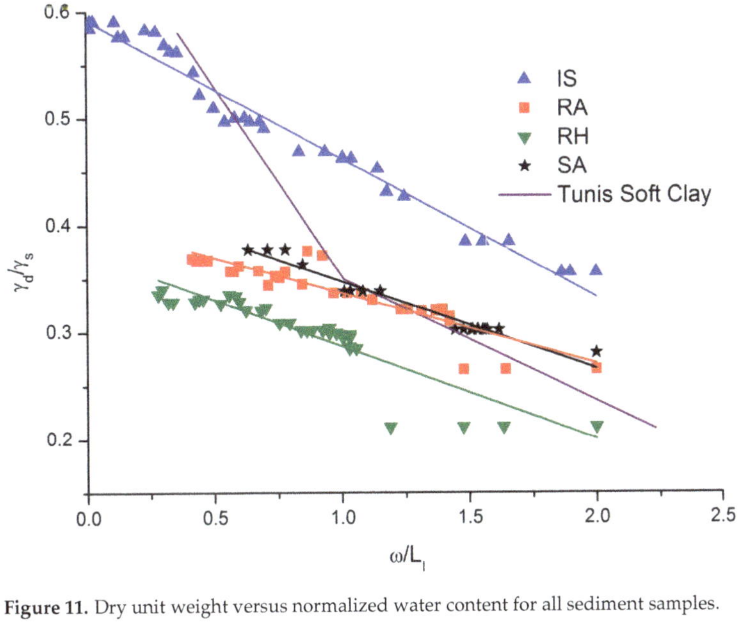

Figure 11. Dry unit weight versus normalized water content for all sediment samples.

Table 8. Possible relationships of dry unit weight vs. normalized water content for sediments.

From Figure 11, $\Upsilon_d/\Upsilon_s = f(\omega/L_L)$ Equation (12): $\Upsilon_d/\Upsilon_s = a(\omega/L_L) + b$ type: Linear, Linear			
Sediment	a	b	r^2
IS	−0.13	0.59	0.97
SA	−0.08	0.43	0.96
RA	−0.07	0.40	0.83
RH	−0.09	0.37	0.83

3.3.2. Shear Strength Relationships

Figure 12 illustrates undrained shear strength evolution with water content leading to the conclusion that shear strength is developed after drainage ending. Maximum values of undrained shear strength for the IS, RA and SA sediments were in the same ranges as shown in Figures 13 and 14. For the SA sediment, the undrained shear strength Su reached 75 kPa in 18 days, followed by the RA sediment, whose shear strength developed faster but not sharper than the IS sediment during the first 22 days, reaching 70 kPa. The IS sediment reached higher values, reaching 130 kPa, whereas the RA sediment only reached 112 kPa at the end of the dewatering; see Figure 14.

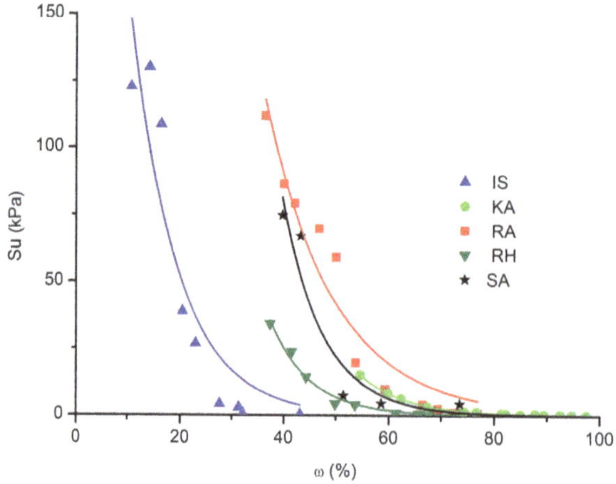

Figure 12. Shear strength evolution versus water content for all samples.

From Figure 12, it can be deduced that the RA sediment, with the highest organic matter content as given in Table 2, could develop high ranges of undrained shear strength for high water content when compared to all the other samples. Considering the initial water content of other sediments, the RA sediment dewatering began from a lower water content, $\omega = 108\%$. This fact also contributed to it developing high undrained shear strength in a shorter time. The RH and SA sediments have comparable organic matter content; however, the SA sediment developed higher undrained shear strength than RH sediment. Considering all the sediments, the IS sediment, which has a low organic matter content, was able to develop undrained shear strength at a low water content. This was also the case for the kaolin which had negligible organic matter content and was the least to develop undrained shear strength since it kept a significant amount of water content during the dewatering process.

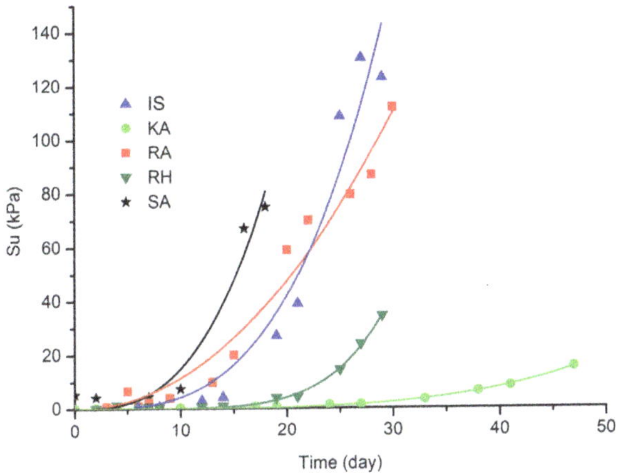

Figure 13. Shear strength evolution versus time for all samples.

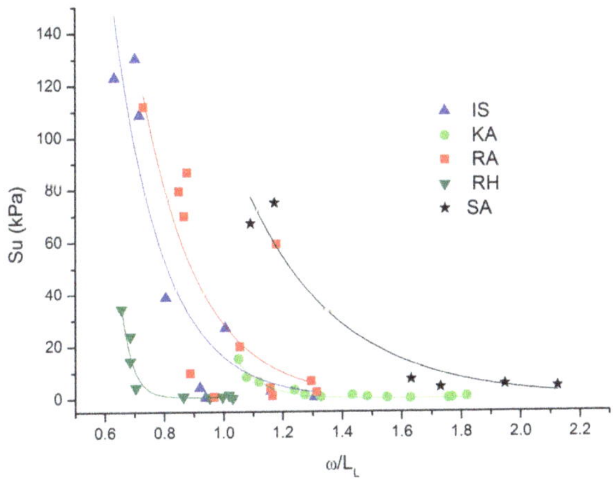

Figure 14. Shear strength evolution versus normalized water content for all samples.

Observing the shape of the curves of the IS and RA sediments in Figure 13, one can note that shear strength is developed at the same rate after the tenth day of dewatering. This observation is also valid for the rate of evolution of the normalized water content to liquid limit when considering Figure 14.

An exponential-type correlation can describe the undrained shear strength evolution with water content for all the samples tested with a minimal regression coefficient of $r^2 = 0.87$; see Equation (13) in Table 9.

Table 9. Possible relationships of Su vs. water content for all samples.

From Figure 12, Su [kPa] = f(ω)[%] Equation (13): Su = abω, type: Exponential, Exp2p					
Sediment	IS	KA	RA	RH	SA
a	492.03	8034.45	1886.19	4485.12	12,051.21
b	0.89	0.89	0.93	0.88	0.88
r2	0.87	0.99	0.94	0.98	0.92

Undrained shear strength evolution versus time was also fitted with an exponential. Equation (14) in Table 10 correctly describes the undrained shear strength evolution with a minimal regression coefficient of $r^2 = 0.94$.

Table 10. Possible relationships of Su vs. time for all samples.

From Figure 14, Su [kPa] = f(t)[day] Equation (14): Su = at^b, type: Power, Allometric1			
Sediment	a	b	r^2
IS	0.00216	3.29537	0.94
RH	1.53×10^{-7}	5.71	0.99
KA	1.77×10^{-6}	4.14	0.99
RA	0.097	2.069	0.99
SA	0.019	2.89	0.94

Undrained shear strength increased faster during the dewatering when attaining the liquid limit for the IS, RA and SA sediments at days 3, 14 and 7, respectively; see Figure 13. Comparing Figures 14 and 15, it is demonstrated that attaining the liquidity limit allows the undrained shear strength to develop. However, maximum values of undrained shear strength are observed when water content nears the plastic limit value, as was the case for the RH sediment. For the latter, the undrained shear strength rose when attaining the plastic limit; see Figure 15.

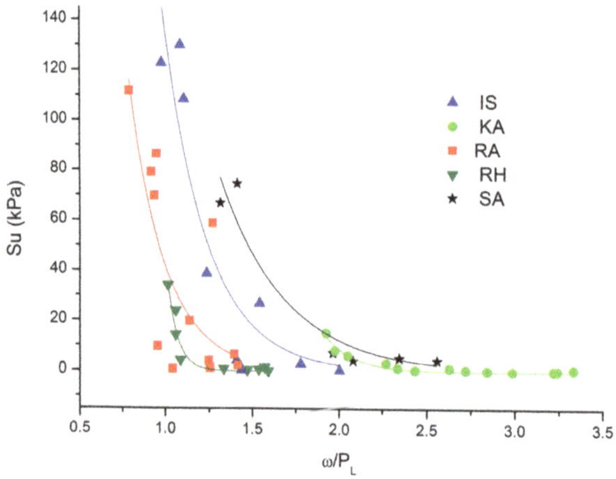

Figure 15. Shear strength evolution versus water content to plastic limit ratio for all samples.

Figures 14 and 15 lead to the conclusion that for sediments having a comparable plasticity index, the undrained shear strength develops a similar trend. This leads to the conclusion that particle bonds begin to develop at a water content equal to the liquidity limit, resulting in undrained shear strength. The undrained shear strength reaches its maximum at the plastic limit.

4. Conclusions

4.1. On the Dewatering Process of Sediments and a Kaolin Clay

In practice, sediment characterization helps to optimize the sediment's recovery process, transport, storage and end use. Sediment characterization plays an important role in sediment management. In practice, it helps to optimize the sediment's recovery process, transport, storage and end use. Sediment management requires a range of data, such as water content, dry unit weight, volume factor and volumetric deformation, which requires shear strength to be quantified. The study of the evolution of the percentage of water drained, ω_2, highlights the correlation of the function of the peak, which varies as a function of time or normalized water content. It was concluded that coarse soil texture and high initial water content contribute to the highest drainage peak, that high organic matter content shortens drainage time and that $CaCO_3$ inhibits the evaporation of already-drained water. The percentage of water evaporated showed a linear trend for all samples tested. While the percentage of water drained followed a sigmoidal trend, based on the normalized water content for all the samples tested Equation (7) in Table 4, a unique and linear correlation was found relating dry unit weight to solid particle weight as a function of water content, as the sediments tested have low clay content. This observation led to the fact that intergranular forces in clays are more dominant when the water content is below the liquid limit, and that below a certain clay content, intergranular forces have no effect on the structure of the material.

4.2. Influence of Sediment Texture

The variation in water content over time for the IS sediment is similar to that for the kaolin clay (KA). It is evident that soil texture and initial water content govern the response of the sediment to dewatering. Similarly, for the RA, RH and SA sediments of the same texture, the evolution of water content was comparable. However, it is worth mentioning that the electrostatic forces in the kaolin clay (KA) compared to the IS sediments helped to prevent the water evaporating and draining. This is demonstrated by the observation of faster evaporation of water in the IS sediment than in the kaolin clay (KA). This could be explained by the large quantities of water continuously reaching the surface of the IS sediment sample due to the phenomenon of suction (finest soil texture) and likely to evaporate. Reaching the liquid limit during dewatering led to the appearance of an inflection point in the evolution of remaining, drained and evaporated waters for all the sediments tested. The result was a decrease in evaporation and drainage rates. In fact, the higher the plasticity index of the sediment, the more perceptible the inflection point, as the variation in the rates of evolution of the different quantities of water began to reach the liquidity limit and continue up to the plasticity limit. Consequently, the inflection point was very distinct in the case of the RH sediment. In contrast to the Tunis soft clays, the evolution of the dry unit weight in relation to the solid particle weight was linear with no variation in slope. One explanation is the low clay content and hence the inability of the sediments to adsorb water. Indeed, before saturation, clay tends to gain water due to electrostatic forces, which is not the case for sediments due to their low clay content. Sediments of comparable texture have solid particles that behave in the same way with water content; consequently, correlations between dry unit weight and water content have been established. The texture of the IS sediment— being similar to that of kaolin clay—its high $CaCO_3$ content and low organic matter content enabled the development of the lowest void ratio and therefore the highest dry unit weight and the achievement of the highest shear strength values [33]. As confirmed by Chen et al. [34], $CaCO_3$ content greater than 5% makes it possible to develop shear strength for particles smaller than 100 μm.

4.3. Development of Undrained Shear Strength during Dewatering

The undrained shear strength values reached 70–130 kPa at the end of dewatering. Correlations between the exponential curves of Su and the normalized water content ω/L_L were found. Undrained shear strength depends on the consistency of the material rather than its water content. Sediments with the lowest water content and highest organic matter content exhibit faster shear strength. Clay texture contributes to undrained shear strength. Materials with a comparable plasticity index develop shear strength according to a similar trend.

Noting the undrained shear strength graphs obtained, it is evident that both the liquid limit and the plastic limit influenced the development and/or evolution of shear strength, which is indeed directly related to soil consistency. More importantly, the fact that the RH sediment only developed shear strength when it reached the plastic limit and, given its high liquid limit, leads to the conclusion that water content also governs the development of shear strength, in addition to soil consistency. This is because the more water the sediment retains, the more voids there are, increasing the distance between particles. This prevents the development of cohesion in the sediment.

This study makes it possible to predict the mechanical properties of sediments during dewatering and their physical characteristics. Consequently, in practice on site, similar conditions need to be provided during sediment dewatering.

Author Contributions: Conceptualization, D.A. and B.B.A.; methodology, D.A. and B.B.A.; software, D.A.; validation, D.A. and D.L.; formal analysis, D.A.; investigation, D.L., D.A. and B.B.A.; writing—original draft preparation, D.A.; writing—review and editing, D.L., D.A. and B.B.A.; visualization, D.A. and B.B.A.; supervision, D.L.; project administration, D.L.; funding acquisition, no funding. All authors have read and agreed to the published version of the manuscript.

Funding: This research received no external funding.

Data Availability Statement: All data results and relative information to this study are available, please contact the corresponding author.

Conflicts of Interest: The authors declare no conflict of interest.

References

1. GEODE—Groupe D'études et D'observation sur les Dragages et L'environnement. Dragages et Immersions en mer et en Estuaire, Revue des Bonne Pratiques Environnementales, 2018, Report, 183p, Annexes 90p. Available online: https://www.cerema.fr/system/files/documents/2018/04/GEODE_BPE_revue_0702018_VF.pdf (accessed on 2 August 2024).
2. Dorleon, G.; Rigaud, S.; Techer, I. Management of dredged marine sediments in Southern France: Main keys to large-scale beneficial re-use. *Environ. Sci. Pollut. Res.* **2024**. [CrossRef] [PubMed]
3. Harrington, J.R.; Murphy, J.; Coleman, M.; Jordan, D.; Debuigne, T.; Szacsuri, G. Economic modelling of the management of dredged marine sediments. *Geol. Geophys. Environ.* **2017**, *42*, 311. [CrossRef]
4. Mymrin, V.; Stella, J.C.; Scremim, C.B.; Pan, R.C.Y.; Sanches, F.G.; Alekseev, K.; Pedroso, D.E.; Molinetti, A.; Fortini, O.M. Utilization of sediments dredged from marine ports as a principal component of composite material. *J. Clean. Prod.* **2017**, *142 Pt 4*, 4041–4049. [CrossRef]
5. Feki, N.; Mbarka, M. Caractérisation physique et environnementale des sédiments du port de pêche de Sfax. In Proceedings of the 3rd Coastal and Maritime Mediterranean Conference, Ferrare, Italy, 25–27 November 2015. [CrossRef]
6. London Convention. Convention on the Prevention of Marine Pollution by Dumping of Wastes and Other Matter, International Maritime Organization (IMO). 1972. Available online: https://www.epa.gov/sites/default/files/2015-10/documents/lc1972.pdf (accessed on 2 August 2024).
7. El Fadili, M.; Messager, M. *Enquête Dragage 2011: Synthèse des Données*; Cerema—Cetmef: Compiègne, France, 2015; 39p.
8. Duan, Z.; Lafhaj, Z.; Bel Hadj Ali, I.; Ducellier, S. Valorisation des sédiments fluviaux traités en vue d'une utilisation en génie civil. *Rev. Paralia* **2013**, *6*, 5.1–5.12. [CrossRef]
9. Goure-Doubi, H.; Lecomte-Nana, G.; Thery, F.; Peyratout, C.; Anger, B.; Levacher, D. Characterization and valorization of dam sediment as ceramic materials. *Int. J. Eng. Innov. Technol.* **2015**, *4*, 84–91.
10. Said, I.; Missaoui, A.; Lafhaj, Z. Reuse of Tunisian marine sediments in paving blocks: Factory scale experiment. *J. Clean. Prod.* **2015**, *102*, 66–77. [CrossRef]
11. Anger, B.; Thery, F.; Levacher, D. Implementation of minimum basic characterization of fine sediments in construction material for practical purposes. In Proceedings of the 11th Geo-Environmental Engineering, Caen, France, 30–31 May 2012.
12. Levacher, D.; Liang, Y. Sediment solidification/stabilization: Similarities differences, difficulties, cost and new developments. In Proceedings of the Geo-Environmental Engineering, Caen, France, 30–31 May 2012.
13. Bel Hadj Ali, I.; Lafhaj, Z.; Bouassida, M.; Said, I. Characterization of Tunisian marine sediments in Rades and Gabes harbors. *Int. J. Sediment Res.* **2014**, *29*, 391–401. [CrossRef]
14. Ben Slama, A.; Feki, N.; Kamoun, A. Etude physique et environnementale des sédiments du port de commerce de Sfax. In Proceedings of the Coastal and Maritime Mediterranean Conference, Split, Croatia, 29 November–1 December 2017; pp. 91–98. [CrossRef]
15. Boutouil, M.; Levacher, D. Traitement et valorisation des vases de dragages par solidification/stabilisation: État de l'art. In Proceedings of the VIth Journées Nationales Génies Civil—Génies Cotier, Caen, France, 17–19 May 2000; pp. 367–375. [CrossRef]
16. Levacher, D.; Sanchez, M. Caractérisation de sédiments marins pour une mise en dépôt à terre et en remblai. *Eur. J. Environ. Civ. Eng.* **2011**, *15*, 167–178. [CrossRef]
17. Samara, M.; Lafhaj, Z.; Chapiseau, C. Valorization of stabilized river sediments in fired clay bricks: Factory scale experiment. *J. Hazard. Mater.* **2009**, *163*, 701–710. [CrossRef] [PubMed]

18. Scordia, P.-Y.; Lafhaj, Z.; Skoczylas, F.; Mongeois, F. Caractérisation et valorisation en technique routière de sédiments fluviaux pollués et traités. *Rev. Eur. Génie Civ.* **2008**, *12*, 1087–1104. [CrossRef]
19. Lafhaj, Z.; Duan, Z.; Bel Hadj Ali, I.; Depelsenaire, G. Valorization of treated river sediments in self-compacting materials. *Waste Biomass Valorization* **2012**, *3*, 239–247. [CrossRef]
20. Zelleg, M.; Said, I.; Hamdi, E.; Lafhaj, Z. Experimental testing for Zarzis port sediments (Tunisia) in road materials. *Geotech. Res.* **2018**, *5*, 13–21. [CrossRef]
21. Dhervilly, P.; Bertrand, M.; Thanneberger, L.; Levacher, D.; Houise, C.; Zoubeir, L. Gestion durable des sédiments: Le projet COVASED. In Proceedings of the XIIIth Journées Nationales Génie Côtier—Génie Civil, Dunkerque, France, 2–4 July 2014; pp. 977–985. [CrossRef]
22. Boullosa Allariz, B. Déshydratation Naturelle et Mécanisée de Sédiments: Étude des Processus mis en jeu et Applications. Ph.D. Thesis, Université de Caen Normandie, Caen, France, 12 December 2018.
23. Boullosa Allariz, B.; Levacher, D. Mechanical dewatering solutions for sediments. In Proceedings of the 15th Geo-Environmental Engineering, Nantes, France, 2–3 June 2016; p. hal-01929615.
24. Boullosa Allariz, B.; Levacher, D.; Thery, F. State parameters during natural dehydration of dam dredged sediments. In Proceedings of the 10th International SedNet Conference, Genoa, Italy, 14–17 June 2017.
25. Gupta, R.; Levacher, D.; Razakamanantsoa, A. Analogy between T-Bar and vane shear test. In Proceedings of the 15th Geo-Environmental Engineering, Nantes, France, 2–3 June 2016; pp. 33–40.
26. Levacher, D.; Boullosa Allariz, B.; Müller, M. Some aspects of sediments dewatering, laboratory testing and useful relationships for dewatering. In Proceedings of the 17th Geo Environmental Engineering, Fukuoka, Japan, 25–26 May 2018; pp. 220–228.
27. Anger, B. Caractérisation de Sédiments Fins de Retenues Hydroélectriques en vue D'une Orientation vers des Filières de Valorisation Matière. Ph.D. Thesis, Université de Caen Basse Normandie, Caen, France, 17 December 2014. Available online: https://normandie-univ.hal.science/tel-01938082 (accessed on 2 August 2024).
28. *D 4648/D4648M-13*; Standard Test Method for Laboratory Miniature Vane Shear Test for Saturated Fine-Grained Clayey Soil. Annual Book of ASTM Standard. ASTM International: West Conshohocken, PA, USA, 2010.
29. USDA, United States Department of Agriculture. Soil Survey Manual, Handbook N°18, Soil Science Division Staff, Issued March 2017, USA, 603p. Available online: https://www.nrcs.usda.gov/sites/default/files/2022-09/The-Soil-Survey-Manual.pdf (accessed on 2 August 2024).
30. *EN 12879: 2000*; Characterization of Sludges—Determination of the Loss of Ignition of Dry Mass. British Standards Institution: London, UK, 2000.
31. Bjerrum, L. Fundamental considerations on the shear strength of soil. *Géotechnique* **1951**, *2*, 209–218. [CrossRef]
32. Touiti, L.; Bouassida, M.; Van Impe, W. Etude de la sensibilité de la vase superficielle de la ville de Tunis. In Proceedings of the Innovative Geotechnical Engineering. International Conference on Geotechnical Engineering, Hammamet, Tunisia, 24–26 March 2008; pp. 379–388.
33. Moore, D.G. Shear strength and related properties of sediments from experimental mohole (Guadalupe site). *J. Geophys. Res.* **1964**, *69*, 4271–4291. [CrossRef]
34. Chen, L.; Chen, X.; Yang, X.; Bi, P.; Ding, X.; Huang, X.; Wang, H. Effect of calcium carbonate on the mechanical properties and microstructure of red clay. *Adv. Mater. Sci. Eng.* **2020**, *2020*, 5298186. [CrossRef]

Journal of
Marine Science and Engineering

MDPI

Article

Provenance Analysis of the Northern Offshore Mud Area of the Shandong Peninsula, China, Spanning the Last 2000 Years

Shuyu Wu [1,2,3,4,5], Jun Liu [2,5,*] and Yongcai Feng [2]

1 School of Engineering, China University of Geoscience (Wuhan), Wuhan 430074, China; hnwushuyu@163.com
2 Ministry of Natural Resources Observation and Research Station of Land-Sea Interaction Field in the Yellow River Estuary, Yantai Center of Coastal Zone Geological Survey, China Geological Survey, Yantai 264000, China; fyc1107@163.com
3 Key Laboratory of Submarine Geosciences, Ministry of Natural Resources, Hangzhou 310012, China
4 Chinese Academy of Geological Sciences, China Geological Survey, Ministry of Natural Resources, Beijing 100037, China
5 Laboratory for Marine Mineral Resources, Qingdao Marine Science and Technology Center, Qingdao 266237, China
* Correspondence: vnlj@163.com; Tel./Fax: +86-0535-2901607

Abstract: The mud area in the northern offshore of the Shandong Peninsula constitutes a dynamic source–sink system in China's continental shelf and is a hotspot for research. However, the provenance of the sediments remains controversial, and the depositional environment is not yet fully understood. This paper performed accelerator mass spectrometry ^{14}C dating, grain-size analysis, clay mineralogic analysis, and geochemical analysis of the ZZ04 sediment core. The results showed that this core primarily comprises silt and clay, reflecting weak sedimentary hydrodynamic conditions and stable deposition. The clay mineral assemblage—illite, smectite, chlorite, and kaolinite—indicates strong physical weathering. The provenance of the mud area was mainly from the Yellow River, Shandong Peninsula rivers, and Yangtze River, highlighting its multi-source characteristics. The smectite/(illite + chlorite) ratio in the ZZ04 core serves as a mineralogical indicator of the East Asian summer monsoon (EASM). Eight significant East Asian winter monsoon (EAWM) intensifications were identified, correlating with global cooling events similar to those at 1.89, 1.4, 1.03, and 0.62 thousand years ago. During the EAWM period, the coastal current and the Yellow Sea warm current play crucial roles in the transportation of matter and heat flux in the mud area. In contrast, during the EASM period, the sediments are predominantly sourced from the Shandong Peninsula rivers, contributing terrigenous materials shaped by chemical weathering.

Keywords: northern offshore mud area of the Shandong Peninsula; mud area; clay minerals; provenance; East Asian monsoon; Yellow River

1. Introduction

The East China shelf marginal sea emerged during the Holocene high-sea-level period, and its sedimentary processes are influenced by the East Asian monsoon. Global sea level fluctuations have led to the formation of several patchy mud areas, which are primarily sourced from sediments carried by major rivers such as the Yellow River (YR) and the Yangtze River. The circulation system, driven by warm and coastal currents, is crucial for sediment transport and deposition [1]. Seven mud areas have been identified on the eastern shelf of China, including the central Bohai Sea (BS) mud area [2], the northern offshore mud area of the Shandong Peninsula [3–6], the eastern Yellow Sea (YS) mud area [7], the central South Yellow Sea (SYS) mud area [8,9], the inland shelf mud area of the East China Sea (ECS) [10,11], the Jeju island mud area [12], and the Beibu Gulf of the South China Sea (SCS) mud area [13]. The northern offshore mud area of the Shandong Peninsula is a significant focus for oceanographic research owing to its rapid sedimentation rates coupled with organic matter burial capabilities and unique geological and geographical attributes [6,14–16]. In the 1980s, Milliman et al. [3] identified an Ω-type subaqueous delta, known as the "Shandong Peninsula mud wedge" [4,17], which has an evolutionary history closely tied to sedimentation from the YR and the Yangtze River. This area provides a detailed record of marine environment changes, including sea level fluctuations, shelf circulation patterns, monsoon evolution, hydrodynamic environment changes, and sediment transport and deposition processes [6,14–16]. Moreover, it serves as an ideal location for studying sediment source–sink dynamics and ocean–land interactions.

Numerous scholars have conducted extensive research on the provenance of the mud area of the Shandong Peninsula. For instance, Wang et al. [18] analyzed the grain-size characteristics of surface sediments in the North Yellow Sea (NYS), reporting that grain-size distribution is primarily influenced by the provenance and sedimentary environment with a significant impact from coastal currents, particularly in the mud area's extension from southwest to northeast. Chen et al. [19] examined the geochemistry of surface sediments in the western part of the NYS, concluding that sediments from the YR are the main source of this region' s sedimentary deposits. Based on ^{210}Pb dating of core sediments, Qi et al. [20] revealed that the western part of the NYS is a multi-source sedimentary area, with YR sediment as the primary source. Liu et al. [21] investigated the sediment provenance in the western part of the NYS since the last glacial maximum using rare-earth elements from the NYS101 borehole sediments. They highlighted that since the Middle Holocene, in addition to the YR sediment, sediments from the Yangtze River and Korean Peninsula rivers have influenced this region. Lan et al. [22] analyzed the major and trace elements of the DLC70-2 borehole sediments, revealing that since the early Late Pleistocene, the YR has been the primary provenance, and the Yalu River exerts a significant influence in the NYS. Xue et al. [6] suggested that the Shandong Peninsula's mud area, rather than forming concurrently with the YR delta, is actually a large mudflat–tidal delta on the seaward side of the Bohai Strait. Furthermore, analyzing the geochemical and clay mineral characteristics of the mud area in the middle of the NYS, Han et al. [23] concluded that the NYS is a multi-source sedimentary area. Its provenance included not only the YR sediments transported by the Shandong coastal current, but also the Liaodong Peninsula river sediments and the Yangtze River sediments carried by the Yellow Sea warm current (YSWC) and atmospheric dust.

Clay minerals serve as a distinctive "fingerprint" for fine-grained sediments of different origins. Owing to their ability to be transported before settling, clay minerals can reveal the sediment source. By analyzing the clay mineralogy of sediments, the transport pathways of suspended matter can be traced. Moreover, the assemblages of clay minerals are intimately connected with the sediment's provenance and climatic evolution, making them a useful tool for reconstructing paleoclimatic and paleoceanographic conditions [24–26]. Given the varied bedrock types and climates across regions, such as the Shandong Peninsula, the Korean Peninsula, and the YR, establishing a correlation between the specific clay mineral assemblages, their sources, and the patterns of their circulation is possible. Monitoring the changes in these assemblages provides a valuable indicator for assessing sedimentary environments and identifying the origins of source materials.

However, the provenance of the sediments of this area remains controversial, and the depositional environment is not yet fully understood. In this study, we utilized a newly acquired core sample, ZZ04, to construct a dating framework using accelerator mass spectrometry (AMS) ^{14}C. Using this framework as a foundation, we conducted grain-size and clay mineral analyses. The study aimed to (1) elucidate the clay mineral assemblages in the mud area and its surroundings, (2) ascertain the provenance of the sediments in the mud area, and (3) investigate the influence of climate change and provenance on the mud area's sedimentary features at different periods.

2. Geological Setting

The NYS is a semi-enclosed marginal sea situated in the Liaodong Peninsula to the northwest, the Shandong Peninsula to the south, and the Korean Peninsula to the east. Its geographical location is approximately between 121°00′ E to 125°10′ E longitude and 37°20′ N to 39°50′ N latitude (Figure 1a). Resembling a half-graben-shaped depression, it opens to the southeast and is characteristic of a typical shallow continental shelf environment [27]. The NYS has an average water depth of 38 m, with the deepest points reaching up to 80 m. It features a shallow plain in the north, where the seabed topography gradually inclines toward the south, and a relatively deep central depression [28].

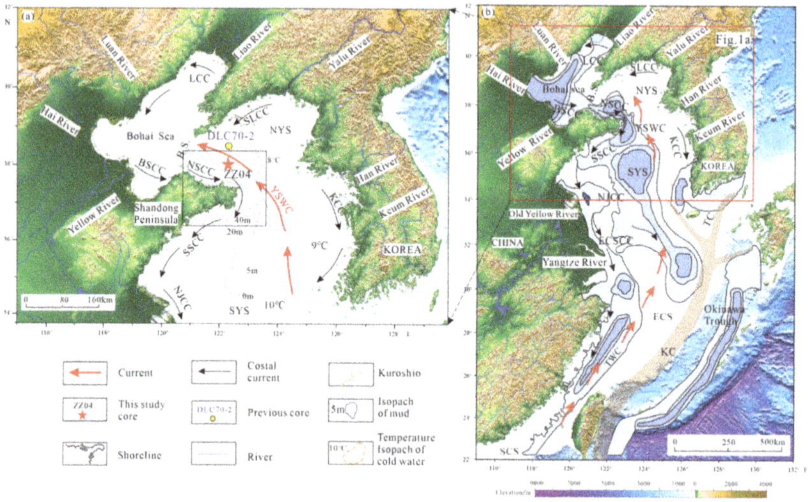

Figure 1. Sketch map showing the oceanographic settings and core location. (**a**) Regional circulation patterns in the Yellow Sea and the Bohai Sea. (**b**) Sketch map showing the circulation patterns in China's eastern marginal sea and the distribution of mud areas. Coastal circulation is indicated by the black arrows, while warm circulation is indicated by the red arrows. The study area is enclosed by the red rectangle. The main currents include the South Liaoning coastal current (SLCC), Liaoning coastal current (LCC), Bohai Sea coastal current (BSCC), North Shandong Peninsula coastal current (NSCC), South Shandong Peninsula coastal current (SSCC), North Jiangsu coastal current (NJCC), Korean coastal current (KCC), East China Sea coastal current (ECSCC), Yellow Sea warm current (YSWC), Taiwan warm current (TWC), and Kuroshio current (KC). The geographical abbreviations include the Bohai Strait (B.S.), North Yellow Sea (NYS), South Yellow Sea (SYS), East China Sea (ECS), and South China Sea (SCS). The main cores studied in this paper are ZZ04 (this study) and a previous study core, DLC70-2 [22].

The basement comprises mainly Archean–Lower Proterozoic metamorphic rocks and Middle and Upper Proterozoic–Cambrian carbonate rocks and slate. From the Middle Jurassic, the basin began to subside, and the sedimentary succession is composed of Middle–Upper Jurassic, Lower Cretaceous, Eocene, Oligocene, and Neogene–Quaternary sediments with terrigenous clastic rock [29]. The NYS receives sediments from several major rivers, including the YR, Yalu River, Korean Peninsula rivers, Shandong Peninsula rivers, and Yangtze River. The YR, with a length of 5463 km, has an average annual runoff of 42.6×10^9 m^3 and transports a sediment discharge of 1089 Mt/year. The Yalu River, with a length of 790 km, has an average annual runoff of 28.9×10^9 m^3 and transports a sediment discharge of 1.13 Mt/year. Approximately 417 km in length, the Han River flows into the SYS in Gyeonggi Province, South Korea. It has an average annual runoff of approximately 19×10^9 m^3 and transports a sediment discharge of approximately 1.84 Mt/year [30].

As a branch of the Kuroshio current (KC), the YSWC flows northwest into the SYS, continuing its journey into the NYS. It extends through the northern part of the Bohai Strait, further reaching into the BS [31,32] (Figure 1b). The YSWC is vigorous in winter yet relatively weak in summer [5,33]. The Shandong coastal current originates from the Bohai Bay, moving eastward along the northern coast of the Shandong Peninsula. Then, it turns southwest at Chengshan Head, following the 40–50 m isobath until it veers south near Laoshan Bay, Qingdao. This coastal current converges with the northbound YSWC, creating a cyclonic circulation system. The path of the Shandong coastal current remains largely stable throughout the year. Near the Chengshan Head, the flow narrows and accelerates to 15 cm/s, peaking at 110 cm/s. The bottom coastal current in the YS is similar to that in the surface layer, albeit with a lower velocity and a more shoreline–proximate flow path [27].

3. Materials and Methods

3.1. Sample Collection

The study samples originated from the ZZ04 core, which was collected by the Yantai Center of Coastal Zone Geological Survey (YCGS) from the northern offshore area of the Shandong Peninsula in 2023 (Figure 1a). The sample site is located at 37.963203° N and 122.370869° E, with a water depth of 44.2 m and a total core length of 3.3 m. The core's lithology primarily comprised gray silty clay and clayey silt.

To ensure the consistency between the river data from the references and our test results, five samples from the YR estuary were collected for concurrent analyses. After collection, these core samples were cut into 2 cm thick sections. Grain-size analysis was conducted at 2 cm intervals, resulting in 163 samples, whereas clay mineralogic and geochemical analyses were performed at 4 cm intervals, resulting in 71 samples, respectively.

3.2. AMS ^{14}C Dating

To establish a dating framework for this core, two samples containing benthic foraminifera were selected and subjected to AMS ^{14}C dating conducted at the Institute of Hydrogeology and Environmental Geology, Chinese Academy of Geological Sciences (Shijiazhuang city, China). The procedure mainly involved three steps. First, the carbon in the sample was converted to CO_2 through combustion. Then, CO_2 was reduced to graphite using a zinc and iron catalyst. Lastly, the sample's age was calculated based on the ^{14}C ratio. The AMS ^{14}C age-dating process was facilitated by a Zn/Fe flame-sealed tube AMS (0.5MV 1.5SDH-1 model, NEC Corporation, Irving, TX, USA), which maintains a long-term $^{14}C/^{12}C$ testing accuracy of higher than 3%. The instrument's background value was Fm = 0.0004, which is equivalent to 63,764 years.

For samples calibrated in the NYS, a correction value of $\Delta R = -276 \pm 52$ years was applied [34]. The ^{14}C age was then calibrated to a calendar age (cal.a B.P., relative to 1950 CE) using the CALIB 7.0.2 software [35].

3.3. Grain-Size Analysis

Grain-size analysis was performed at the YCGS Testing Center using the PSA 1190LD laser grain-size analyzer (Anton Paar, Ashland, VA, USA). This instrument has a measurement range of 0.98–2500 μm and ensures a relative error of less than 1% for repeated measurements. Briefly, after adding 0.1 mol/L HCl to the samples to remove carbonate and 10% H_2O_2 to eliminate organic materials, 0.5 mol/L $Na(PO_3)_6$ was added and thoroughly mixed. Then, the samples were diluted to an appropriate concentration after removing the supernatant and allowing the samples to stand for 24 h. Prior to measurement, additional disaggregation was carried out using an ultrasonic cleaner for 1 min. Parameters, including the mean particle size, sorting coefficient, skewness, and kurtosis, of the sediment fractions were calculated using Folk's method [36].

3.4. Clay Mineralogic Analysis

Clay mineralogic analysis was performed at the Qingdao Sibada Company using the Bruker D8 X Ray diffractometer (Bruker AXS, Karlsruhe, Germany). First, the sample was placed in a 50 mL centrifuge tube and immersed in 10% H_2O_2 to remove organic material. Residual H_2O_2 was neutralized by adding pure water, followed by centrifugation at 3500 rpm. Next, 50% glacial acetic acid was added to the sample to remove carbonate minerals. Then, it was centrifuged and washed with water several times until it reached a neutral pH. The suspended particles of the upper clay fraction (less than 2 μm) were transferred to a new centrifuge tube containing two drops of 26% saturated NaCl. After centrifugation, the supernatant was extracted and used to prepare a slide. The slide was allowed to air-dry naturally before being placed in a dryer containing ethylene glycol at the bottom. It was then baked at 60 °C for 6 h prior to X-ray diffraction (XRD) testing.

Each sample underwent three evaluations: (1) In dry air conditions, a $\Delta 2\theta$ scan was performed from 3° to 30.1°, with a step size of 0.021°. (2) Under ethylene glycol saturation, a $\Delta 2\theta$ scan was conducted from 3° to 30.1° with the same step size. (3) Under ethylene glycol saturation, a $\Delta 2\theta$ scan was performed from 24° to 26°, with a smaller step size of 0.011°. Additionally, some of the ethylene glycol saturation samples were heated at 550 °C for 2 h to further identify kaolinite and chlorite and to confirm the presence of smectite.

In the analysis of clay minerals, we employed the Al_2O_3 standard spectrum as a benchmark. The relative abundances of the clay minerals were determined according to Biscay's method [37], and the ethylene glycol saturation curves were drawn using the Jade 7.5 software. The diffraction peak areas for smectite (17 Å), illite (10 Å), and kaolinite–chlorite (7 Å) were multiplied by their respective intensity factors of 1, 4, and 2. The proportion of chlorite to kaolinite was derived from their diffraction peak areas at 3.54 Å and 3.58 Å. These characteristic peak areas of the four clay minerals were normalized to a total of 100%.

The illite chemical index (ratio of the 5 Å to 10 Å peak areas in the glycolated curve) is often used as an intuitive indicator of provenance characteristics [38,39]. According to Esquevin's classification [40], illite is considered Fe-Mg rich when its chemical index is less than 0.5, indicating strong physical weathering. Conversely, an index greater than 0.5 suggests Al-rich illite, reflecting strong chemical weathering [41,42]. The 10Å FWHM is defined as the full width at half-maximum of the 10Å mica peak.

Cluster analysis is an unsupervised learning method that reveals the intrinsic structure and patterns of data by grouping similar data points into clusters. In source classification, cluster analysis can be used to identify and differentiate various data, thereby providing a basis for the source tracer [43]. In this paper, we used the ternary diagram and crossplot methodologies to achieve these analytical objectives.

3.5. Geochemical Analysis

Major and trace element analysis was conducted at the YCGS Testing Center using the Axios X fluorescence spectrometer (Panaco Company, Almelo, The Netherlands). Samples were initially dried, pelletized, and digested using a mixture of 4 mL nitric acid (HNO_3) and 1 mL perchloric acid ($HClO_4$), followed by a mixture of 4 mL hydrofluoric acid (HF) and 1 mL perchloric acid ($HClO_4$). The procedure was completed by adding 10 mL of nitric acid (HNO_3).

X-ray fluorescence spectroscopy was utilized to measure the concentration of major elements, including Si, Al, Ca, Mg, K, and Na, and trace elements, including Cu, Pb, Zn, Cr, Ni, Co, Cd, Rb, Sr, and Ba. The MgO/Al_2O_3 ratio in marine sediments indicates the extent of terrigenous material input but does not account for grain-size effects in regions where terrigenous clastic materials predominate. Generally, a higher MgO/Al_2O_3 ratio indicates a lower terrigenous material input [44]. Changes in the Sr/Ca ratio serve as an indicator of the biological activity of phytoplankton in the marine surface waters. As microscopic plants grow, they incorporate calcium ions (Ca^{2+}) into their skeletal structures while taking up comparatively lower amounts of strontium ions (Sr^{2+}). Given the chemical similarities between Sr^{2+} and Ca^{2+}, their competitive uptake can establish a correlation between the

Sr/Ca ratio and the level of marine productivity. An elevated productivity signifies a higher absorption of Ca^{2+} by the organisms, which subsequently leads to a decreased Sr/Ca ratio. During periods of lower productivity, the Sr/Ca ratio is relatively higher, reflecting less competition for calcium during phytoplankton growth [45].

4. Results

4.1. Grain-Size Analysis

The lithology of core ZZ04 predominantly consisted of gray clayey silt and silty clay (Figure 2a). The sediments were primarily composed of silt (4–63 μm), clay (<4 μm), and sand (>63 μm), whose contents ranged from 75.7–81.09% (average: 77.94%, Figure 2b), 17.88–24.19% (averaged 21.64%, Figure 2c) and <1.5% (Figure 2d), respectively. The sediments primarily consisted of silt, according to the Shepard classification (Figure 3a). The grain-size curves indicated a clear stratification into two sections at a depth of 1.83 m (~1.63 cal.ka B.P.).

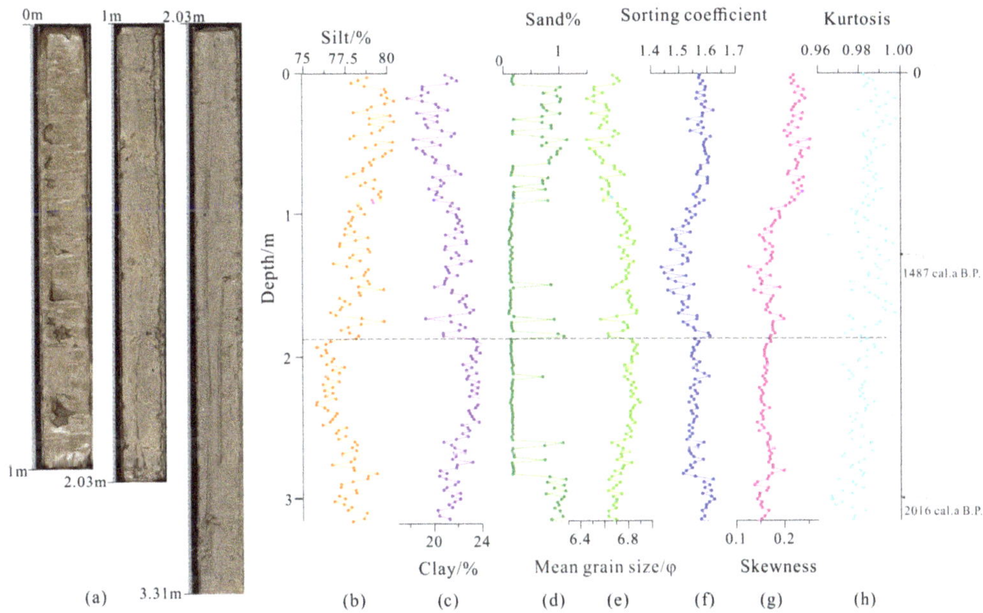

Figure 2. Sediment grain-size curves in the ZZ04 core. From left to right: (**a**) photographs of the cores, (**b**) silt content (4–63 μm), (**c**) clay content (<4 μm), (**d**) sand content (>63 μm), (**e**) mean particle size, (**f**) sorting coefficient, (**g**) skewness, and (**h**) kurtosis. Black arrows show the calendar ASM ^{14}C-age of the benthic foraminifera.

In the lower section (1.83–3.37 m), from the bottom to the top depth, the silt content decreased, and the clay content increased (Figure 2b,c). The mean particle size slightly increased from 6.63 Φ to 6.87 Φ, indicating a transition from coarser to finer sediments (Figure 2e). The sorting coefficient and skewness curves exhibited minor fluctuations around 1.55 and 0.15, respectively, suggesting a moderate sorting degree and a positive bias distribution (Figure 2f,g). The kurtosis showed a medium level and an overall upward trend from 0.96 to 0.99 (Figure 2h).

In the upper section (0–1.83 m), there was a transition from finer to coarser sediments. As the primary component, the silt content increased from 76% to 81% (Figure 2b). Concurrently, the clay content decreased from 22% to 18% (Figure 2c), whereas the sand content remained minimal, fluctuating between 0 and 1% (Figure 2d). The mean particle size showed moderate decreases from 6.87 Φ to 6.46 Φ (Figure 2e). The sorting coefficient initially decreased and then increased, indicating a medium sorting level (Figure 2f). Skewness was stable below 1 m, with a value of 0.15, then began to rise above 1 m, showing an overall positive bias (Figure 2g). The kurtosis fluctuated at approximately 0.9, indicating a medium level (Figure 2h).

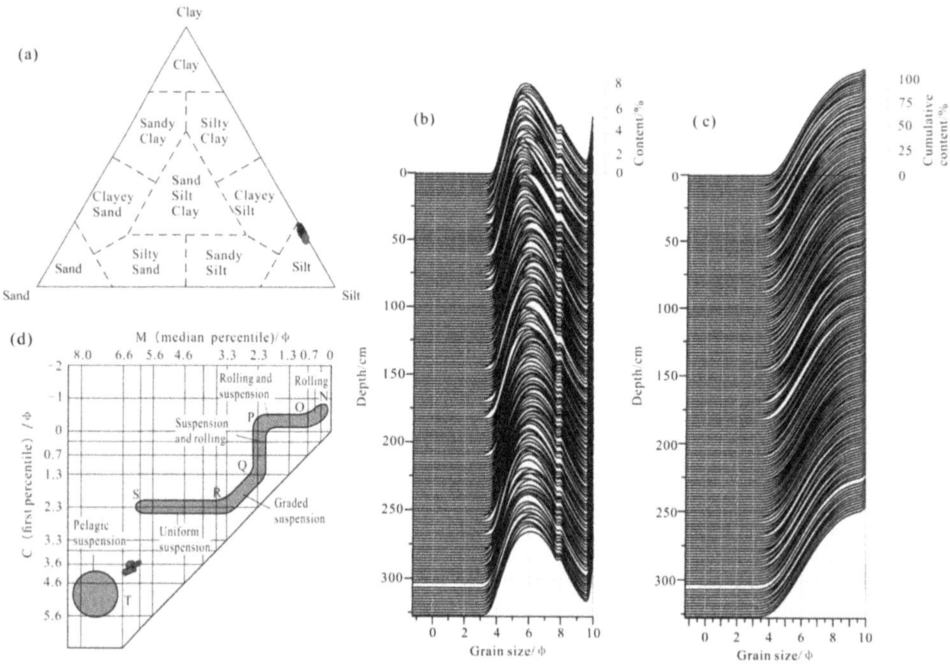

Figure 3. Characteristics of sediment grain-size of core ZZ04. (**a**) Lithology ternary diagrams based on the Shepard classification. (**b**) Grain-size distribution, (**c**) grain-size probability distribution, and (**d**) C-M diagrams.

The grain-size distribution of the ZZ04 core showed a unimodal pattern, with the main peak mode between 5Φ and 6Φ, suggesting a relatively singular provenance (Figure 3b). The grain-size probability curve showed a two-stage distribution, highlighting the predominance of suspension components and scarcity of saltation components (Figure 3c). The C-M diagram showed a sedimentation mode between marine and uniform suspensions, implying weak sedimentary hydrodynamic conditions (Figure 3d).

4.2. Sediment Chronology

The AMS ^{14}C dating results are presented in Table 1 and Figure 2. The age of each sampling interval was established by linear interpolation.

Table 1. AMS ^{14}C and calendar ages of core ZZ04.

Sample ID	Depth (cm)	Material	AMS^{14} C Age (a B.P.)	Calendar Age (cal.a B.P.)	1σ Error Range (cal.a B.P.)
ZZ04-135	134	Mixed benthic foraminifera	1810 ± 35	1487	1385–1576
ZZ04-315	314	Mixed benthic foraminifera	2250 ± 30	2016	1911–2119

The ZZ04 samples collected in 2023 indicated a calendar age of −73 cal.a B.P. at 0 m. The linear sedimentation rates (LSRs) of the sediment layer showed a lower mean value

(0.086 cm/a) above 134 cm and a higher mean value (0.34 cm/a) between 134 cm and 314 cm. Moreover, the ZZ04 core sediments covered a time span of −73 to 2050 cal.a B.P., providing a continuous sediment record.

4.3. Clay Minerals

The XRD pattern of core ZZ04 revealed that it was predominantly composed of clay minerals, including illite, smectite, chlorite, and kaolinite. Additionally, the sample contained some clastic minerals, such as quartz and feldspar (Figure 4).

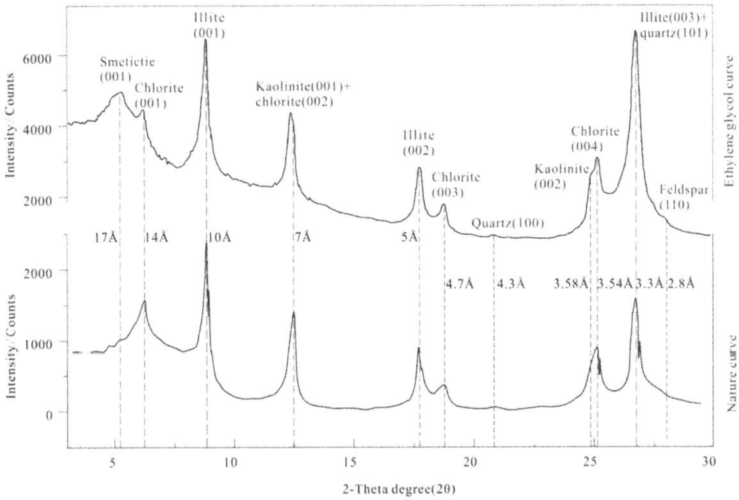

Figure 4. Ethylene glycol curves (up) and natural curves (down) of the X-ray diffraction of clay minerals from core ZZ04.

The clay minerals in core ZZ04 were primarily composed of illite, smectite, chlo rite, and kaolinite, whose content ranged from 52.93–69.78% (average: 62.1%, Figure 5a), 12.78–26.4% (average: 18.11%, Figure 5b), 7.95–16.53% (average: 11.7%, Figure 5c), and 5.27–10.39% (average: 7.9%, Figure 5d), respectively. The clay mineral assemblage was illite–smectite–chlorite–kaolinite. The distribution pattern of the clay mineral assemblage revealed a stratification into two sections at a depth of 1.83 m. The clay mineral content remained relatively stable in the lower section. However, the smectite content exhibited a declining trend. Meanwhile, in the upper section, the illite content increased, whereas the chlorite content decreased (Figure 5a,d). The kaolinite content showed a positive correlation with that of chlorite (with a correlation coefficient of 0.66). In contrast, there was a negative correlation between the contents of kaolinite and illite (correlation coefficient of −0.65; Table 2).

Table 2. Correlation analysis of the clay minerals.

	Illite	Smectite	Chlorite	Kaolinte
Illite	1	−0.722	−0.602	−0.650
Smectite	−0.722	1	−0.071	0.068
Chlorite	−0.602	−0.071	1	0.660
Kaolinte	−0.650	0.068	0.660	1

In the ZZ04 core, the 10Å FWHM ranged from 0.2° to 0.32° (2θ), with an average of 0.24° (2θ) (Figure 5e). The illite chemical index in the core ZZ04 samples varied from

0.23–0.47 (average: 0.34), placing it between Fe-Mg types. This is indicative of strong physical weathering (Figure 5f).

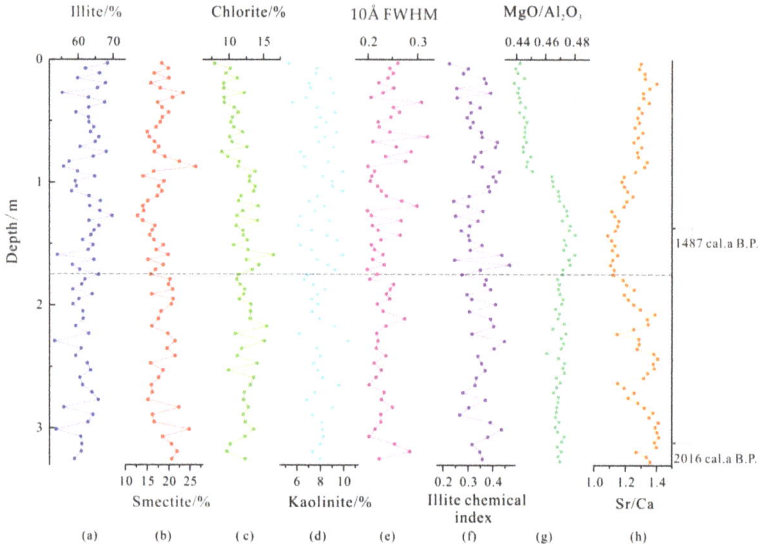

Figure 5. Clay mineral and geochemical curves of core ZZ04. (**a**) Illite content, (**b**) smectite content, (**c**) chlorite content, (**d**) kaolinite content, (**e**) 10Å FWHM, (**f**) illite chemical index, (**g**) MgO/Al_2O_3 ratio, and (**h**) Sr/Ca ratio. Black arrows show the calendar ASM ^{14}C age of the benthic foraminifera.

4.4. MgO/Al_2O_3 and Sr/Ca Ratios

The MgO/Al_2O_3 ratio ranged between 0.44 and 0.48, with an average value of 0.46. Below 0.95 m, this ratio remained stable, averaging approximately 0.47. Above 0.95 m, the ratio had a lower value of 0.44 (Figure 5g).

The Sr/Ca ratio ranged from 1.09 to 1.42, with an average value of 1.27. In the lower section (1.83–3.37 m), the Sr/Ca ratio decreased from 1.4 to 1.1, whereas in the upper section (0–1.83 m), it increased from 1.1 to 1.4 (Figure 5h).

5. Discussion

5.1. Provenance Identification of the Mud Area

The clay minerals in this area predominantly originated from a terrestrial source. These minerals were transported and deposited by the surrounding rivers [46,47]. The NYS receives inputs potentially from the YR, Yangtze River, Yalu River, Shandong Peninsula rivers, and Korean Peninsula rivers (including the Han, Keum, and Yeongsan rivers). A comparative analysis of the clay mineral composition from these rivers enables an examination of how sediments from various rivers influence the distribution patterns of clay minerals (Table 3, Figure 6).

The sediments transported by the YR are largely influenced by the characteristics of the Loess Plateau. The loess is predominantly composed of evaporites and carbonates and has an alkaline-to-weakly-alkaline pH. The YR basin is in a cold and dry climate and is strongly affected by physical weathering processes. The clay mineral compositions reported in the references and those from our results were consistent, which attests to the reliability of the test outcomes. The YR's clay minerals were notably high in smectite (12–23%) and relatively low in kaolinite (8–13%). Additionally, the illite/smectite ratio was less than 6 (Table 3) [2,48–54].

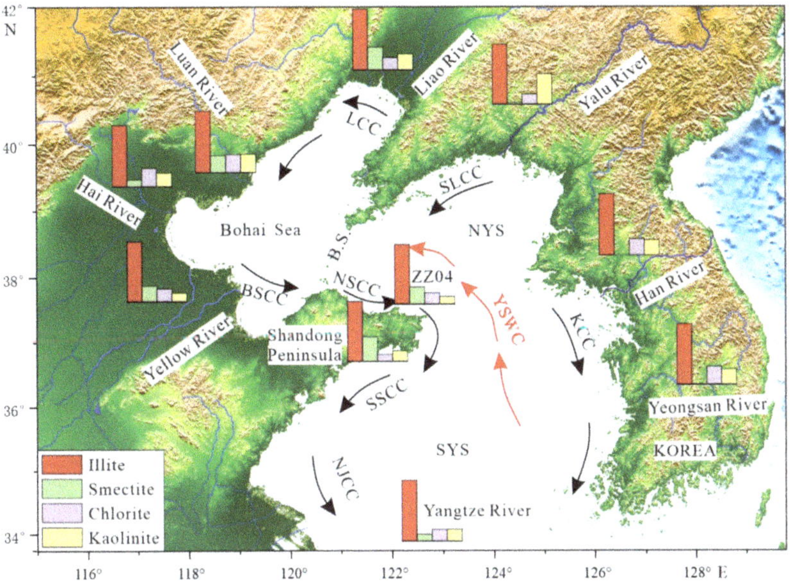

Figure 6. Diagram of the clay mineral composition in the study area and around the YS and BS. The main currents include the South Liaoning coastal current (SLCC), Liaoning coastal current (LCC), Bohai Sea coastal current (BSCC), North Shandong Peninsula coastal current (NSCC), South Shandong Peninsula coastal current (SSCC), North Jiangsu coastal current (NJCC), Korean coastal current (KCC), and Yellow Sea warm current (YSWC). Geographical abbreviations include the Bohai strait (B.S.), North Yellow Sea (NYS), South Yellow Sea (SYS), and East China Sea (ECS).

The Yangtze River's drainage spans the South China orogenic belt, the Yangtze platform, the Qinling–Dabie Mountain orogenic belt, and the Sanjiang Paleo-Tethys orogenic belt. The river's drainage area contains extensive areas of carbonate rocks and terrigenous clastic rocks, with a significant presence of medium-acid intrusive rocks, schists, and gneiss [39,50]. The Yangtze River basin is in a warm and humid climate, and the soil is mostly weakly acidic to acidic, and the chemical weathering is stronger than that in the YR basin [39,50]. The clay mineral assemblage indicates moderate levels of kaolinite at approximately from 8–15%, low levels of smectite ranging from 2–10%, and an illite/smectite ratio exceeding 8 (Table 3) [48,53,55–57].

The Shandong Peninsula's surface is characterized by exposure to Precambrian metamorphic rocks and Yanshanian granite [47]. The composition of clay minerals in this regional river resembles that in the YR; however, the smectite content is notably higher (20–25%) (Table 3) [47,58], which is likely attributed to the extensive weathering of basic volcanic rocks [59].

The Korean Peninsula rivers—the Han, Keum, and Yeongsan rivers—are situated in the Yangtze Plate and are mainly composed of Precambrian gneiss, Jurassic and Cretaceous granites, and Quaternary loose alluvial deposits [32,59–61]. Carbonate outcrops are scarce, with the Jurassic Daebo granite and Late Cretaceous Bulguksa granite as the primary source rocks for the Keum and Yeongsan Rivers, respectively [60]. Such geological conditions can increase the contents of illite, chlorite, and kaolinite while hindering the development of smectite, resulting in a relatively low smectite content (typically less than 2%) (Table 3) [52,62–66].

Situated within the Sino-Korean Plate, the Yalu River predominantly consists of metamorphic rocks, such as amphibole, granulite, and biotite gneiss, along with various pyroxenite groups. The developed rock joints and fissures lead to pronounced spherical and network weathering patterns [67]. The distribution of clay minerals in the Yalu River is similar to that in the Korean Peninsula rivers, which favors the formation of illite and kaolinite but not that of smectite (whose content is typically below 2%) (Table 3) [52,61].

The ZZ04 core showed a clay mineral assemblage comprising illite, smectite, chlorite, and kaolinite. The relatively high smectite content (average value is 18%, Table 3) suggests a substantial influence from the YR materials. By contrast, the relatively low kaolinite content (average value is 8%, Table 3) indicates that physical weathering predominantly affected this study area. Based on this analysis, the composition of the clay minerals in core ZZ04 is similar to that of the YR and Shandong Peninsula rivers.

Table 3. Clay mineral assemblages and their relative content of river sediments around the NYS.

Rivers	Illite/%	Smectite/%	Chlorite/%	Kaolinite/%	References
Yellow River	61	21	9	9	Liu et al., 2010 [2]
	63	15	13	10	Fan et al., 2001 [48]
	65	14	12	9	Milliman et al., 1985 [49]
	56	21	13	10	Xu et al., 2009 [50]
	59	23	9	9	Xu, 1985 [51]
	67	13	12	8	Ren, 1986 [52]
	62	16	12	10	Yang, 1988 [53]
	62	12	16	10	Yang et al., 2003 [54]
	60	10	18	12	Yang et al., 2003 [54]
	55	13	18	13	This paper
	60	15	13	11	This paper
	56	17	16	11	This paper
	57	17	15	11	This paper
	64	13	14	9	This paper
Yangtze River	71	7	13	9	Fan et al., 2001 [48]
	65	10	11	14	Yang, 1988 [53]
	66	2	17	15	He et al., 2013 [55]
	68	6	14	13	Xu, 1983 [56]
	71	7	14	8	Lu et al., 2015 [57]
Shandong Peninsula rivers	58	25	10	8	Liu et al., 2022 [47]
	58	24	7	11	Liu et al., 2022 [47]
	59	24	7	10	Liu et al., 2022 [47]
	64	19	9	8	Liu et al., 2022 [47]
	57	12	15	16	Hu et al., 2018 [58]
Han River	57	0	23	20	Park et al., 1991 [62]
	70	1	13	17	Park et al., 1992 [63]
	60	2	21	17	Cho et al., 2015 [64]
Keum River	59	1	10	30	Ren et al., 1986 [52]
	55	1	24	21	Cho et al., 2015 [64]
	64	0	17	19	Choi, 1981 [65]
Yeongsan River	64	2	20	15	Cho et al., 2015 [64]
	64	0	19	17	Kim, 1980 [66]
Yalu River	59	1	10	30	Ren et al., 1986 [52]
	68	2	12	18	Yang et al., 2003 [61]
Average of Yellow River (N = 14)	60	16	14	10	
Average of Yangtze River (N = 5)	68	6	14	12	
Average of Shandong Peninsula rivers (N = 5)	59	21	10	11	
Average of Han River (N = 3)	62	1	19	18	
Average of Keum River (N = 3)	59	1	17	23	
Average of Yeongsan River (N = 2)	64	1	19	16	
Average of Yalu River (N = 2)	64	2	11	24	
Average of core ZZ04 (N = 71)	62	18	12	8	

The analysis of clay minerals present in the surface sediments indicates that this study area arose mainly from the YR's sediments. We applied cluster analysis to the ZZ04 core and to samples from adjacent rivers to further elucidate the provenance of the mud area in the Shandong Peninsula. This analysis included a ternary clay mineral diagram (smectite–chlorite–illite + kaolinite; Figure 7a) and a crossplot (illite/smectite and chlorite/kaolinite ratios, Figure 7b). In addition, the clay mineral of the core DLC70-2 in the northern part of this study area (see Figure 1a for location) was analyzed. The clay minerals in both the ZZ04 and DLC70-2 cores were within the range of sediments from the YR, the Shandong Peninsula rivers, and, in part, the Yangtze River. The influence of the Yangtze River may be attributed to the YSWC, which transports sediments from the south.

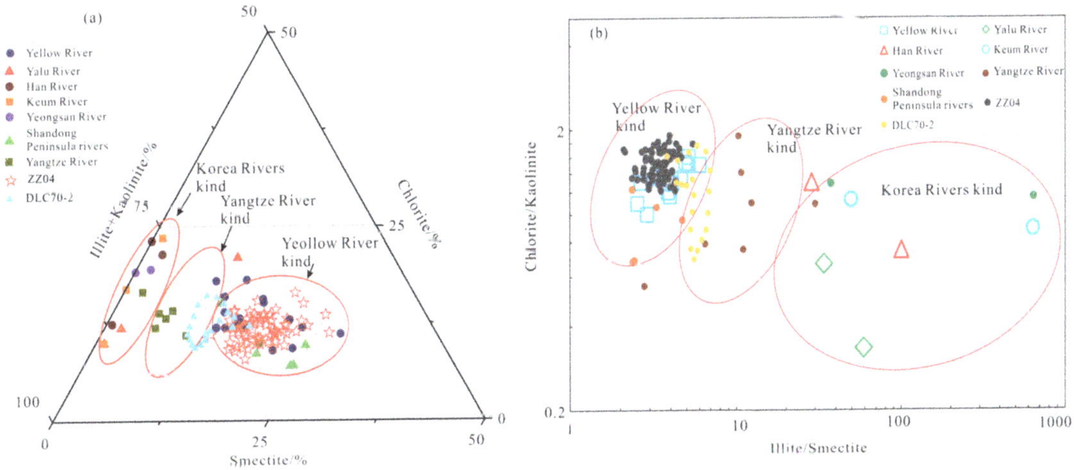

Figure 7. Cluster analysis of clay minerals in the cores of ZZ04 and DLC70-2, as well as river sediments around the NYS. (**a**) Ternary diagram of Smectite–chlorite–illite + kaolinite, (**b**) Crossplot of illite/smectite and chlorite/kaolinite.

5.2. Climate Change in the Mud Area over the Last 2000 Years

The formation of clay minerals is influenced by various factors, including river flows, warm currents, coastal currents, and climate. Generally, higher illite and chlorite levels suggest intensified physical weathering, meaning a drier and colder climate, whereas higher smectite and kaolinite levels indicate intensified chemical weathering, implying a warmer and more humid climate. A previous study has established that an increasing smectite/(illite + chlorite) ratio correlates with a warming climate trend [25,39].

In this study, we made a specific attempt to analyze the paleoclimate dynamics in China relative to global patterns, taking into account the influence of the EAM history. Our research extended to assessing the EAM's influence on the sediment sources within the NYS, including an evaluation of the terrestrial sources and the marine productivity. A comprehensive comparative analysis of the smectite/(illite + chlorite), Sr/Ca, and MgO/Al_2O_3 ratios in the ZZ04 core and the $\delta^{18}O$ data from the Dunde ice core [68], the Greenland GISP2 ice core [69], and Hongyuan peat [70] over the past 2000 years is presented in Figure 8. The climatic index, smectite/(illite + chlorite) ratio, was negatively correlated with both the Sr/Ca ratio and the Dunde ice core $\delta^{18}O$ values (proxy index of the East Asian winter monsoon; EAWM) [68]. This suggests that the smectite/ (illite + chlorite) ratio serves as a proxy for the East Asian summer monsoon (EASM) in the NYS. By analyzing the variations in these indicators, we identified four distinct climate change phases over the last 2000 years. They are discussed below in detail.

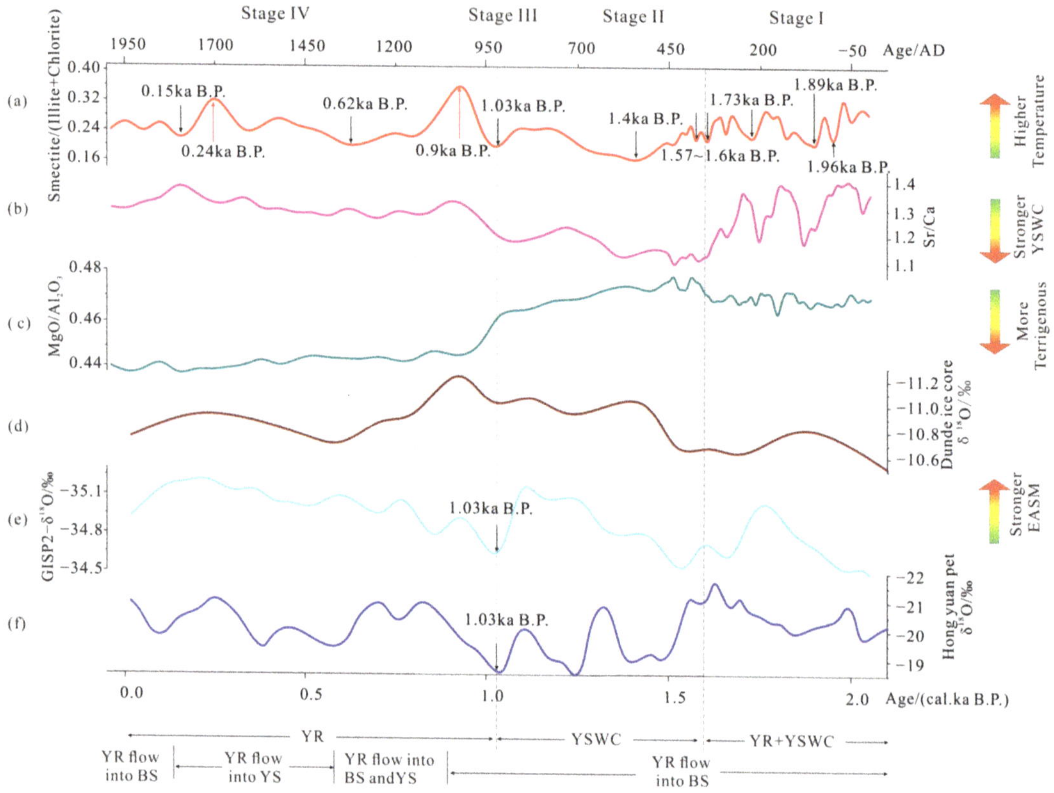

Figure 8. Comparison of clay mineral and geochemical element ratios in core ZZ04 sample with EAM curves over the past 2000 years. (**a**) The smectite/(illite + chlorite) ratio of the core ZZ04, (**b**) the Sr/Ca ratio of core ZZ04, (**c**) the MgO/Al_2O_3 ratios of the core ZZ04, (**d**) the $\delta^{18}O$ record of the Dunde ice core [68], (**e**) the $\delta^{18}O$ record of the GISP2 ice core from Greenland [69], and (**f**) the $\delta^{18}O$ record of the Hongyuan peat [70].

Stage I (2–1.6 cal.ka B.P.), characterized by moderate-to-high-frequency temperature fluctuations, aligns with the late Western Han Dynasty to the early Eastern Jin Dynasty in Chinese history (−50 to 350 AD) (Figure 8a). The climate during this period was marked by fluctuating intensities of the EAWM and EASM, with three notable cooling events at 1.96, 1.89, and 1.73 cal.ka B.P. These events correspond to the EAWM intensification periods from 2.2 cal.ka B.P. to 1.9 cal.ka B.P. and 1.7 cal.ka B.P. (Figure 8a), as suggested by Zhu et al. [71], and 1.88 cal.ka B.P., as suggested by Xiao et al. [72]. The grain-size of sediments in Stage I gradually became finer over time (Figure 2b,d). In terms of the clay mineral composition, the chlorite levels diminished, whereas kaolinite levels had a more pronounced variation (Figure 5c,d). Physical weathering predominantly occurred, although temperatures fluctuated. During the EAWM, sediments were predominantly sourced from the YR, with fine sands transported to the study area via the Bohai and Shandong coastal currents. Conversely, during the EASM, sediments were mainly sourced from the Shandong Peninsula rivers, contributing terrigenous materials shaped by chemical weathering processes.

Stage II (1.6–1.3 cal.ka B.P.) marks a cooling period that aligns with the low-temperature era of the Southern and Northern Dynasties in Chinese history (350–540 AD) (Figure 8a) [73,74]. This stage was characterized by the prevalence of the EAWM. A minor ice age occurred between 1.6 and 1.57 cal.ka B.P., followed by a significant cooling event at 1.4 cal.ka B.P.

(Figure 8a), which corresponds to the 1.4 cal.ka B.P. ice floe event in the North Atlantic [75]. Sediment grain-size analysis indicates a gradual increase in silt content and a corresponding decrease in clay content (Figure 2c,d). In terms of the clay mineral composition, there was a notable rise in illite content but a decline in smectite and chlorite contents (Figure 5a,c), indicative of a cold and dry climate. A high MgO/Al_2O_3 ratio and a low Sr/Ca ratio (Figure 8b,c) suggest a lower terrigenous input and higher marine productivity, respectively. The primary source of sediment during this period was the Yangtze River, with the YSWC facilitating the transport and deposition of sediments in the study area.

Stage III (1.3–0.8 cal.ka B.P.) represents a warming period in Chinese history, spanning from the Sui and Tang Dynasties (581–907 AD) to the Southern Song and Yuan Dynasties (1127–1320 AD). A brief cooling event occurred during the Five Dynasties to the Northern Song Dynasty (907–1127 AD) (Figure 8a) [73], during which the sediment's silt content continued to rise, whereas the clay content diminished (Figure 2c,d). The clay mineral composition showed a decrease in illite content and increases in the smectite and kaolinite contents (Figure 5b,d), signaling a significant paleoclimatic warming. Extreme weather events, i.e., a cold event at 1.03 cal.ka B.P. and a hot event at 0.9 cal.ka B.P., were recorded (Figure 8a). These cold events align with the Little Ice Age, as documented in the Dunde ice core [68], the GISP2 ice core [69], and Hongyuan peat [70] (Figure 8d–f). From 1128 AD, the YR's course was altered, flowing into both the BS and YS until 1494 AD, after which it exclusively flowed into the YS from Yancheng City, Jiangsu province, until 1579 AD [76]. After the cold event at 1.03 cal.ka B.P., there was a decline in the MgO/Al_2O_3 ratio and an increase in the Sr/Ca ratio (Figure 8b,c), indicating an intensification of the terrigenous material influx, which may be associated with the YR's course change. In Stage III, the EAWM's influence waned, whereas that of the EASM was strengthened (Figure 8a). At the beginning of this stage, the Yangtze River sediments, which were carried by the YSWC, were the main source significantly affecting the study area. As the EAWM weakened, the EASM became dominant, and the study area became increasingly influenced by the Shandong Peninsula rivers. During the warming climate, heavy precipitation events led to a decrease in the illite content and an increase in the smectite content, which is attributed to the leaching of potassium from illite and its subsequent transformation into smectite. This period concluded with a temperature decline at the end of the Yuan Dynasty.

Lastly, Stage IV (0.8 cal.ka B.P. to present), which was characterized by a stable cooler climate with a strengthened EAWM, coincided with the Little Ice Age and corresponded with the cold period during the Ming and Qing Dynasties in China (1321–1920 AD) [73,74] (Figure 8a). This stage included two Little Ice Age events at 0.62 cal.ka B.P. and 0.15 cal.ka B.P. The 0.62 cal.ka B.P. (Figure 8a), which was consistent with the climate events recorded in the Dunde ice core (1420–1520 AD) [77] and Hongyuan peat (1370–1400 AD) [70] (Figure 8d,f). The lithological trend of increasing silt content and decreasing clay content continued from the previous stages (Figure 2c,d). Additionally, decreases in chlorite and kaolinite contents (Figure 5c,d) reflected the predominantly cold climate. During this period, the MgO/Al_2O_3 ratio reached its minimum (Figure 8c), whereas the Sr/Ca ratio peaked at its maximum (Figure 8b), signifying the predominant influence of terrigenous material sources. The significant strengthening of the EAWM and the changes in the YR's course led to old YR-derived terrigenous clastic sediments, transported by the YSWC, becoming the primary source. The YR returned to the BS until 1855 AD, with minimal changes observed.

6. Conclusions

The main conclusions of this study are as follows:

(1) The ZZ04 core sample, acquired from the mud area in the northern offshore area of the Shandong Peninsula, predominantly consists of silt. The clay mineral composition is an assemblage of illite–smectite–chlorite–kaolinite. Illite, enriched with Fe-Mg, suggests strong physical weathering. Comparative analysis of the clay mineral composition of samples from the surrounding rivers, along with ternary diagram

and crossplot analyses, confirm that the sample's clay minerals are mainly sourced from the YR, the Shandong Peninsula rivers, and the Yangtze River, highlighting the multi-source characteristics of the sediments.

(2) The formation of the mud area was influenced by shifts in the East Asian monsoon, alterations in the YR's course, and dynamics of the YSWC. During the EAWM period, the coastal current's deposition rate and transport capacity led to a proportional increase in the terrigenous material input. Meanwhile, the YSWC was an important contributor in the transportation of matter and heat flux from the south. During the EASM period, sediments were mainly sourced from the Shandong Peninsula rivers, contributing terrigenous materials shaped by chemical weathering.

Author Contributions: Writing—original draft preparation, S.W.; supervision and writing, J.L.; generating the original data and performing the analytic calculations, Y.F. All authors have read and agreed to the published version of the manuscript.

Funding: This research was jointly funded by the Science and Technology Innovation Fund of the Command Center of Natural Resources Comprehensive Survey entitled "Temporal and spatial distribution of palaeochannel and origin of organic carbon burial in the Western Bohai sea since 2.28 Ma" (grant number KC20220011), the project entitled "Characterization of Carboniferous-Early Permian heterogeneous porous carbonate reservoirs and hydrocarbon potential analysis in the central uplift of the South Yellow Sea Basin" (grant number KLSG2304) funded by the Key Laboratory of Submarine Science, Ministry of Natural Resources, and the project entitled "1:25000 Marine Regional Geological Survey in Weihai Sea Area, North Yellow Sea (grant number DD20230412)", supported by the China Geological Survey.

Institutional Review Board Statement: Not applicable.

Informed Consent Statement: Not applicable.

Data Availability Statement: All the data and materials used in this paper are available from the corresponding authors upon request.

Acknowledgments: The authors would like to express their gratitude to the Yantai Center of Coastal Zone Geological Survey, China Geological Survey for providing the basic data and the Ph.D. students Jian Liu and X. Kong from the Qingdao Institute of Marine Geology for their help in this study. The authors are also grateful to the editors and reviewers who provided sincere comments and assisted in the writing of this manuscript. We are deeply grateful for the valuable suggestions provided by the three anonymous reviewers and the editorial assistance by the editors.

Conflicts of Interest: The authors declare no conflicts of interest.

References

1. Lv, W.Z. Sedimentary Processes and Mechanism of the Mud Areas on the China Sea Continental Shelf during the Middle to Late Holocene. Ph.D. Thesis, East China Normal University, Shanghai, China, 2022.
2. Liu, J.G.; Li, A.C.; Chen, M.H. Environmental evolution and impact of the Yellow River sediments on deposition in the Bohai Sea during the last deglaciation. *J. Asian Earth Sci.* **2010**, *38*, 26–33. [CrossRef]
3. Milliman, J.D.; Qin, Y.-S.; Ren, M.; Saito, Y. Man's influence on the erosion and transport of sediment by Asian rivers; the Yellow River (Huanghe) example. *J. Geol.* **1987**, *95*, 751–762. Available online: https://www.jstor.org/stable/30063822 (accessed on 2 August 2024). [CrossRef]
4. Liu, J.P.; Milliman, J.D.; Gao, S. The Shandong mud wedge and post-glacial sediment accumulation in the Yellow Sea. *Geo-Mar. Lett.* **2001**, *21*, 253–254. [CrossRef]
5. Liu, J.; Saito, Y.; Wang, H.; Yang, Z.G.; Nakashima, R. Sedimentary evolution of the Holocene subaqueous clinoform off the Shandong Peninsula in the Yellow Sea. *Mar. Geol.* **2007**, *236*, 165–187. [CrossRef]
6. Xue, C.T.; Qin, Y.C.; Ye, S.Y.; Laws, E.A.; Wang, Z.B. Evolution of Holocene ebb-tidal clinoform off the Shandong Peninsula on East China Sea shelf. *Earth-Sci. Rev.* **2018**, *177*, 478–496. [CrossRef]
7. Park, S.C.; Lee, H.H.; Han, H.S.; Lee, G.H.; Kim, D.C.; Yoo, D.G. Evolution of late Quaternary mud deposits and recent sediment budget in the southeastern Yellow Sea. *Mar. Geol.* **2000**, *170*, 271–288. [CrossRef]
8. Yang, Z.S.; Liu, J.P. A unique Yellow River-derived distal subaqueous delta in the Yellow Sea. *Mar. Geol.* **2007**, *240*, 169–176. [CrossRef]

9. Qiu, J.D.; Liu, J.; Saito, Y.; Yang, Z.G.; Yue, B.J.; Wang, H.; Kong, X.H. Sedimentary evolution of the Holocene subaqueous clinoform off the southern Shandong Peninsula in the Western South Yellow Sea. *J. Ocean Univ. China* **2014**, *13*, 747–760. [CrossRef]
10. Milliman, J.D.; Huang-Ting, S.; Zuo-Sheng, Y.; Meade, R.H.; Milliman, J.D.; Jin, Q. Transport and deposition of river sediment in the Changjiang Estuary and adjacent continental shelf. *Cont. Shelf Res.* **1985**, *4*, 37–44. [CrossRef]
11. Liu, J.P.; Li, A.C.; Xu, K.H.; Velozzi, D.M.; Yang, Z.S.; Milliman, J.D.; De Master, D.J. Sedimentary features of the Yangtze River-derived along-shelf clinoform deposit in the East China Sea. *Cont. Shelf Res.* **2006**, *26*, 2141–2156. [CrossRef]
12. Xiang, R.; Yang, Z.S.; Saito, Y.; Guo, Z.G.; Fan, D.J.; Li, Y.H.; Xiao, S.B.; Shi, X.F.; Chen, M.H. East Asia Winter Monsoon changes inferred from environmentally sensitive grain-size component records during the last 2300 years in mud area southwest off Cheju Island, ECS. *Sci. China Ser. D* **2006**, *49*, 604–614. [CrossRef]
13. Li, G.; Miao, L.; Yan, W. Holocene evolution of the shelf mud deposits in the north-western South China Sea. *Front. Mar. Sci.* **2022**, *9*, 937616. [CrossRef]
14. Alexander, C.R.; DeMaster, D.J.; Nittrouer, C.A. Sediment accumulation in a modern epicontinental-shelf setting; the Yellow Sea. *Mar. Geol.* **1991**, *98*, 51–72. [CrossRef]
15. Zhu, Y.T.; Xu, Y.C.; Liu, M.Y.; Lin, Z.B.; Zhu, L.H. Geochemistry and Holocene Sedimentary Environment Evolution of Subaqueous Clinoform off Shandong Peninsula (Yellow Sea). *Minerals* **2021**, *11*, 1209. [CrossRef]
16. Chang, X.; Liu, X.; Li, T.; Xiong, Z.; Duan, B.; Huang, J.; Liu, J.P.; Zhang, M.; Wang, A.; Wang, H. Late Quaternary marine transgressions off the Shandong Peninsula inferred from paleosalinity indicators: Implications for Holocene mud wedge formation. *Chem. Geol.* **2024**, *658*, 122117. [CrossRef]
17. Liu, J.P.; Milliman, J.D.; Gao, S.; Cheng, P. Holocene development of the Yellow River's subaqueous delta, North Yellow Sea. *Mar. Geol.* **2004**, *209*, 45–67. [CrossRef]
18. Wang, W.; Li, A.C.; Xu, F.J.; Huang, P.; Li, Y. Distribution of Surface Sediments and Sedimentary Environment in the North Yellow Sea. *Oceanol. Limnol. Sin.* **2009**, *40*, 525–531.
19. Chen, X.H.; Li, T.G.; Zhang, X.H.; Li, R.H. A Holocene Yalu River-derived fine-grained deposit in the southeast coastal area of the liaodong Peninsula. *Chin. J. Oceanol. Limnol.* **2013**, *31*, 636–647. [CrossRef]
20. Qi, J.; Li, F.Y.; Song, J.M.; Gao, S.; Wang, G.Z.; Peng, C. Sedimentation rate and flux of the North Yellow Sea. *Mar. Geol. Quat. Geol.* **2004**, *24*, 9–14.
21. Liu, J.; Saito, Y.; Kong, X.; Wang, H.; Zhao, L. Geochemical characteristics of sediment as indicators of post-glacial environmental changes off the Shandong Peninsula in the Yellow Sea. *Cont. Shelf Res.* **2009**, *29*, 846–855. [CrossRef]
22. Lan, X.H.; Mi, B.B.; Chen, X.H.; Li, R.H.; Wang, Z.B.; Lu, K. Tracing of Rare Elements in Late Quanternary Sediments from Central North Yellow Sea. *J. Chin. Soc. Rare Earths* **2015**, *33*, 241–252.
23. Han, Z.; Wang, Y.; Sun, Y.; Ai, X.; Wu, X.; Yang, Y. Composition of minerals in surface sediments of the Yellow Sea and their provenance. *Mar. Geol. Front.* **2022**, *38*, 10–19.
24. Dou, Y.G.; Li, J.; Zhao, J.T.; Wei, H.L.; Yang, S.Y.; Bai, F.L.; Zhang, D.L.; Ding, X.; Wang, L.B. Clay mineral distributions in surface sediments of the Liaodong Bay, Bohai Sea and surrounding river sediments: Sources and transport patterns. *Cont. Shelf Res.* **2014**, *73*, 72–82. [CrossRef]
25. Guo, R.S.; He, L.; Ye, S.Y.; Zhao, L.H. Mineral Characteristics, Provenance and Climatic Significance of Wetland Sediments from Dalinghe River Estuary in Liaohe Delta Since Late Pleistocene. *Geoscience* **2020**, *34*, 154–165.
26. Yang, S.X.; Ye, S.Y.; He, L.; Yuan, H.M.; Zhao, G.M.; Ding, X.G.; Pei, S.F.; Lu, J.F. Geochemical and clay mineral characteristics of the Holocene sediments on the west coast of Bohai Bay and their implications for environmental and climatic changes. *Mar. Geol. Quat. Geol.* **2021**, *41*, 75–87.
27. Xu, D.Y.; Liu, X.Q.; Zhang, X.H.; Li, T.G.; Chen, B.Y. *Offshore Geology of China*; Geological Publishing House: Beijing, China, 1997; p. 8.
28. Qin, Y.S.; Zhao, Y.Y.; Chen, L.R. *Geology of Yellow Sea*; Maritime Press: Beijing, China, 1989; p. 249.
29. Wu, S.Y.; Liu, J.; Chen, J.W.; Wu, H.R. Characteristics of Milankovitch cycles recorded in Eocene strata in the eastern depression of North Yellow Sea Basin, North China. *China Geol.* **2021**, *4*, 274–287. [CrossRef]
30. Shi, X.F. *China Offshore Oceanography-Seabed Matter*; Maritime Press: Beijing, China, 2016; pp. 25–139.
31. Guan, B.X. Patterns and Structures of the Currents in Bohai, Huanghai and East China Seas. *Oceanol. China Seas.* **1994**, *1*, 17–26.
32. Kim, K.H. Geochemical Study of Some Mesozoic Granitic Rocks in South Korea. *J. Korean Inst. Mining Geol.* **1992**, *25*, 435–446.
33. Hu, D.X. Upwelling and sedimentation dynamics. *China J. Oceanol. Limnol.* **1984**, *2*, 12–19.
34. Southon, J.; Kashgarian, M.; Fontugne, M.; Metivier, B.; Yim, W.W. Marine reservoir corrections for the Indian Ocean and southeast Asia. *Radiocarbon* **2002**, *44*, 167–180. [CrossRef]
35. Reimer, P.J.; Bard, E.; Bayliss, A.; Beck, J.W.; Blackwell, P.G.; Ramsey, C.B.; Buck, C.E.; Cheng, H.; Lawrence, E.R.; Michael, F.; et al. IntCal13 and Marine13 Radiocarbon Age Calibration Curves 0–50,000 Years cal BP. *Radiocarbon* **2013**, *55*, 1869–1887. [CrossRef]
36. Folk, R.L. The Distinction between Grain-size and Mineral Composition in Sedimentary-Rock Nomenclature. *J. Geol.* **1954**, *62*, 344–359. [CrossRef]
37. Biscaye, P.E.; Eittreim, S.L. Suspended particulate loads and transports in the nepheloid layer of the abyssal Atlantic Ocean. *Mar. Geol.* **1977**, *23*, 155–172. [CrossRef]
38. Ehrmann, W. Implications of late Eocene to early Miocene clay mineral assemblages in McMurdo Sound (Ross Sea, Antarctica) on paleoclimate and ice dynamics. *Palaeogeogr. Palaeoclimatol. Palaeoecol.* **1998**, *139*, 213–231. [CrossRef]

39. Liu, Z.F.; Trentesaux, A.; Clemens, S.C.; Colin, C.; Wang, P.X.; Huang, B.Q.; Boulay, S. Clay mineral assemblages in the northern South China Sea: Implications for East Asian monsoon evolution over the past 2 million years. *Mar. Geol.* **2003**, *201*, 133–146. [CrossRef]
40. Esquevin, J. Influence De La Composition Chimique Des Illites Sur Leur Cristallinite. *Bull. Cent. Rech. Pau-SNPA* **1969**, *3*, 147–153.
41. Gingele, F.X. Holocene climatic optimum in Southwest Africa—Evidence from the marine clay mineral record. *Palaeogeogr. Palaeoclimatol. Palaeoecol.* **1996**, *122*, 77–87. [CrossRef]
42. Liu, Z.F.; Tuo, S.T.; Colin, C.; Liu, J.T.; Huang, C.Y.; Selvaraj, K.; Chen, C.A.; Zhao, Y.L.; Siringan, F.P.; Boulay, S.; et al. Detrital fine-grained sediment contribution from Taiwan to the northern South China Sea and its relation to regional ocean circulation. *Mar. Geol.* **2008**, *255*, 149–155. [CrossRef]
43. Liu, S.F.; Shi, X.F.; Liu, Y.G.; Wu, Y.H.; Qiao, S.Q.; Yang, G. Recent 2000 a climatic record of mud area on the inner shelf of the East China Sea. *Acta Oceanol. Sin.* **2011**, *33*, 85–93.
44. Lund, B.; Ma, J. A review of cluster analysis techniques and their uses in library and information science research: K-means and k-medoids clustering. *Perform. Meas. Metr.* **2021**, *22*, 161–173. [CrossRef]
45. Stoll, H.M.; Schrag, D.P. Coccolith Sr/Ca as a new indicator of coccolithophorid calcification and growth rate. *Geochem. Geophys. Geosystems* **2000**, *1*, 1006. [CrossRef]
46. Li, Y.; Li, A.C.; Huang, P.; Xu, F.J.; Zheng, X.F. Clay minerals in surface sediment of the north Yellow Sea and their implication to provenance and transportation. *Cont. Shelf Res.* **2014**, *90*, 33–40. [CrossRef]
47. Liu, J.Q.; Liu, Y.L.; Yin, P.; Gao, F.; Cao, K.; Chen, X.Y. Composition, Source and Environmental Indication of Clay Minerals in Sediments from Mud Deposits in he Southern Weihai Offshore, Northwestern Shelf of the South Yellow Sea, China. *J. Ocean Univ. China* **2022**, *21*, 1161–1173. [CrossRef]
48. Fan, D.J.; Yang, Z.S.; Mao, D.; Guo, Z.G. Clay Minerals and Geochemistry of the Sediments from the Yangtze and Yellow Rivers. *Mar. Geol. Quat. Geol.* **2001**, *21*, 7–12.
49. Milliman, J.D.; Beardsley, R.C.; Zuo-Sheng, Y.; Limeburner, R. Modem Huanghe-derived muds on the outer shelf of the East China Sea: Identification and potential transport mechanisms. *Conitinental Shelf Res.* **1985**, *4*, 175–188. [CrossRef]
50. Xu, K.H.; Milliman, J.D.; Li, A.C.; Liu, J.P.; Kao, S.J.; Wan, S.M. Yangtze and Taiwan derived sediments on the inner shelf of East China Sea. *Cont. Shelf Res.* **2009**, *29*, 2240–2256. [CrossRef]
51. Xu, D.Y. Mud sedimentation on the east China sea continental shelf. *Mar. Geol. Quat. Geol.* **1985**, *5*, 17–26.
52. Ren, M.; Shi, Y. Sediment discharge of the Yellow River (China) and its effect on the sedimentation of the Bohai and the Yellow Sea. *Cont. Shelf Res.* **1986**, *6*, 785–810. [CrossRef]
53. Yang, Z.S. Mineralogical assemblages and chemical characteristics of clays from sediments of the huanghe, changjiang, zhujiang rivers and their relationship to the climate environment in their sediment source areas. *Oceanol. Limnol. Sin.* **1988**, *19*, 336–346.
54. Yang, S.Y.; Jung, H.S.; Lim, D.I.; Li, C.X. A review on the provenance discrimination of sediments in the Yellow Sea. *Earth-Sci. Rev.* **2003**, *63*, 93–120. [CrossRef]
55. He, M.Y.; Zheng, H.B.; Huang, X.T.; Jia, J.T.; Li, L. Yangtze River sediments from source to sink traced with clay mineralogy. *J. Asian Earth Sci.* **2013**, *69*, 60–69. [CrossRef]
56. Xu, D.Y. Mud sedimentation on the East China Sea shelf. In Proceedings of the International Symposium on Sedimentation on the Continental Shelf with Special Reference to the East China Sea, Hangzhou, China, 12–16 April 1983.
57. Lu, J.; Li, A.C.; Huang, P.; Li, Y. Mineral distributions in surface sediments of the western South Yellow Sea: Implications for sediment provenance and transportation. *Chin. J. Oceanol. Limnol.* **2015**, *33*, 510–524. [CrossRef]
58. Hu, G.; Xu, K.; Clift, P.D.; Zhang, Y.; Li, Y.; Qiu, J.; Kong, X.; Bi, S. Textures, provenances and structures of sediment in the inner shelf south of Shandong Peninsula, western South Yellow Sea. *Estuar. Coast. Shelf Sci.* **2018**, *212*, 153–163. [CrossRef]
59. Lan, X.H.; Zhang, X.J.; Liu, X.B.; Li, R.H.; Zhang, Z.X. Distribution pattern of clay minerals in surface sediments of South Yellow Sea and their provenance. *Mar. Geol. Quat. Geol.* **2011**, *31*, 11–16. [CrossRef]
60. Seo, K.W.; Chi, J.M.; Jang, Y.H. Geochemical relationship between shore sediments and near terrestrial geology in Byunsan-Taean area, west coast of korea. *Econ. Environ. Geol.* **1998**, *31*, 69–84. Available online: https://koreascience.kr/article/JAKO199823034627177.do (accessed on 2 August 2024).
61. Yang, S.Y.; Li, C.X.; Lee, C.B. Rare earth element geochemistry and sediment source tracing in rivers around the Yellow Sea. *Chin. Sci. Bull.* **2003**, *48*, 1233–1236. [CrossRef]
62. Park, C.K.; Oh, J.K. A Study on the Clay Minerals in the Han River Estuary and the Kyonggi Bay Areas. *J. Korea Soc. Oceanogr.* **1991**, *26*, 313–323. Available online: https://koreascience.kr/article/JAKO199111920647699.do (accessed on 2 August 2024).
63. Park, Y.A.; Khim, B.K. Origin and dispersal of Recent clay minerals in the Yellow Sea. *Mar. Geol.* **1992**, *104*, 205–213. [CrossRef]
64. Cho, H.G.; Kim, S.; Kwak, K.Y.; Choi, H.; Khim, B. Clay mineral distribution and provenance in the Heuksan mud belt, Yellow Sea. *Geo-Mar. Lett.* **2015**, *35*, 411–419. [CrossRef]
65. Choi, J.H. Recent Clay Minerals in the Kunsan Estuary and the Adjacent Continental Shelf. Master's Thesis, Seoul National University, Seoul, Republic of Korea, 1981.
66. Kim, D.C. Recent Clay Minerals of the Yeongsan Estuary and the Adjacent Continental Shelf. Master's Thesis, Seoul National University, Seoul, Republic of Korea, 1980.
67. Gao, J.H.; Li, J.; Wang, Y.P.; Bai, F.L.; Li, J.S.; Cheng, Y. Heavy mineral distributions and their implications for sediment dynamics in the Yalu Estuary and its adjacent sea area. *Acta Oceanol. Sin.* **2009**, *31*, 84–94.

68. Shi, Y.F.; Kong, Z.C.; Wang, S.M.; Tang, L.Y.; Wang, F.B.; Yao, C.D.; Zhao, X.T.; Zhang, P.Y.; Shi, S.H. Climatic fluctuations and important events during the Holocene Great Warm Period in China. *Sci. China Ser. B* **1992**, *12*, 1300–1308.
69. Stuiver, M.; Grootes, P.M.; Braziunas, T.F. The GISP2 δ18O climate record of the past 16,500 years and the role of the sun, ocean, and volcanoes. *Quaternary Res.* **1995**, *44*, 341–354. [CrossRef]
70. Xu, H.; Hong, Y.T.; Lin, Q.H.; Hong, B.; Jiang, H.B. Temperature changes from 6ka to present as indicated by oxygen isotope of cellulose in Hongyuan Peat. *Chin. Sci. Bull.* **2002**, *47*, 1181–1186.
71. Zhu, Y.M.; Tian, Y.; Yin, P.; Duan, X.Y.; Cao, K.; Liu, D.Y. Response of high-resolution sedimentary records to East Asian winter monsoon in the inner shelf of the East China Sea over the past 8000 years. *Mar. Geol. Front.* **2023**, *39*, 1–10.
72. Xiao, S.B.; Li, A.C.; Chen, M.H.; Liu, J.P.; Jiang, F.Q. Recent 8ka Mud Records of the East Asian Winter Monsoon from the Inner Shel of the East China Sea. *Earth Sci. J. China Univ. Geosci.* **2005**, *30*, 573–581.
73. Ge, Q.S.; Zhu, H.Y. Changes of the physical and human geographical environment in China during the past 2000 years. *Acta Geogr. Sin.* **2021**, *76*, 3–14.
74. Ge, Q.S.; Fang, X.Q.; Zheng, J.Y. Learning from the historical impacts of climatic change in China. *Adv. Earth Sci.* **2014**, *29*, 23–29.
75. Bond, G.; Showers, W.J.; Cheseby, M.; Lotti, R.; Almasi, P.; DeMenocal, P.; Priore, P.; Cullen, H.; Hajdas, I.; Bonani, G. A pervasive millennial-scale cycle in North Atlantic Holocene and glacial climates. *Science* **1997**, *278*, 1257–1266. [CrossRef]
76. Xue, C.T.; Liu, J.; Kong, X.H. Channel shifting of lower Yellow River in 1128-1855AD and its influence to the sedimentation in Bohai, Yellow and east China Seas. *Mar. Geol. Quat. Geol.* **2011**, *31*, 25–36. [CrossRef]
77. Yao, C.D.; Xie, Z.C.; Wu, X.L.; Thompson, L.G. Little Ice Age climate records in the Dunde Ice Cap. *Sci. China Ser. B* **1990**, *11*, 1196–1201.

Journal of
Marine Science and Engineering

MDPI

Article

Analysis of Downstream Sediment Transport Trends Based on In Situ Data and Numerical Simulation

Yuxi Wu [1,2], Xiwen Li [1,*], Enjin Zhao [2,3,4,*], Yang Wang [1], Shiyou Zhang [1], Zhiming Xu [5], Qinjun Wang [5], Dongxu Jiang [1] and Zhuang Xing [1]

1 Haikou Marine Geological Survey Center, China Geological Survey, Haikou 571127, China; yuxiwu@cug.edu.cn (Y.W.); wangyang01@mail.cgs.gov.cn (Y.W.); 17608918675@163.com (S.Z.); 13078929611@163.com (D.J.); 17683345561@163.com (Z.X.)
2 College of Marine Science and Technology, China University of Geosciences, Wuhan 430074, China
3 Shenzhen Research Institute, China University of Geosciences, Shenzhen 518057, China
4 Shandong Provincial Key Laboratory of Marine Environment and Geological Engineering, Ocean University of China, Qingdao 266100, China
5 Unit 91656 of the Chinese People's Liberation Army, Shanghai 200231, China; 18606629153@163.com (Z.X.); wyx959022104@icloud.com (Q.W.)
* Correspondence: author: lxw1818168@163.com (X.L.); zhaoej@cug.edu.cn (E.Z.)

Abstract: This study conducted an in-depth analysis of the sediment dynamics in the lower reaches of the Changhua River and its estuary on Hainan Island. Through field collection of topographic data and sediment sampling, combined with advanced computational techniques, the study explored the transport pathways and depositional patterns of sediments. The grain size trend analysis (GSTA) method was utilized, in conjunction with the Flemming triangle diagram method, to classify the dynamic environment of the sediments. Furthermore, hydrodynamic modeling results were integrated to further analyze the transport trends of the sediments. The study revealed that the sediment types in the research area are complex, primarily consisting of gravelly sand and sandy gravel, indicating a generally coarse sedimentary environment in the region. The sediments in the lower reaches of the Changhua River generally transport towards the south and southwest (in the direction of Beili Bay). The net sediment transport directions inferred from the GSTA model are largely consistent with the Eulerian residual flow patterns, especially in the offshore area, where discrepancies are observed in the nearshore zone. The nearshore transport is influenced by the combined effects of alongshore currents, residual flows, and river inputs, while the offshore transport exhibits a shift from the northwest to southwest directions, reflecting the regional circulation patterns.

Keywords: sediment grain size characteristics; sediment transport trends; Gao–Collins method; hydrodynamic; Changhua River estuary

1. Introduction

In recent years, with the advancement in marine scientific research [1–6], sediment grain size analysis techniques have been extensively applied [7–12]. Grain size parameters, such as mean grain size, sorting coefficient, and skewness, have been established as effective indicators for reflecting the provenance, transportation, and depositional environment of sediments [13–15]. Concurrently, the continuous refinement of sediment dynamic zoning and sediment transport trend analysis methods has provided robust tools for uncovering sedimentary processes [16,17]. The transport process of sediments not only reflects the complexity and variability of the regional hydrodynamic environment, but also significantly impacts regional geomorphological evolution, ecological environments, and resource development and utilization [18]. The grain size characteristics and distribution of sediments are key parameters for revealing sediment transport patterns, and detailed grain size analysis can provide an in-depth understanding of the sources, transport pathways, and depositional mechanisms of sediments [19]. It is against this technological backdrop that we are able to more meticulously explore the transport patterns of sediments in specific areas, such as the lower reaches of the Changhua River, and their environmental impacts.

In the exploration of sediment transport directions in the lower reaches of the Changhua River, the issue of sediment accumulation, as one of the core challenges in the field of water conservancy projects, cannot be overlooked due to its complex causes and the urgency of its management [20]. The Gao–Collins method, a two-dimensional sediment grain size trend analysis model, has been widely applied in various marine environments to infer net sediment transport patterns from the spatial distribution of grain size parameters [21–23]. Yet, its application in the Changhua River estuary has been sparse, and the results have not been fully synthesized with the local hydrodynamic conditions and sediment sources [12,24]. With the rapid development of computational technology, sediment transport models, particularly the GSTA (grain size trend analysis) model, have provided a powerful tool for accurately simulating and predicting the dynamic processes of river sediments [25–27]. The GSTA model, an evolution of the Gao–Collins method, incorporates advanced computational techniques to enhance the accuracy and resolution of sediment transport predictions [28–30]. The GSTA model not only integrates classical hydrodynamics theory with modern numerical simulation techniques, but also innovatively considers the coupled effects of topographical changes, flow conditions, sediment grain size distribution, and riverbed morphology, providing a more comprehensive and accurate reflection of sediment transport and deposition processes within river systems.

The lower reaches of the Changhua River in Hainan Island and its estuary, a critical transition zone between terrestrial and marine systems, have been subject to complex hydrodynamic conditions shaped by fluvial inputs, tidal actions, and wave energies [12,24,31]. These conditions are further influenced by the presence of dams upstream, which regulate the river's flow and significantly impact sediment transport patterns. The water discharge data used in our hydrodynamic model, sourced from the nearest Baoqiao Hydrological Station, captures the regulatory role of these dams, ensuring that our analysis reflects the real-world conditions influenced by their presence. The lower reaches of the Changhua River, characterized by low-lying terrain, suffer from severe sediment accumulation, which not only restricts the river's flood discharge capacity, increasing the risk of flood disasters, but also directly threatens the safety of the lives and property of surrounding residents and agricultural production [32,33]. Additionally, the land use patterns within the basin significantly influence the sediment dynamics. According to recent research, the predominant land use types in the Changhua River basin include arable land, forest land, and urban areas, with significant changes observed over the past two decades [34,35]. Arable land, primarily distributed in coastal areas, and forest land, mainly in the central mountainous regions (Figure 1), have experienced fluctuations that reflect the interplay between natural conditions and human activities. These changes directly affect the generation and transport of sediment, exacerbating the sediment accumulation issues in the lower reaches. Therefore, in-depth research on the transport direction of sediments in the lower reaches of the Changhua River is crucial for understanding the spatiotemporal distribution characteristics of sediment accumulation, evaluating the effectiveness of river channel regulation projects, and formulating scientific management measures.

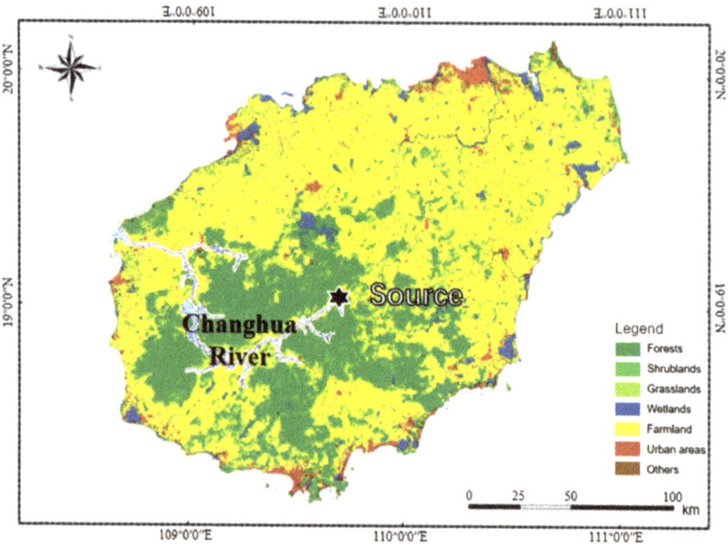

Figure 1. Land use patterns in Hainan province with a focus on the Changhua River Basin, the black star refers to the source of the Changhua River (Origin: International Research Center of Big Data for Sustainable Development Goals. DOI: 10.12237/casearth.63eb24d7819aec795e09af68. [36]).

The application of the GSTA model will enable us to more accurately simulate the sediment transport processes in the lower reaches of the Changhua River under different seasonal and flow conditions, revealing the transport patterns of sediments in the river channel and predicting potential future accumulation trends. This will provide a scientific basis and technical support for river channel regulation, flood control planning, navigation safety, and ecological environment protection in the lower reaches of the Changhua River. Additionally, through the simulation analysis of the GSTA model, we can evaluate the effectiveness of different management measures, optimize resource allocation, and achieve sustainable development in the lower reaches of the Changhua River. Furthermore, the incorporation of hydrodynamic modeling in this study allows for a detailed analysis of sediment transport trends by simulating the complex interactions between flow velocity and direction, which are crucial for understanding the net transport patterns in the lower reaches of the Changhua River.

Building on the achievements of previous research, this study, in conjunction with the actual conditions of the lower reaches of the Changhua River and the nearby sea area of the river mouth, employs the Flemming triangle diagram method and the GSTA model. Focusing on the surface sediments of the lower reaches of the Changhua River and the nearby sea area of the river mouth, the study aims to explore the transport direction and dcontrolling factors of sediments in the area through grain size analysis, sediment type classification, and dynamic environmental zoning. Additionally, it will utilize the residual current from hydrodynamic modeling to further analyze the transport of sediments, providing a more comprehensive understanding of the sediment dynamics in the study area.

2. Study Area and Data

The study area is situated in the western part of Hainan Island, mainly encompassing the lower reaches of the Changhua River and its estuary (Figure 2). The approximate coordinates range from 108°36′ E to 108°50′ E and 19°15′ N to 19°22′ N. The study area covers a large part of the region from Chahe Town to the estuary of the Changhua River, including Xiantiancun, Dangchangcun and Jiuxiancun, among others. A total of 40 offshore sampling points were investigated and collected by the Haikou Geological Survey Center. Sediment data for the open sea were obtained through historical data [24,32,33,37].

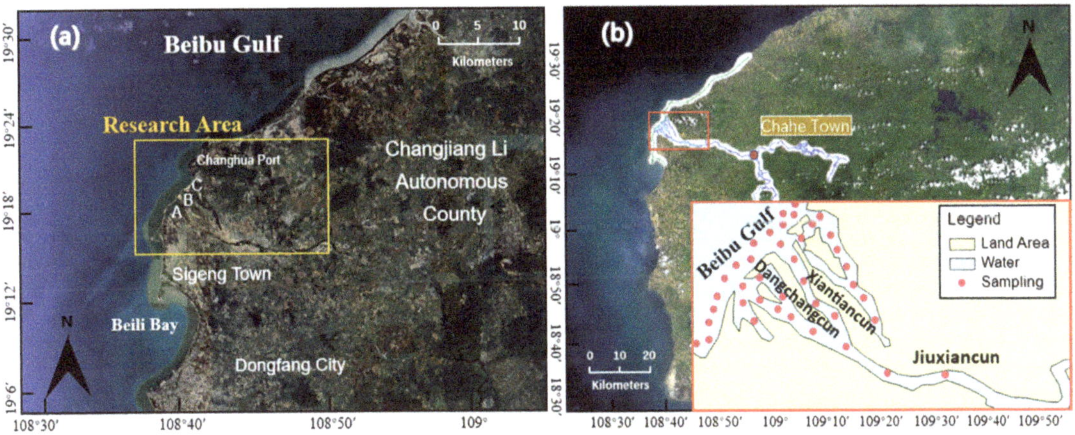

Figure 2. Geographical extent of the study area in the lower reaches of the Changhua River and its estuary, highlighting the key sampling locations and topographical features that influence sediment transport dynamics: (**a**) Orange Box Denotes the Study Area with 3 Major Estuaries (A, B, C), and Changhua Port Located at Estuary C. (**b**) Red Box Encompasses Villages and 40 Sampling Points Within the Study Area.

The hydrological regime of the lower Changhua River is characterized by a distinct seasonal variability, with a pronounced flood season typically occurring from May to October and a dry season from November to April [38]. The river's discharge is highly variable, with the flood season accounting for approximately 77% of the annual net flow [39–43]. The estuary, shaped by the river's dynamic interaction with tidal and wave forces, exhibits a weak tidal regime with a mean tidal range of less than 2 m, indicating the dominance of fluvial inputs over tidal influences [32,44]. The river's sediment is predominantly coarse-grained, reflecting its strong erosional capacity, particularly during the flood season when high discharges lead to significant sediment mobilization and transport [32,35]. Figure 3 illustrates the long-term dynamics of the relationship between water discharge and sediment transport in the lower reaches of the Changhua River. This figure presents the annual discharge, sediment concentration, and sediment transport rate, highlighting the seasonal variability and the dominant factors influencing sediment movement. The data reveal a pronounced flood season from May to October, during which approximately 77% of the annual net flow occurs, significantly impacting sediment mobilization and transport. The dry season, from November to April, shows a marked decrease in both water discharge and sediment transport, indicating the critical role of seasonal hydrological patterns in shaping sediment dynamics.

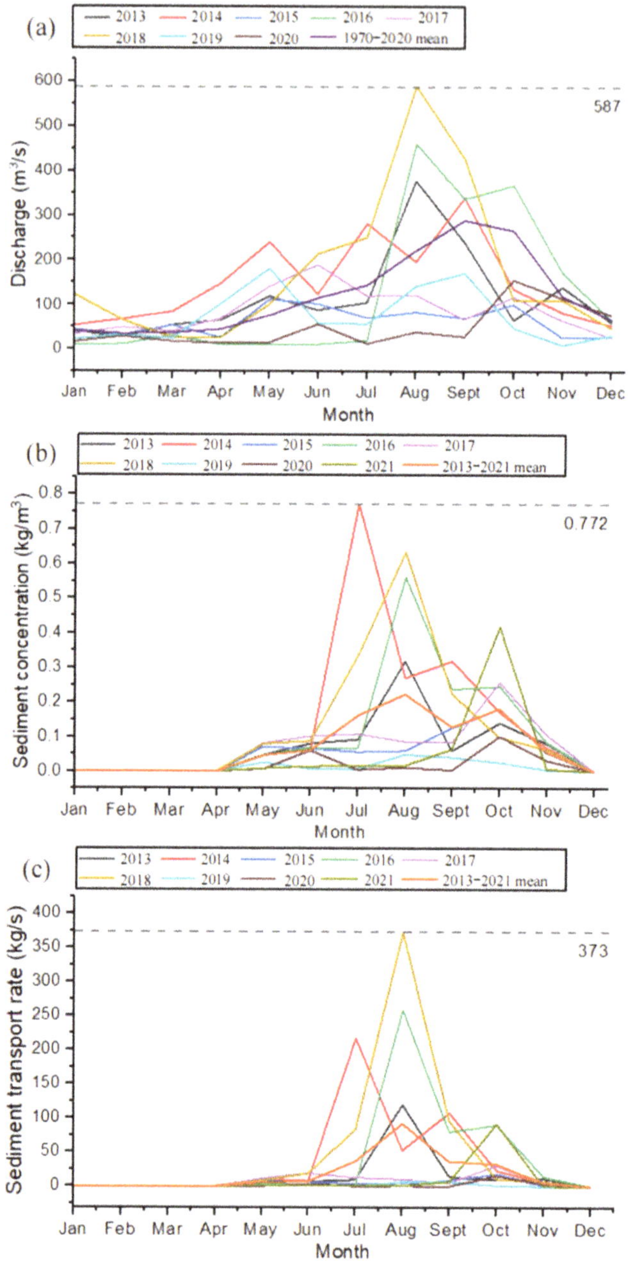

Figure 3. (**a**) Annual Discharge in the Lower Reaches of the Changhua River: The maximum value of 587 m^3/s occurred in August 2018. (**b**) Annual Sediment Concentration in the Lower Reaches of the Changhua River: The maximum value of 0.772 kg/m^3 was recorded in July 2014. (**c**) Annual Sediment Transport Rate in the Lower Reaches of the Changhua River: The maximum value of 373 kg/s was reached in August 2018.

3. Materials and Methods

3.1. Sediment Sampling

Sediment grain size analysis was conducted at the Hainan Provincial Research Center for Geological Testing. Initially, an appropriate amount of the original sample was thoroughly mixed and uniformly dried at 80 °C. For samples with particle sizes less than 2000 μm, a suitable amount of the sample was treated with 10 mL of hydrogen peroxide and hydrochloric acid solutions to remove organic matter and calcareous biomaterials, respectively. Subsequently, 0.05 N sodium hexametaphosphate was added, and the mixture was allowed to stand in a beaker for 24 h. Afterward, it was sonicated for 15 min using an ultrasonic oscillator and then analyzed using a Master-sizer 2000 laser particle size analyzer. The measurement range of this analyzer is 0.02 to 2000 μm, with a relative error of less than 2%, a particle size resolution of 0.01Φ, and a repeatability relative error of less than 3%. The data from both parts were merged using a laser diffraction particle size analyzer simulation program to obtain a complete particle size distribution.

To evaluate the sensitivity of the sample point selection, we conducted additional analyses. By varying the number of sample points, we found that the results remained robust within a certain range: we set up multiple groups with different quantities of sample points and observed the changes in the Flemming triangle diagram. When the number of sample points was reduced by 20%, the results exhibited minimal variation, consistent with those obtained using the full set of sample points in terms of the dynamical partitioning. This indicates that our sample point selection is representative to a certain extent and that the research findings possess high robustness. Figure 4 illustrates three scenarios: all sampling points, a 20% reduction in sampling points, and a 40% reduction in sampling points.

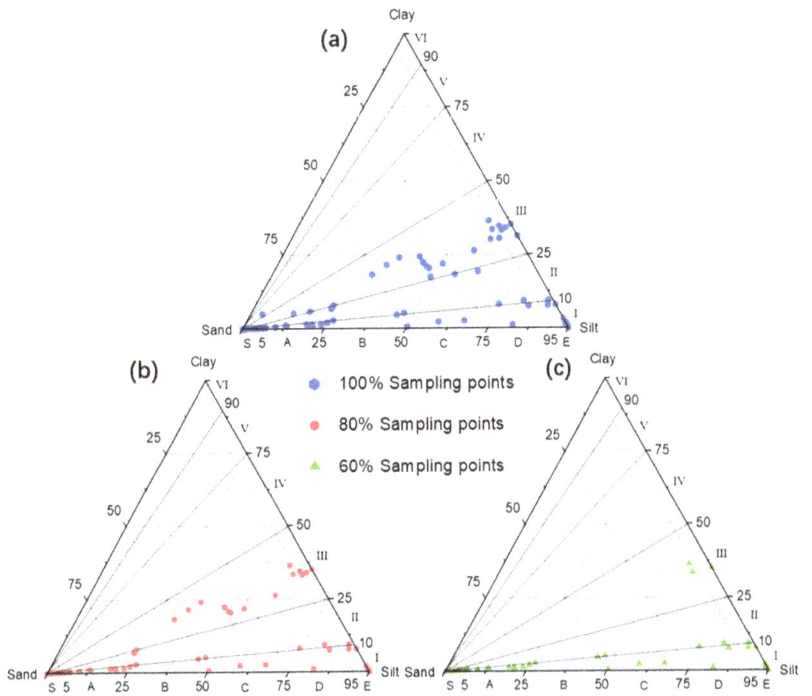

Figure 4. The Flemming Triangle Diagram under Different Numbers of Sample Points: (**a**) The blue hexagons represent the distribution of the original sampling points, serving as the control group. (**b**) The red dots indicate the distribution of 80% of the sampling points. (**c**) The green triangles show the distribution of 60% of the sampling points.

The sample particle sizes were expressed using the Udden–Wentworth scale in phi (Φ) units. The mean particle size (Mz), sorting coefficient (δ), and skewness (Sk) were calculated using the Folk–Ward graphical method with corresponding formulae [13].

3.2. GSTA Model for Sediment Transport Trend Analysis

The planar distribution trend of sediment grain size parameters is known as the grain size trend. By comparing the grain size parameters at a given point with those of its surrounding points, the possible direction of the net transport can be determined. In the case of any two adjacent sampling points, based on sediment grain size parameters such as the mean grain size, sorting coefficient and skewness, a grain size trend vector can be defined, which is a unit vector from the sampling point to the point of interest [29].

A characteristic distance is used to determine whether two sampling points are adjacent. The characteristic distance is usually defined by the maximum sampling interval (or geostatistical distance) and should first be determined to assess the nearest neighbors of a sampling site [22,23]. In this paper, the characteristic distance chosen is the maximum sampling interval. Based on the aforementioned grain size trend vector conditions, all grain size trend vectors between the sampling point and its adjacent sampling points can be obtained. Then, by summing all the grain size trend vectors at this point, the resultant vector for this sampling point can be derived. The length only indicates the significance of this particle size trend, not its magnitude. Its direction represents the net transport direction of sediments at the sampling point [28].

3.3. Hydrodynamic Modeling

In the study, bathymetric data are derived from ETOPO1 global seafloor topography data and in situ measurements using ADCP. The spatial resolution of ETOPO1 data is $1/60^\circ \times 1/60^\circ$, which is insufficient for the research requirements. ADCP depth measurements have higher density in nearshore areas and provide actual measured data with higher accuracy.

The model's open boundary conditions are defined by the forced tidal water level, incorporating eight primary tidal components: M2, S2, K1, O1, N2, K2, P1, and Q1. The open boundary water level is dynamically adjusted to match simulation outcomes. The model's closed boundary aligns with the terrestrial boundary, where the normal velocity of ocean currents is set to zero, precluding any exchange of temperature and salt between land and seawater. The model also integrates the impact of wind fields, with data sourced from ECMWF at a resolution of $1/8^\circ \times 1/8^\circ$. This dataset encompasses the u (east–west) and v (north–south) components of the wind vector, along with the sea level pressure.

An unstructured grid, finite volume, regional ocean model FVCOM [45] was used to simulate the hydrodynamic background and hydrological features. It has been widely used for the study of coastal oceanic and estuarine circulation [46–48]. To enhance computational accuracy and reduce computation time, the density of boundary nodes gradually decreases from nearshore to offshore. In the offshore region, the grid density is lower, with a resolution of 0.2 km, while the nearshore part of the open boundary has a higher grid resolution. In the main research area near the river channel, the grid resolution is highest, reaching 50 m. The entire study area grid comprises a total of 25,921 computational nodes (Figure 5). Since the primary focus is on the surface-related hydrodynamic information, the vertical dimension is discretized into five layers. The hydrodynamic model used in this study has been previously validated for its ability to simulate current velocity and direction [49]. In that work, we compared the model outputs with in situ measurements and found a high degree of correspondence, confirming the reliability of the model for our region of interest.

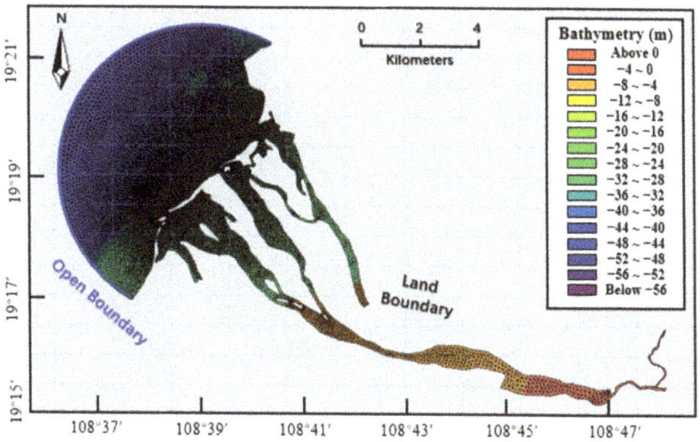

Figure 5. The study and simulation area consisted of a mesh file (mesh triangulation and boundary conditions) and applied in the model: The grid resolution varies from 0.2 km in the offshore region to 50 m in the main research area near the river channel.

4. Results

4.1. Sediment Types

Granulometric analysis results were utilized to construct a distribution map of surficial sediment types (Figure 6d). The study area, encompassing the lower reaches of the Changhua River and the adjacent marine area, exhibits a diverse array of surficial sediment types. These include gravelly sand, silty gravel, sandy silt, silt, sandy gravel, sand, gravelly silt, gravel sand, silty sand, and gravelly muddy sand, among which gravelly sand and gravelly sand are predominant. The overall sediment particle size is relatively coarse, with gravelly sand constituting 30% and gravel sand 17.5% of the sediment types. Silty sand and sand account for 7.5% each, while silt, gravelly muddy sand, and silty gravel each make up 10%. The sediment types in the study area display a patchy distribution pattern, dominated by gravelly sand and gravelly muddy sand. In contrast, the offshore sea area exhibits simpler sediment type distributions with larger distribution areas. The northern sea area (the Beibu Gulf) is primarily characterized by fine-grained sediments such as mud and silty sand; the central area is dominated by coarse-grained sediments like gravelly sand and gravel sand; and the southern sea area is mainly composed of sandy silt.

4.2. Distribution Characteristics of Sediment Grain Size Parameters

The isopleth distribution of the mean grain size of sediments is shown in Figure 6a. The mean grain size of sediment samples in the study area ranges from 0.02Φ to 7.18Φ, with an average value of 3.17Φ. It can be observed that the distribution of the mean grain size of sediments in the study area shows a fine-coarse-fine change from north to south, which is consistent with the trend in the sediment type distribution. This reflects that the hydrodynamic energy in the study area gradually transitions from a low-energy environment in the northern bay to a high-energy environment at the estuary, and then shifts to a low-energy environment in the southernmost part (the Bay direction). Within a 3 km range of the downstream river channel to the estuary, there is a coarse-grained area. Most of the mean grain size of sediments is less than 2Φ, dominated by coarse-grained sediments, and the isopleth lines of 0.4–1.6Φ are distributed in a SE–NW strip, reflecting a general southeast to northwest sediment transport direction. From the estuary extending 3 km outward to the open boundary of the ocean, there is another area of coarse-grained sediments. From the southeast to the northwest, from the downstream river channel to the estuary extending nearly straight to the far sea, the sediments are mainly coarse-grained, reflecting the strong average dynamic force of sediment transport in this area, which is the manifestation of wave convergence and transportation force dominated by waves and wave-generated longshore currents.

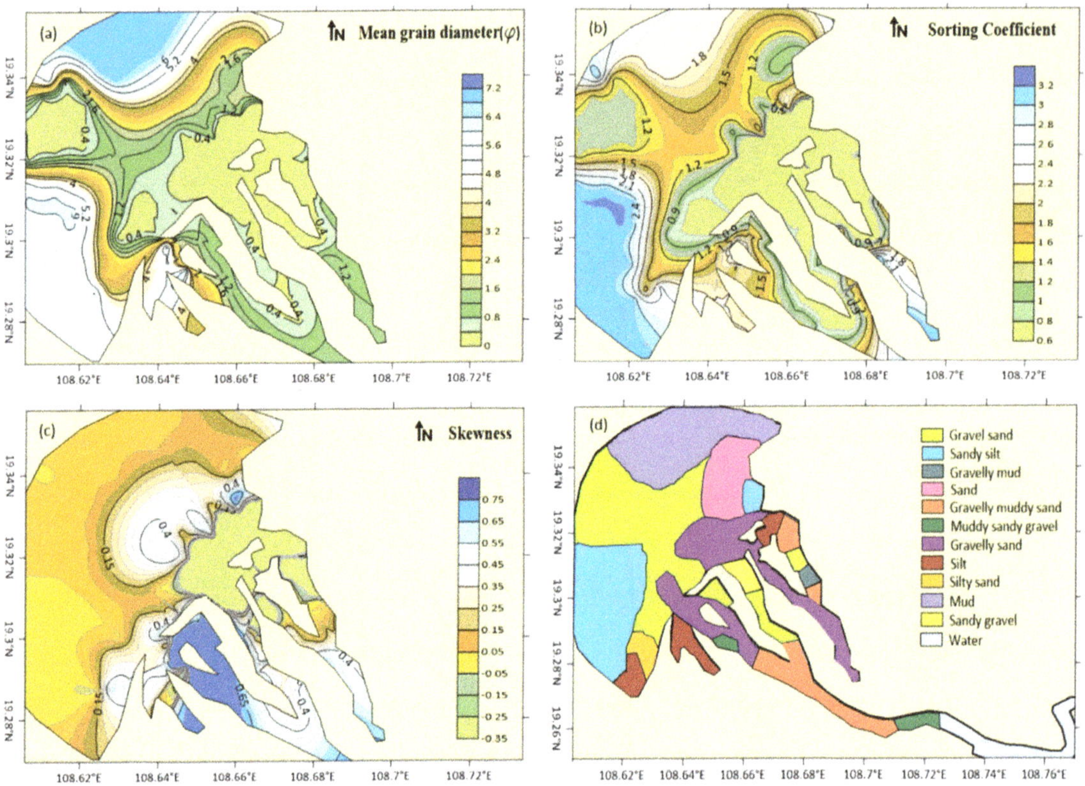

Figure 6. (**a**) Spatial distributions of the mean diameter of the surface sediment. (**b**) Spatial distributions of the sorting coefficient. (**c**) Spatial distributions of the skewness. (**d**) Spatial distributions of the sediment types in the study area.

It can be seen from Figure 6b that the sorting of mud and sand in the study area is between 0.6 and 3.2, which is poor. The sorting coefficient in the main river channel of the lower reaches of the Changhua River is as high as 3.0, and the degree of sorting is very poor. This is because the action of the current is supported by the tide, and the mud and sand are mixed and settle directly without sorting. The sorting coefficient in the southern part of the study area (the Beili Bay direction) is generally greater than 2.4. The region is affected by the current, waves, and tides, with strong hydrodynamic conditions, which can transport particles of different grain sizes together, resulting in a mixture of coarse and fine sediments and poor sorting. The sorting coefficient in the northern part of the study area (the Beibu Gulf) is greater than 1.5, and the degree of sorting is relatively poor, which is better than that in the southern part of the study area. The best sorting area is located in the nearshore area and the far northwest sea area, with a sorting coefficient between 0.6 and 1.2, and the degree of sorting is moderate. Overall, the study area has a high overall sorting coefficient and poor sorting, reflecting the complex material source and hydrodynamic conditions of the area.

It can be seen from Figure 6c that the range of skewness in the study area is from −0.66 to 0.75. The southern part of the study area (Beili Bay) and the B and C estuaries are all negatively skewed, with the median at the finer end and the grain size enriched at the finer end. Most of the remaining areas are positively skewed, with sediments enriched at the coarser end, and the grain size is mainly enriched at the coarser end. The study area's skewness is extremely negatively skewed to extremely positively skewed, and most of the outer sea of the study area is nearly symmetrical. There is an area with extremely positive skewness at the C estuary.

4.3. Sediment Dynamic Zoning

The dynamic environmental characteristics of surficial sediments in the research area were analyzed using the Flemming triangle diagram method [50], which integrates sediment grain size features and the strength of hydrodynamic forces they represent to delineate sediment dynamic environments and sub-environments. The sediment grain size classification standards in the dynamic zoning map are as follows: sand sediment particles with a diameter of 0.063–2.000 mm, silt from 0.004 to 0.063 mm, and clay less than 0.004 mm. The structural classification lines in the diagram are based on the mass fraction of sand as nodes, at 5%, 25%, 50%, 75%, and 95%, creating six components from S to E, indicating a gradual decrease in sediment particle size from S to E. The fine-grained components (clay and silt) in the sediments are used as classification lines for the triangular structure, at 10%, 25%, 50%, 75%, and 90%, dividing into six hydrodynamic zones from I to VI. The closer to the clay and silt end-members, the weaker the hydrodynamic force; from I to VI, hydrodynamic forces gradually weaken. Thus, the Flemming triangle diagram method divides the triangle into 25 areas, each representing different sediment dynamic environments. Consequently, in the Flemming triangle diagram, the sediment grain size increases from S to E; hydrodynamic environments weaken from I to VI.

Based on grain size parameters, topography, sediment sources, hydrodynamics, and sediment transport, the study area is divided into four sedimentary regions (Figure 7a,b): the Northern Offshore Depositional Zone, Central Offshore Depositional Zone, Southern Offshore Depositional Zone, and Downstream River Channel Depositional Zone.

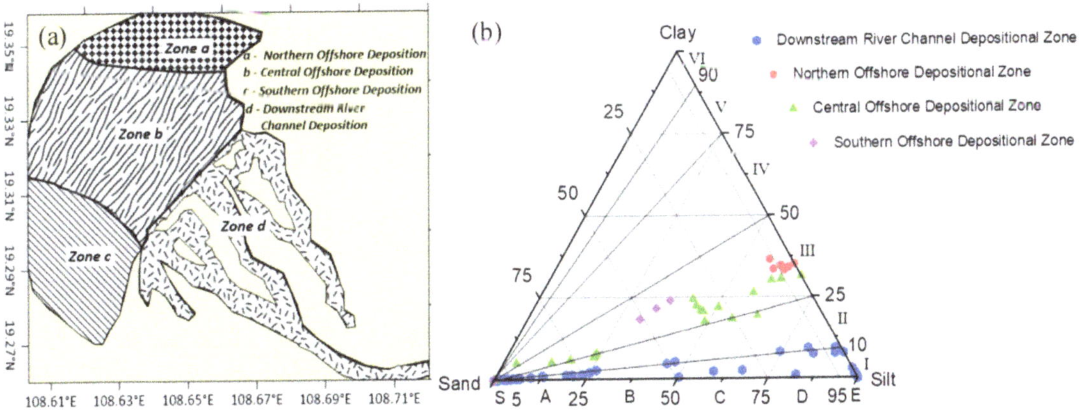

Figure 7. (**a**) Distribution characteristics of sediment classification in the study area. (**b**) Sediment classification according to Flemming triangle diagram: Dividing the study area into four sedimentary regions: Northern Offshore Depositional Zone (red dots), Central Offshore Depositional Zone (green triangles), Southern Offshore Depositional Zone (purple rhombuses), and Downstream River Channel Depositional Zone (blue hexagons).

4.4. Sediment Transport Trend Analysis

In this paper, the GSTA model is used to calculate the grain size transport trends of surficial sediments in the study area based on the spatial distribution of average grain size, sorting coefficient, and skewness (Figure 8). The maximum sampling interval of 2 km is used as the characteristic distance. The vector arrows in the figure represent the net transport direction of sediments, and the length of the vectors indicates the significance of sediment transport, not the rate of sediment transport.

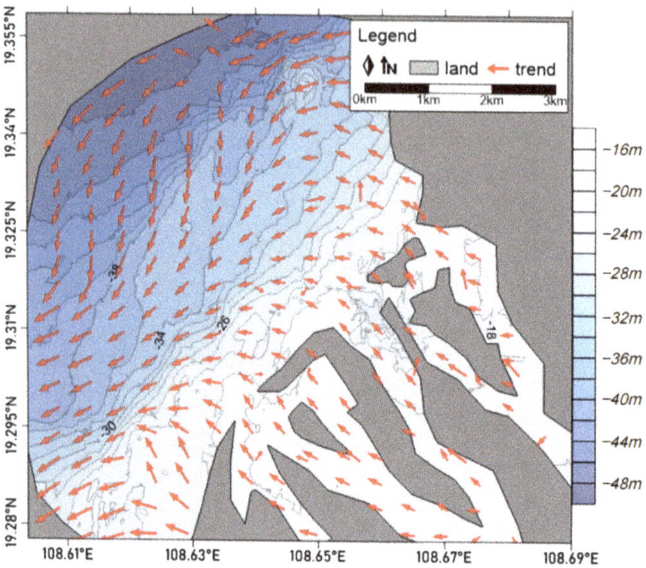

Figure 8. Trends in sediment transport in the Changhua river estuary and downstream areas: The base map is a bathymetric map of the seabed, and the red arrows represent the direction of sediment transport.

The transport trend in surficial sediments from the Changhua River to the nearshore sea area in the study area generally shows a trend from the river to the near sea and then to the far sea. The transport trend at the river outlet is generally strong, mainly moving northwestward after entering the ocean. The transport trend in the near sea is influenced by the combined effects of alongshore currents, residual currents, and rivers. The alongshore current flows in the NE–SW direction, the nearshore residual current flows in the northeast direction, mainly affecting the direction of sediments, and the river mainly flows in the northwest direction. At the southern outlet, affected by both the alongshore current and the river, the sediments gradually change from northwest to southwest; at the central river outlet, affected by the comprehensive influence of seawater, the transport trend changes from northwest to west; at the northern river outlet, affected by the influence of the bay's horn, it moves westward, and after reaching the far sea, the transport direction gradually changes to southwest, consistent with the direction of sediments from other outlets. The overall transport characteristics are northwestward near the sea and southwestward in the far sea. The transport trends of the several distributaries of the Changhua River are the same, all moving towards the ocean. The transport trend in the far sea changes from northwest to south to southwest from north to south. The northwestward transport in the north, together with the circulation of the Beibu Gulf carrying material, will form a deposition center [24,32,33], converging fine-grained sediments, which is consistent with the muddy sediments shown in Figure 2. Under the strong hydrodynamic action of waves, alongshore currents, and runoff, the southwestward transport in the southern far sea will transport coarse particles to Beihai Bay.

5. Discussion

5.1. Implications of Sediment Grain Size Parameters

The fine-coarse-fine change in the mean grain size of sediments from north to south, as observed in the study area, is a direct reflection of the varying hydrodynamic energy. The transition from a low-energy environment in the northern bay to a high-energy environment at the estuary, followed by a shift to a low-energy environment in the southernmost part, highlights the importance of hydrodynamic forces in shaping sediment distribution. The coarse-grained area within 3 km of the downstream river channel to the estuary, together with the SE–NW strip distribution of isopleth lines, indicates a general southeast to northwest sediment transport direction, which is a significant finding for understanding sediment dynamics in the area.

The poor sorting observed in the study area, with a high overall sorting coefficient, suggests a complex material source and hydrodynamic conditions. The very poor sorting in the main river channel of the lower reaches of the Changhua River, due to the mixing and settling of mud and sand without sorting, is a notable characteristic that impacts sediment transport and deposition processes.

The skewness range in the study area, from extremely negatively skewed to extremely positively skewed, indicates varying sediment enrichment patterns. The negative skewness in the southern part and the B and C estuaries, with enrichment at the finer end, contrasts with the positive skewness in most remaining areas, where sediments are enriched at the coarser end. This variation in skewness provides insights into the sediment transport mechanisms and the influence of different hydrodynamic forces on sediment distribution.

5.2. Sediment Dynamic Zoning and Its Significance

The Flemming triangle diagram method provides a comprehensive framework for analyzing the dynamic environmental characteristics of surficial sediments. The division of the study area into four sedimentary regions based on grain size parameters, topography, sediment sources, hydrodynamics, and sediment transport offers a clear understanding of the sediment dynamic environments and sub-environments. The distribution of sampling points across different zones in the Flemming triangle diagram indicates a dispersed sediment grain size distribution and a predominantly strong overall hydrodynamic force in the study area.

The Northern Offshore Depositional Zone, with its fine sediment particles and low-energy hydrodynamic environment, is likely influenced by the fine-grained material transport of the Beibu Gulf circulation. The Central Offshore Depositional Zone, with its mixture of clay, silt, and sand, suggests moderate hydrodynamic conditions and a broad distribution of sediment grain sizes, influenced by strong tides and seasonal storm tides. The Southern Offshore Depositional Zone, dominated by sandy sediments, indicates a higher-energy hydrodynamic environment and possible influence from distant material inputs. The Downstream River Channel Depositional Zone, with its complex and variable sediment grain size compositions, reflects the diversity of river flow velocity, water depth, and sediment supply sources, showing a strong nature of transition areas.

5.3. Interpretation of Sediment Transport Trends

The sediment transport trends in the study area can be divided into three paths: one from the downstream river channel to the estuary to the southern part of the study area's outer sea; one from the downstream river channel to the estuary to the northern part of the study area's outer sea and then to the southern part of the outer sea; and one from the northern part of the study area's outer sea to the southern part of the outer sea. The first two sediment sources are from the lower reaches of the Changhua River, and the last one is from the Beibu Gulf. This is consistent with previous studies on the sediment transport trends in the lower reaches of the Changhua River [33,35,38,45].

In the estuarine and downstream regions of the Changhua River, climate and water levels exhibit distinct seasonal variations [24,33,38,39,41,42,45], which significantly influence the direction of sediment transport. During the flood season, which corresponds to the summer months, the discharge of the Changhua River increases markedly due to the substantial rise in rainfall, enhancing the river's sediment transport capacity [32,33,42]. During this period, the river carries a substantial load of sediment, leading to increased sediment deposition in the estuary and downstream areas. The faster water flow velocities characteristic of the flood season facilitate the easier transportation and dispersion of sediment particles near the estuary. Against the backdrop of the prevailing southerly winds in summer, this rapid water flow, in conjunction with the impact of waves and the southwest–northeast coastal current, induces a shift in the sediment transport direction from northwest to southwest. Particularly in the estuarine region, this change is closely related to the intensity of riverine sediment transport and wave action [32,33,45]. Moreover, the high discharge during the flood season may also alter the estuarine topography, creating new erosional channels and depositional areas, which further affect the distribution and transport pathways of the sediment [35,38,45].

In the dry season, which corresponds to spring, the flow of the Changhua River diminishes, leading to a decrease in the river's sediment transport capacity. During this period, the river carries less sediment, and the sediment deposition rate in the estuary and downstream areas is correspondingly reduced. The slower water flow velocities typical of the dry season result in sediment particles being more prone to deposition on the riverbed, leading to more localized and limited sediment transport patterns [32]. Additionally, the low flow during the dry season may enhance the relative significance of wave action in the estuary, exerting a greater influence on the transportation and deposition of sediment. These seasonal variations in water flow and wave action result in different sediment transport characteristics across different seasons. Especially in the estuarine area, the shift in sediment transport direction from northwest to southwest reflects the intensity of riverine sediment transport and wave action during the flood season, as well as the relative enhancement of wave action during the dry season.

To discuss and verify the rationality of the GSTA model calculation results, this paper calculates the surface residual current within 15 synodic days according to the vector synthesis and decomposition method based on the hydrodynamic model results of sea currents, and then uses the drawing tool kit in the Surfer 11 software to draw the residual current distribution map (Figure 9).

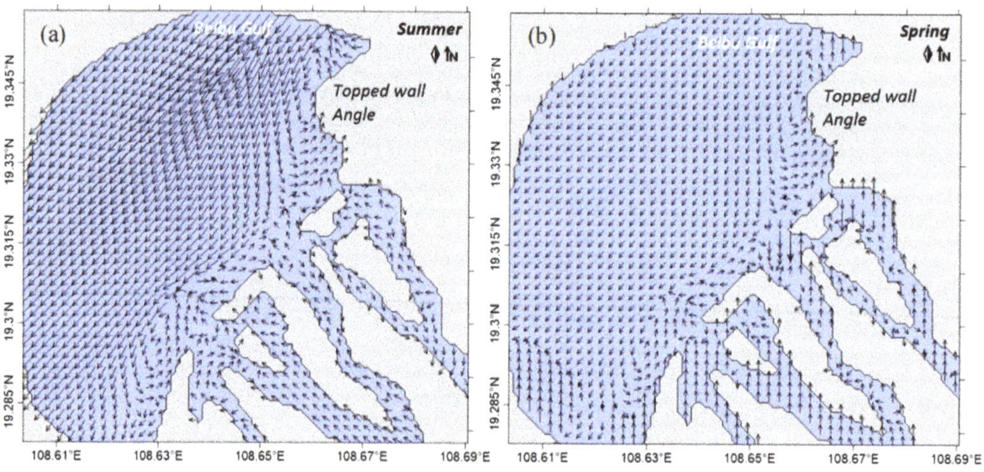

Figure 9. (**a**) Eulerian residual current fields during summer (flood season) (**b**) Eulerian residual current fields during spring (dry season).

There are two vortices in the residual current of the study area, a large counterclockwise vortex in the middle nearshore area, and a small counterclockwise vortex in the southern area. The residual current near the shore flows northeastward, deflects after passing below the topped wall angle, and changes to west–southwestward, moving together with the northwestward residual current from the Beibu Gulf [38]. Most of the residual currents in the far sea are southward and southwestward, and the residual current directions in the river channels all point to the ocean.

Overall, the residual current direction is consistent with the net sediment transport direction of the GSTA model, indicating that the analysis of the sediment transport trend map of the GSTA model is feasible. The main difference is near the shore, where the residual current direction is parallel to the shore and flows northeastward, while the sediment transport direction is parallel to the river channel and flows northwestward. This may be the effect of terrain guidance. The river channel terrain may guide the water flow, causing the water to flow in a specific direction in the river channel. This terrain guidance effect may mask the direct impact of the residual current on sediment transport, causing the sediment transport direction to be inconsistent with the residual current direction. Secondly, the seasonal variation characteristics of the lower reaches of the Changhua River will also lead to inconsistencies between sediment transport and the Euler residual flow field. Although seasonal changes are short-term, they can have a significant impact on long-term sediment transport patterns through long-term cumulative effects. During the flood season, the water flow of the Changhua River increases, carrying a large amount of sediment into the river mouth. These sediments, under the action of strong water flow, may be more easily transported along the river channel direction (or the direction affected by the terrain) and are deposited at the river mouth, thus affecting the morphology and position of the sand spit. During the dry season, the water flow weakens and the amount of sediment carried decreases, but the action of waves and alongshore currents relatively strengthens. These forces may re-transport the already deposited sediment to other locations, causing the sediment transport direction to be inconsistent with the Euler residual current direction.

6. Conclusions

The following conclusions summarize the key findings of our study on sediment transport dynamics in the lower reaches of the Changhua River and its estuary:

Three Main Paths of Sediment Transport: One from the downstream river channel to the estuary to the southern part of the study area's outer sea; one from the downstream river channel to the estuary to the northern part of the study area's outer sea and then to the southern part of the outer sea; and one from the northern part of the study area's outer sea to the southern part of the outer sea. The first two sediment sources are from the lower reaches of the Changhua River, and the last one is from the Beibu Gulf.

Sediment Transport Direction: Sediments in the lower reaches of the Changhua River are primarily transported towards the south and southwest, a trend consistent with Eulerian residual flow patterns, particularly in the offshore areas. Nearshore sediment transport is influenced by the combined effects of alongshore currents, residual flows, and river inputs, leading to complex sediment transport directions.

Mixed Sediment Origins: The sediments in the outer southern part of the study area originate from both the fine-grained materials transported by the circulation of Beili Bay in the north and the coarse-grained materials from the lower reaches of the Changhua River, resulting in a mixed distribution of sediment types in the outer southern part.

Tidal Currents and Sediment Transport: Tidal currents are a significant factor affecting sediment transport, with residual current directions in the outer sea and downstream river channel generally aligning with the net sediment transport directions. However, discrepancies in the nearshore area suggest that the guiding effect of the river mouth sand spit may override the influence of residual currents on sediment transport, causing sediment transport directions to diverge from the residual current flow.

Limitations: In order to assess the sensitivity of the sampling point selection, we conducted additional analyses. By altering the position and quantity of the sampling points, we observed that the results remained robust within a certain range. This suggests that our selection of sampling points is representative to a certain extent, and the study outcomes exhibit a high degree of robustness. However, we also recognize that increasing the number of sampling points, particularly in certain critical areas, may further enhance the representativeness of the results. Future studies could consider augmenting the sampling points in these areas to further validate our findings.

Author Contributions: Conceptualization, Y.W. (Yuxi Wu) and E.Z.; Methodology, Z.X. (Zhiming Xu); Software, Y.W. (Yang Wang) and Q.W.; Validation, X.L., Q.W., D.J. and Z.X. (Zhuang Xing); Formal analysis, Y.W. (Yuxi Wu); Investigation, Y.W. (Yuxi Wu), X.L., Y.W. (Yang Wang), S.Z., Z.X. (Zhiming Xu) and Z.X. (Zhuang Xing); Resources, Y.W. (Yuxi Wu), E.Z., S.Z. and D.J.; Data curation, Y.W. (Yuxi Wu), Y.W. (Yang Wang), S.Z. and Z.X. (Zhiming Xu); Writing—original draft, Y.W. (Yuxi Wu) and E.Z.; Writing—review and editing, Y.W. (Yuxi Wu) and E.Z.; Visualization, Y.W. (Yuxi Wu); Supervision, X.L., E.Z. and Q.W.; Project administration, X.L. All authors have read and agreed to the published version of the manuscript.

Funding: This research was funded by the National Natural Science Foundation of China (Grant Nos. 52371295, 52001286), Guangdong Basic and Applied Basic Research Foundation (Grant No. 2022A1515240002), Hubei Provincial Natural Science Foundation of China (Grant No. 2023AFB576), the Open Fund of Shandong Provincial Key Laboratory of Marine Environment and Geological Engineering (Grant No. MEGE2024006), the China Geological Survey Project (Grant No. DD20220956), and the Fundamental Research Funds for the Central Universities, China University of Geosciences (Wuhan) (Grant Nos. 138-162301212657, 138-162301212679).

Institutional Review Board Statement: Not applicable.

Informed Consent Statement: Not applicable.

Data Availability Statement: Data were obtained from the Haikou Marine Geological Survey Center and are available from the authors with the permission of the Haikou Marine Geological Survey Center.

Conflicts of Interest: The authors declare no conflicts of interest.

References

1. Jiang, F.Y.; Zhao, E.J. Damage mechanism and failure risk analysis of offshore pipelines subjected to impact loads from falling object, considering the soil variability. *Mar. Struct.* **2024**, *93*, 103544. [CrossRef]
2. Sun, Q.L.; Wang, Q.; Shi, F.Y.; Alves, T.; Gao, S.; Xie, X.N.; Wu, S.G.; Li, J.B. Runup of landslide-generated tsunamis controlled by paleogeography and sea-level change. *Commun. Earth Environ.* **2022**, *3*, 244. [CrossRef]
3. Zhao, E.J.; Dong, Y.K.; Tang, Y.Z.; Sun, J.K. Numerical investigation of hydrodynamic characteristics and local scour mechanism around submarine pipelines under joint effect of solitary waves and currents. *Ocean. Eng.* **2021**, *222*, 108553. [CrossRef]
4. Zhao, E.J.; Qu, K.; Mu, L. Numerical study of morphological response of the sandy bed after tsunami-like wave overtopping an impermeable seawall. *Ocean. Eng.* **2019**, *186*, 106076. [CrossRef]
5. Zhao, E.J.; Sun, J.K.; Tang, Y.Z.; Mu, L.; Jiang, H.Y. Numerical investigation of tsunami wave impacts on different coastal bridge decks using immersed boundary method. *Ocean. Eng.* **2020**, *201*, 107132. [CrossRef]
6. Zhao, E.J.; Wu, Y.X.; Jiang, F.Y.; Wang, Y.; Zhang, Z.Y.; Nie, C.H. Numerical investigation on the influence of the complete tsunami-like wave on the tandem pipeline. *Ocean. Eng.* **2024**, *294*, 116697. [CrossRef]
7. Brooks, H.L.; Steel, E.; Moore, M. Grain-Size Analysis of Ancient Deep-Marine Sediments Using Laser Diffraction. *Front. Earth Sci.* **2022**, *10*, 820866. [CrossRef]
8. Dawelbeit, A.; Jaillard, E.; Eisawi, A. Grain size analysis of the latest Quaternary Kordofan Sand of Central Sudan: Depositional environment and mode of transportation. *Aeolian Res.* **2022**, *55*, 100785. [CrossRef]
9. Eltijani, A.; Mohammed, M.A.A.; Abuobida, Y.; Yousif, I.M. Integrating CoDA and PCA for enhanced characterization of fluvial depositional processes: A case study of the Shendi formation, Sudan. *Discov. Geosci.* **2024**, *2*, 10. [CrossRef]
10. Pszonka, J.; Sala, D. Application of the mineral liberation analysis (MLA) for extraction of grain size and shape measurements in siliciclastic sedimentary rocks. *E3S Web Conf.* **2018**, *66*, 02002. [CrossRef]
11. Spychala, Y.T.; Ramaaker, T.A.B.; Eggenhuisen, J.T.; Grundvåg, S.A.; Pohl, F.; Wróblewska, S. Proximal to distal grain-size distribution of basin-floor lobes: A study from the Battfjellet Formation, Central Tertiary Basin, Svalbard. *Depos. Rec.* **2021**, *8*, 436–456. [CrossRef]
12. Syvitski, J.P.; Vorosmarty, C.; Kettner, A.J. Impact of Humans on the Flux of Terrestrial Sediment to the Global Coastal Ocean. *Science* **2005**, *308*, 376–380. [CrossRef] [PubMed]
13. Folk, R.L.; Ward, W.C. A Study in the Significance of Grain-Size Parameters. *J. Sediment. Petrol.* **1957**, *27*, 3–26. [CrossRef]
14. Pszonka, J.; Schulz, B. SEM Automated Mineralogy applied for the quantification of mineral and textural sorting in submarine sediment gravity flows. *Gospod. Surowcami Miner.—Miner. Resour. Manag.* **2022**, *38*, 105–131. [CrossRef]
15. Shideler, G.L.; Ślączka, A.; Unrug, R.; Wendorff, M. Textural and mineralogical sorting relationships in Krosno Formation (Oligocene) turbidites, Polish Carpathian Mountains. *J. Sediment. Petrol.* **1975**, *45*, 44–56.
16. McCave, I.N. Size sorting during transport and deposition of fine sediments: Sortable silt and flow speed. *Dev. Sedimentol.* **2008**, *60*, 121–142. [CrossRef]

17. Paterson, G.A.; Heslop, D. New methods for unmixing sediment grain size data. *Geochem. Geophys. Geosyst.* **2015**, *16*, 4494–4506. [CrossRef]
18. Pszonka, J.; Godlewski, P.; Fheed, A.; Dwornik, M.; Schulz, B.; Wendorff, M. Identification and quantification of intergranular volume using SEM automated mineralogy. *Mar. Pet. Geol.* **2024**, *162*, 106708. [CrossRef]
19. Milliman, J.D.; Syvitski, J.P. Geomorphic/Tectonic Control of Sediment Discharge to the Ocean: The Importance of Small Mountainous Rivers. *J. Geol.* **1992**, *100*, 525–544. [CrossRef]
20. Qi, Y.L.; Yu, Q.; Gao, S.; Li, Z.Q.; Fang, X.; Guo, Y.H. Morphological evolution of river mouth spits: Wave effects and self-organization patterns. *Estuar. Coast. Shelf Sci.* **2021**, *262*, 107567. [CrossRef]
21. Gao, S.; Collins, M.B. Analysis of Grain Size Trends, for Defining Sediment Transport Pathways in Marine Environments. *J. Coast. Res.* **1994**, *10*, 70–78. Available online: https://www.jstor.org/stable/4298194 (accessed on 10 October 2024).
22. Gao, S.; Collins, M.; McLaren, P.; Bowles, D. A critique of the "McLaren Method" for defining sediment transport paths; discussion and reply. *J. Sediment. Res.* **1991**, *61*, 143–147. [CrossRef]
23. Pedreros, R.; Howa, H.L.; Michel, D. Application of grain size trend analysis for the determination of sediment transport pathways in intertidal areas. *Mar. Geol.* **1996**, *135*, 35–49. [CrossRef]
24. Xiao, X.; Shi, Y.H.; Feng, X.L.; Xu, Y.Q. Surface sediment characteristics and dynamics in beibu gulf. *Period. Ocean. Univ. China* **2016**, *46*, 83–89. [CrossRef]
25. Kawakami, G.; Nishina, K.; Poizot, E. Dominant updriftward sediment transport on the updrift-side of a modern deflected delta, Ishikari coast, Hokkaido, Japan. *Mar. Geol.* **2021**, *436*, 106480. [CrossRef]
26. Paladino, I.M.; Mengatto, M.F.; Mahiques, M.M.; Noernberg, M.A.; Nagai, R.H. End-member modeling and sediment trend analysis as tools for sedimentary processes inference in a subtropical estuary. *Estuar. Coast. Shelf Sci.* **2022**, *278*, 108126. [CrossRef]
27. Duman, M.; Eronat, A.H.; Talas, E. Interplay of natural and anthropogenic factors in sediment dynamics and trace element distribution in Güllük Gulf, western Türkiye: A comprehensive geochemical and hydrodynamic analysis. *Cont. Shelf Res.* **2024**, *282*, 105332. [CrossRef]
28. Gao, S.; Collins, M. Net sediment transport patterns inferred from grain-size trends, based upon definition of 'Transport Vector'. *Sediment. Geol.* **1992**, *81*, 47–60. [CrossRef]
29. Gao, S. A FORTRAN program for grain-size trend analysis to define net sediment transport pathways. *Comput. Geosci.* **1996**, *22*, 449–452. [CrossRef]
30. Gao, S. Grain size trend analysis: Principle and applicability. *Acta Sedimentol. Sin.* **2009**, *27*, 826–836.
31. Wright, L.D.; Friedrichs, C.T. Gravity-driven sediment transport on continental shelves: A status report. *Environ. Sci. Geol.* **2006**, *26*, 2092–2107. [CrossRef]
32. Wang, X.M.; Qu, H.B.; Xiong, Y.K.; Lu, L.; Hu, K. Grain-size characteristics and transport trend of bottom sediments at the estuary of Changhua River in Hainan. *Geoscience* **2022**, *36*, 88–95. [CrossRef]
33. Xiao, M.; Wu, J.; Chen, Q.; Jin, M.; Zhang, Y. Dynamic change of land use in changhua downstream watershed based on ca-markov model. *Trans. Chin. Soc. Agric. Eng.* **2012**, *28*, 231–238. [CrossRef]
34. Dai, S.P.; Luo, H.X.; Hu, Y.Y.; Zheng, Q.; Li, H.L.; Li, M.F.; Yu, X. Dynamic Land Use Change of Hainan Island in Recent 20 Years Based on GLC30 Data. *Agric. Eng.* **2021**, *11*, 61–69. [CrossRef]
35. Lu, L. Research of Modern Sediment Transport Model of Changhua River Estuary in Hainan Province. Master's Thesis, China University of Geosciences (Beijing), Beijing, China, 2021. [CrossRef]
36. Hainan Provincial Academy of Environmental Sciences. *Dataset of Hainan Island Ecological System Pattern from 2000 to 2019 (30 m) [Data Set]*; International Research Center for Big Data on Sustainable Development Goals: Beijing, China, 2022. [CrossRef]
37. Xu, D. Sedimentary Records Since Last Deglaciation and the Formation of Modern Sedimentary Pattern in Eastern Beibu Gulf. Ph.D. Thesis, University of Chinese Academy of Sciences, Beijing, China, 2014.
38. Mao, L.M.; Zhang, Y.L.; Bi, H. Modern pollen deposits in coastal mangrove swamps from northern Hainan Island, China. *J. Coast. Res.* **2006**, *22*, 1423–1436. [CrossRef]
39. Wu, J.Q.; Xiao, M.; Yang, J.T.; Xiao, X.B.; Tang, W.H. Study on distribution characteristics of soil erosion in the lower reaches of Changhua River in Hainan. *Technol. Soil Water Conserv.* **2012**, *2012*, 12–15. [CrossRef]
40. Zhang, P.; Ruan, H.M.; Dai, P.D.; Zhao, L.R.; Zhang, J.B. Spatiotemporal river flux and composition of nutrients affecting adjacent coastal water quality in Hainan Island. *J. Hydrol.* **2020**, *591*, 125293. [CrossRef]
41. Zhang, W.Y.; Xiong, P.; Meng, Q.C.; Dudzinska-Nowak, J.; Chen, H.; Zhang, H.; Zhou, F.; Miluch, J.; Harff, J. Morphogenesis of a late Pleistocene delta off the south-western Hainan Island unraveled by numerical modeling. *J. Asian Earth Sci.* **2020**, *195*, 104351. [CrossRef]
42. Zhao, L.; Cai, G.Q.; Zhong, H.X.; Li, B.; Zou, L.Q.; Li, S.; Han, Y.F. Grain-size characteristics and sedimentary environment of surface sediments in the shallow sea in the southeast of Hainan Island. *Mar. Geol. Quat. Geol.* **2021**, *41*, 64–74. [CrossRef]
43. Zhu, L.R.; Liu, Y.H.; Ye, C.Q. Runoff change and influencing factors of Changhua River in arid area of Hainan Island. *Ecol. Sci.* **2020**, *39*, 183–189.
44. Gao, J. Study on Sediment Transport Model in Changhua River Estuary of Hainan Province Based on Remote Sensing Analysis. Master's Thesis, China University of Geosciences (Beijing), Beijing, China, 2014.
45. Chen, C.; Liu, H.; Beardsley, R.C. An unstructured grid, finite-volume, three-dimensional, primitive equations ocean model: Application to coastal ocean and estuaries. *J. Atmos. Ocean. Technol.* **2003**, *20*, 159–186. [CrossRef]

46. Chen, C.; Xue, P.; Ding, P.; Beardsley, R.C.; Xu, Q.; Mao, X.; Gao, G.; Qi, J.; Li, C.; Lin, H.; et al. Physical mechanisms for the offshore detachment of the Changjiang Diluted Water in the East China Sea. *J. Geophys. Res. Ocean.* **2008**, *113*, C02002. [CrossRef]
47. Jiang, L.; Xia, M. Dynamics of the Chesapeake Bay outflow plume: Realistic plume simulation and its seasonal and interannual variability. *J. Geophys. Res. Ocean.* **2016**, *121*, 1424–1445. [CrossRef]
48. Lai, W.; Pan, J.; Devlin, A.T. Impact of tides and winds on estuarine circulation in the Pearl River Estuary. *Cont. Shelf Res.* **2018**, *168*, 68–82. [CrossRef]
49. Wu, Y.; Zhao, E.; Li, X.; Zhang, S. Application of Wave-current coupled Sediment Transport Models with Variable Grain Properties for Coastal Morphodynamics: A Case Study of the Changhua River, Hainan. *EGUsphere* **2024**, preprint. [CrossRef]
50. Flemming, B.W. A revised textural classification of gravel free muddy sediments on the basis of ternary diagrams. *Cont. Shelf Res.* **2000**, *20*, 1125–1137. [CrossRef]

Journal of
*Marine Science
and Engineering*

MDPI

Article

Atypical Linear Tectonic Block of the Intraplate Deformation Zone in the Central Indian Ocean Basin

Vsevolod V. Yutsis [1,*], Oleg V. Levchenko [2], Alexander V. Tevelev [3], Yulia G. Marinova [2], Ilia A. Veklich [2] and Abraham Del Razo Gonzalez [1]

1 Geosciences Division, Potosino Institute of Scientific and Technological Research, San Luis Potosí 78216, Mexico; abraham.delrazo@ipicyt.edu.mx
2 Shirshov Institute of Oceanology, Russian Academy of Sciences, Moscow 119991, Russia; olevses@mail.ru (O.V.L.); marinova.ocean@gmail.com (Y.G.M.); veklich.ilya11@mail.ru (I.A.V.)
3 Geological Faculty, Moscow State University, Moscow 119899, Russia; atevelev@yandex.ru
* Correspondence: vsevolod.yutsis@ipicyt.edu.mx; Tel.: +52-444-2033174

Abstract: The Central Indian Ocean Basin (CIOB) is distinguished by unusually high tectonic activity, setting it apart from all other passive oceanic basins. Within the interior of the Indo-Australian lithospheric plate lies a unique area of intraplate deformation. This region is characterized by the highest recorded intraplate oceanic seismicity, with earthquake magnitudes reaching up to M = 8, abnormally high heat flow—measured to be two to four times higher than background levels for the ancient oceanic lithosphere of the Cretaceous age—and, most notably, intense folding and faulting of sediments and the basement, which are typically associated only with boundary zones of lithospheric plates. This anomalously tectonically active intraplate area was studied during regular research cruises in the 1970s–1980s, after which new conclusions were mainly drawn from satellite data modeling. Substantially new geophysical data were obtained in 2017 after a long gap. Bathymetric surveys using multibeam echosounders during the 42nd cruise of the R/V (Research Vessel) Akademik Boris Petrov and the SO258/2 cruise of the R/V Sonne provided full coverage of a large portion of the intraplate deformation area in the CIOB. This confirmed the mosaic-block structure of the intraplate deformation zone in the Central Indian Ocean Basin, consisting of numerous isometrically deformed tectonic blocks. A linear block at 0.2–0.6° S, which has a branch-like shape in plain view, is morphologically distinct from these blocks. It represents a system of structural elements of different scales (folds, flexures, ruptures), which constitute a structural paragenesis formed in the mechanical environment of a dextral transpressive tectonic setting.

Keywords: Indian Ocean; intraplate deformation; Central Indian Ocean Basin; tectonic block; shear; swell; folding; transpression

1. Introduction

The deep-water basins of the world ocean are characterized by a generally flat sub-horizontal bottom with rare individual seamounts rising above the surface. This structure is caused by the tectonic passivity of the oceanic lithospheric plates far from their boundaries, where the deep-water basins are located. The Central Indian Ocean Basin, which shows unusually high tectonic activity in its northern part, is an exception to this general picture. This is the area with the highest oceanic intraplate seismicity [1]. High young tectonic activity is also evidenced by anomalously high heat flux, with measured values two to four times higher than background values for the ancient Cretaceous oceanic lithosphere [2,3]. Continuous seismic reflection profiling (CSP) in this area has revealed intense fold-rupture deformation of the sedimentary cover and basaltic basement forming large bedding irregularities [2,4,5]. The complex of observed unique tectonic structures characterized by expressive geophysical anomalies makes the area of Intraplate Deformation of the Indian Ocean Lithosphere (IDIOL) in the Central Indian Ocean Basin one of the most deformed zones in the world ocean. This significantly distinguishes it from the known manifestations of intraplate activity known within the continental lithosphere in various areas of the Earth [6–9].

Although IDIOL has been studied for quite some time [2,4,10–24], its structure has not yet been elucidated in detail, and its nature is still largely debatable. This is mainly due to the fact that, unfortunately, the most important full-scale geological and geophysical studies of IDIOL were carried out before the 1990s at the end of the last century, during the cruises of research vessels that were regularly conducted in the 1970s–1980s. Moreover, in them, the main method was Continuous Seismic Profiling (CSP), with a rather powerful pneumatic source, according to the data of which intraplate deformations of the basement and sedimentary cover were studied. And the references in the bibliography emphasize this. Then followed a long pause, until 2017, when we received new data during the 42nd cruise of the R/V Akademik Boris Petrov and the SO258 cruise of the R/V Sonne (Figure 1). There was multibeam bathymetry, but unfortunately, there was no more CSP (except for a small single profile of multichannel seismic profiling of the common-midpoint method (CMP) in the "Sonne" cruise), and there was shallow high-resolution seismo-acoustic profiling (Parasound), which, in the presence of a thick sediment cover, was not very informative for studying tectonics.

The seismic profiles show that the basement, together with the overlying sediments, is being "squeezed" upward with the formation of large uplifts up to 1–2 km high and up to 100–200 km in size, which are complicated by numerous faults and folds with amplitudes of tens to hundreds of meters [2,11,13–15,19–21]. Deep-sea drilling (Ocean Drilling Programm LEG 116 sites 717–719) has established that these faults and folds are about 8 Ma old, i.e., late Miocene, which determines the time of the onset of intraplate deformation of the Indian Ocean lithosphere [25]. It is also suggested that some of this deformation started earlier than ~14–15 Ma [21]. The geological and geophysical data collected in the IDIOL area until the early 1990s and the results of their processing and interpretation are summarized and analyzed in the monograph "Intraplate deformation in the Central Indian Ocean Basin" [19].

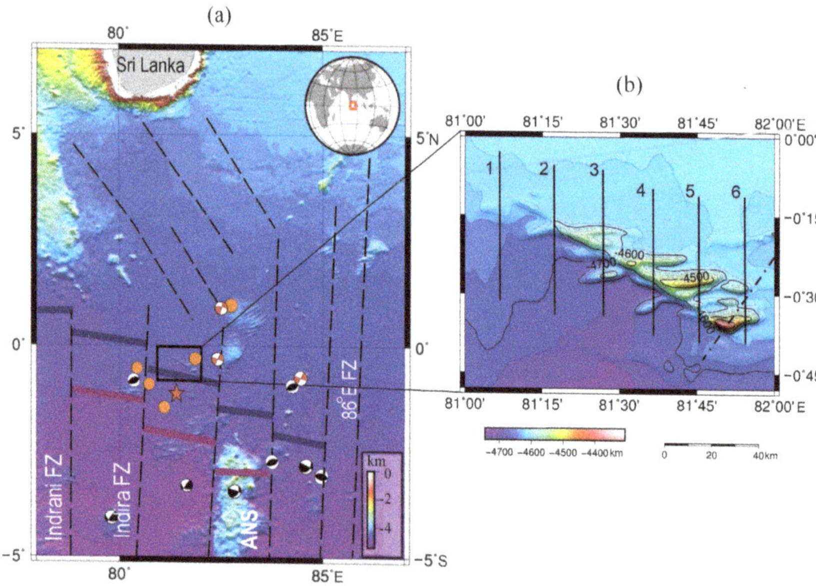

Figure 1. General tectonic setting of the Central Indian Ocean Basin underlain by satellite-derived bathymetry (**a**). Bottom relief map of the CIOB according to [26], with the addition of new multibeam

echo-sounder data from the 42nd cruise of the R/V "Akademik Boris Petrov [27] and from R/V Sonne SO258/2 [28]. The map shows the focal mechanisms of recent earthquakes with a magnitude greater than 5. Black and white beach-balls are thrusts; red and white—shear; and orange circles—mechanism undetermined [29–31]. Fracture zones (FZ) are shown by dash lines [32,33]. Magnetic anomalies are shown according to [32,34]—dark blue line is a 34 anomaly with an age of 83 Ma; violet line—33 anomaly with an age of 79 Ma [24]. ODP LEG 116 sites are marked by a red star [25]. ANS—Afanasy Nikitin seamount. The black box notes the study area shown in Figure (**b**). Black solid lines—seismic profiles (lines 1–6) from the SO258/2 cruise of the R/V Sonne [35]; black dot-dashed line—seismic profile from the 22nd cruise of the R/V "Professor Shtokman" [11,19].

Modern remote sensing techniques have made some contribution to the study of the intraplate deformation zone. A bathymetric map has been constructed from satellite altimetry (ETOPO1) and shipboard data [36,37], which clearly shows the complex structure of the Central Indian Ocean Basin, with variations in the elevation of the ocean floor up to almost four kilometers within the study area (Figure 2a).

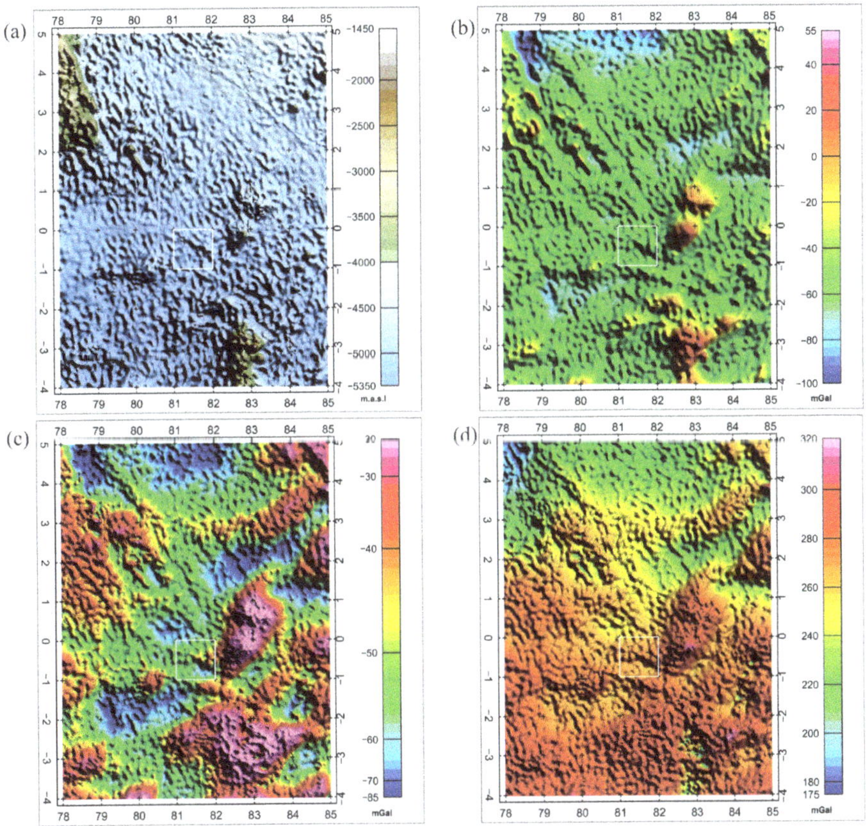

Figure 2. Geophysics of the Central Indian Ocean Basin. (**a**) Bottom topography; (**b**) Free air gravity; (**c**) Airy isostatic gravity; (**d**) Complete Bouguer Gravity Anomaly. The white rectangle indicates the position of the detailed survey polygon.

Gravity anomaly maps describe variations in Earth's gravity that result from differences in the density of the Earth's crust. These variations can reveal the distribution of subsurface masses and help understand the structure and composition of the Earth's crust, as well as the geological processes that have formed it. The free-air and complete Bouguer gravity anomalies, which we obtained from satellite/shipborne data (EIGEN-6C4), reflect the differentiated density structure associated with the anomalous region of the lithosphere (Figure 2b,d) [38,39]. Within the polygon studied in detail (Figure 1; 0.2–0.6° S and 81–82° E), the complete Bouguer anomaly is characterized by values of 260–270 mGal and a general northwest trend. The morphological structure itself is weakly expressed in the gravity field. However, the residual gravity in this area corresponding to the morphological uplift reveals a narrow negative anomaly up to −90 mGal. In addition, the gravimetric isostatic anomaly from WGM2012 provides insight into the differential vertical motions of the crust and lithosphere under unstable isostatic equilibrium (Figure 2c). Isostasy refers to the gravitational balance between the densest and least dense parts of the Earth's crust, and theoretically, at the center of a level plateau, both gravity anomalies should be approximately equal. In the studied local area, isostatic anomalies are negative and have values of the order of −50 mGal.

In order to better understand the nature of the studied morphostructure, we analyzed the magnetic field of the area using the EMAG2v3 model. Regionally (Figure 3a), the magnetic anomaly reduced to Ecuador can be divided into three areas (magnetic domains), the Northeast, Central, and Southwest domains. The NE and SW domains are characterized by positive anomalies isometric in plan with values from 0 to115 nT. The central one is dominated by the NW trend, with narrow, elongated, sinuous, variable anomalies from +100 nT to −110 nT.

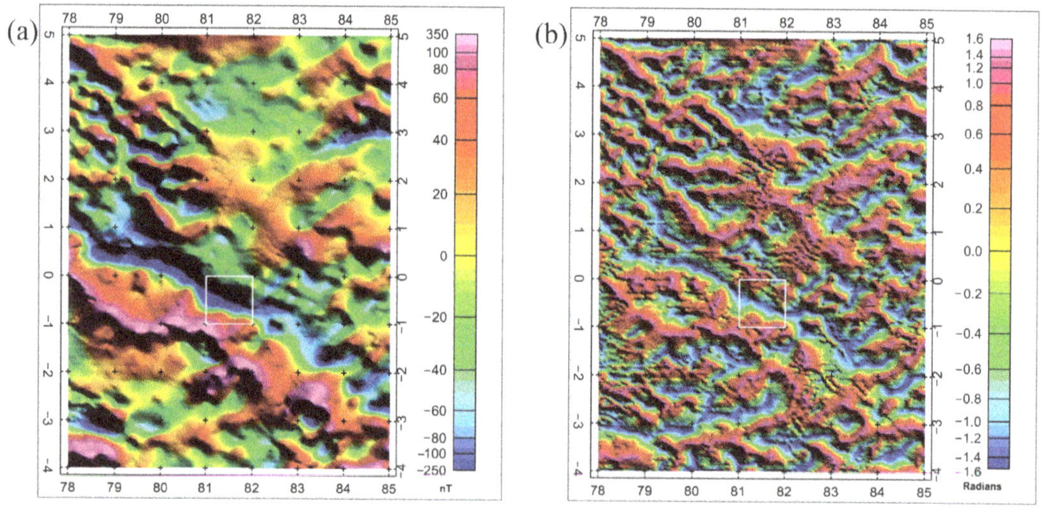

Figure 3. Magnetic field of the Central Indian Ocean Basin. (**a**) Magnetic anomaly reduced to Ecuador; (**b**) Magnetic field tilt derivative. The white rectangle indicates the position of the detailed survey polygon.

Finally, a relatively thin sedimentary layer overlying a complexly constructed geologic basement provided an excellent opportunity to demonstrate the effectiveness of tilt derivative techniques applied to magnetic satellite, ship, and airborne data (EMAG2v3) as an interpretative tool in an area of subtle magnetic signals (Figure 3b) [40–42]. This is a simple and fast method for locating vertical contacts where zero radian contours delineate the spatial location of the magnetic source edges; the method in its simplest form assumes that the source structures have vertical contacts, that there is no remanent magnetization, and that the magnetization is vertical. This map shows that magnetic bodies are oriented in a preferential NW-SE direction and a complementary ENE-WSW direction. The tilt-depth method only depends on mapping specific contours of magnetic inclination angles, while the depth to the source is the distance between the zero contour and the −45° or +45° contour or their average.

Based on the rare seismic profiles and relatively low-detail satellite measurements of the gravity field, it has been suggested that meridional compressional stress within the Indo-Australian plate as a result of the continental collision of India with Eurasia caused the crust of the Central Indian Ocean Basin to buckle, forming a series of extended conjugate latitudinal ridges and troughs [2]. The oceanic lithosphere of the Central Indian Ocean Basin was approximated as a homogeneous thin elastic slab. Under the influence of meridional horizontal compression, long-period folds—"ridges and troughs"—oriented in a sublatitudinal direction were formed within this plate. This type of long-wavelength regular folding of the lithosphere in the region of intraplate deformation in the CIOB served as the basis for simplified models of this process [18,43–45]. Such concepts of the corrugated structure of the lithosphere in the IDIOL region were conditioned by the prevailing views at that time about the simple structure of the deep-sea basin floors of the World Ocean, which were associated with their poor exploration and the low level of detail of regional geophysical data at the end of the last century.

These first schematic views of the tectonic structure and the lateral distribution of deformed uplifts based on rare seismic profiles and satellite measurements of the gravity field have survived up to the present day (Figure 2) [21]. However, the first detailed geophysical survey during the 31st cruise of the R/V "Dmitri Mendeleev" in 1984 corrected these ideas. It was found that the highly deformed tectonic block in the area of 4° S and 80° E has an isometric rhomboid shape, bounded on all sides by faults of two generations [11,12,19,46]. It is bounded in the west and east by ancient meridional paleotransformations—the Indrani at 79° E and the unnamed one at 80.5° E—and from the north and south by a series of sublatitudinal echelon faults of the late Miocene age. Instead of the wavy structure of the IDIOL, created by a series of extended latitudinal rises and troughs, a mosaic-block structure of the intraplate deformation region was proposed, created by chaotically distributed isometric deformed tectonic blocks [27,46]. Physical modeling of the compressive deformation of the Indo-Australian plate during the continental collision of India with Eurasia demonstrated that this is precisely how the plate deforms in the presence of meridional weakened zones of transform faults [47].

The geophysical survey carried out later in the 22nd cruise of the R/V "Professor Shtokman" (1989) at three polygons in the CIOB corroborated the isometric shape of isolated deformed uplifts and the mosaic distribution of such tectonic blocks [10,11,19]. Simultaneously, this survey revealed that these blocks differ from each other in morphology and exhibit highly individual tectonic structures [11]. Continuous seismic profiling revealed the structure of the sedimentary cover in this area within the IDIOL to have a thickness of 2.5–3 km [10,11,19,21]. The data indicate that the sediments and basement are affected by significant conformational folding and fracture disturbances, which are characteristic of the entire IDIOL in the Central Indian Ocean Basin [11].

In cruise SO258/2 of the R/V Sonne (2017), geophysical work was carried out in the Central Indian Ocean Basin spanning the band from 6° N to 3.5° S from 81° to 84° E [35]. Eight meridional profiles of 600 miles each were completed in a corridor of approximately 140 km in width. Two NW-SE profiles (250 miles each), two SW-NE profiles (100 miles each), and several short linking profiles were also completed. A new detailed map of the bottom topography, based on the results of a bathymetric survey with a multibeam echosounder, revealed a clear, chaotic mosaic distribution of isometric rises in the CIOB [27]. This distribution differs from the previously depicted extended latitudinal ridges [2,48]. However, this original scheme of conjugate latitudinal ridges and troughs is still accepted by numerous researchers [21], which renders our paper pertinent. This paper discusses the results of the SO258/2 cruise survey on an atypical linear deformed IDIOL linear tectonic block at 0.5° S, which exhibits a morphology and deformation character that differs markedly from all other isometric tectonic blocks (Figure 1). We refer to this block as "the branch".

2. Materials and Methods

The continuous geophysical survey conducted on the R/V Sonne during cruise SO258/2, which included multibeam bathymetry, seismo-acoustic profiling, magnetometry, and gravimetry, was carried out along the entire ship's route at a speed of 10–13 knots [35]. Six of the eight meridional profiles crossed the considered atypical linear tectonic block IDIOL, which is the subject of this study (Figure 1).

The bathymetric survey was acquired with a Kongsberg EM122 shipboard deep-sea multibeam echosounder, which was equipped with a 16 × 8 m antenna with up to 288 beams focused. The longitudinal and transverse beam widths were 0.5° and 1.0°, respectively, with a viewing angle of 140°. This approach ensured high data density and resolution, as well as a large coverage width in a six-depth water band (up to 30 km).

The seismo-acoustic survey was conducted using a shipboard Atlas Parasound DS3 (P70) narrow-beam parametric profiler. This device emits acoustic signals with a power of 70 kW, which provides a maximum penetration depth of up to 200 m in weakly consolidated sediments. In the course of the SO258/2 cruise, the penetration depth did not exceed 100 m at a signal frequency of 4 kHz.

In order to characterize the structure of the entire sedimentary section, the results of continuous pneumatic source seismic profiling, which were previously obtained in this area during the 22nd cruise of the R/V "Professor Shtokman", were utilized. The aforementioned data were recorded in analog form.

The multibeam echosounder data were processed in GenericMapping Tools software, with the visualization performed in GlobalMapper-22 and Golden Software Surfer-21. The seismo-acoustic data were processed using the RadExPro-2019 software package, with Kingdom Suite 2017 used for data interpretation. The archived seismic data recorded in analog form were interpreted in CorelDRAW-22 software.

A new high-resolution bathymetric map of the "branch" study area, volumetric models, and a series of seismic profiles from different years were used as the data for morphostructural analysis. Detailed morphostructural analysis became possible only after the creation of high-precision relief models showing even the smallest features.

3. Results

3.1. Morphostructural Analysis

The studied area represents a hilly elevation with a highly dissected bottom relief and appears as branch-like in plan (Figure 4a). The northern edge of the entire structure is raised by 70–80 m relative to the south along a narrow–elongated ridge extending in the southeast–northwest (120°) direction, with a height of over 100 m. Along its entire length, a series of short sub-latitudinal ridges adjoin it from both sides, with the largest located approximately equidistantly (7–10 km). These regular short ridges are clearly manifested in the northern block, where they are the same height—over 100 m (Figure 4b). In the southern block, its own system of ridges has formed, which is practically independent of the northern system. This new interpretation of the bottom relief is radically different from the previous one, which depicted three large latitudinal ridges instead of a series of small latitudinal ridges [19,49], with no signs of an oblique ridge.

The studied block has a complex morphology, and a detailed analysis allows us to consider it as a structural paragenesis [50,51] of structural elements (folds and various types of faults) at different scales. As mentioned earlier, the defining structural element of the block is the complexly constructed Main Ridge, which has a strike azimuth of about 120° and is located on a hypsometric step, where the northern block is raised by 70–80 m relative to the southern block.

From a structural point of view, the Main Ridge is a long narrow Main Anticline of a cylindrical type with a sharply asymmetric structure. Its northern limb is a gentle slope, while the southern limb is complicated by a left-stepping series of short faults along the entire length of the fold. These faults are morphologically expressed as deep furrows and represent tension cracks oriented at an angle of about 45° (±10°) to the strike of the anticline (azimuth 335–355°). All tension cracks are arranged in a series of left-stepping en echelon arrays and form three domains.

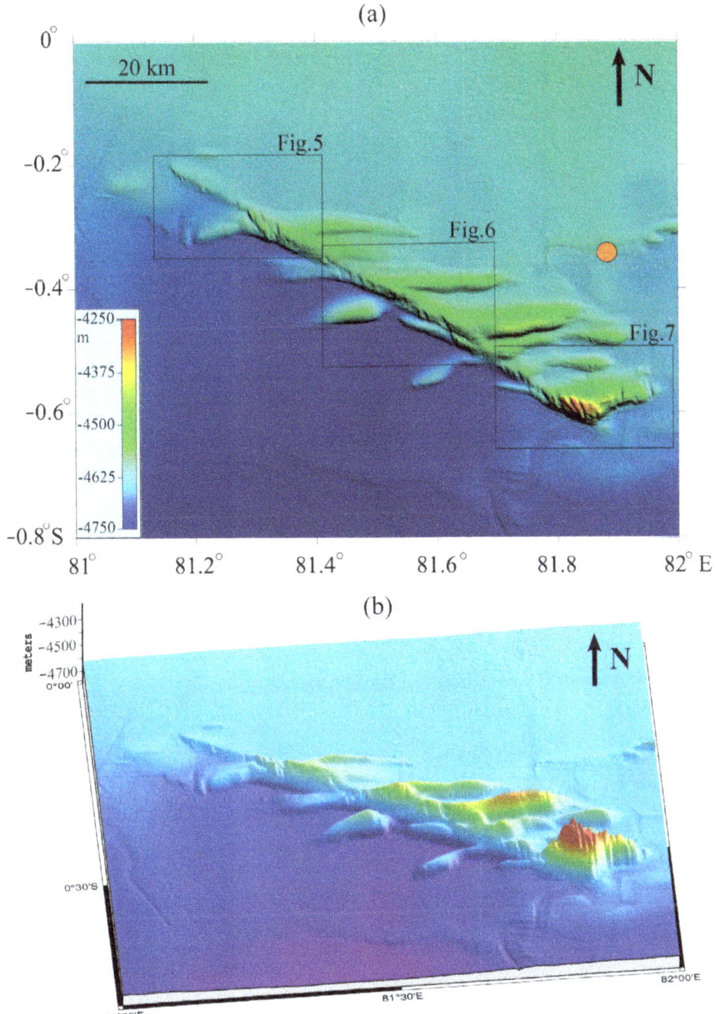

Figure 4. Bottom relief map (**a**) and 3D-model (**b**) of the atypical linear tectonic block "branch". The location of the nearest earthquake epicenter is shown by the orange circle.

The first domain is distinguished in the western part of the Main Anticline, where it is significantly narrower than in other parts (Figure 5). The tension cracks here are oriented at the steepest angles to the strike of the fold (about 35°). In places, they are accompanied by fractures parallel to the strike of the anticline, most likely shear fractures (strike-slip faults). Such a combination of morphostructural elements is typically formed in a dextral simple shear environment. In several places between the tension cracks, relative depressions in the topography are observed. As a result, structures of the pull-apart type arise, bounded by pairs of opposite normal faults and fractures. The hinge point of the fold is represented by a narrow pericline with a gentle plunge of the hinge to the west–northwest.

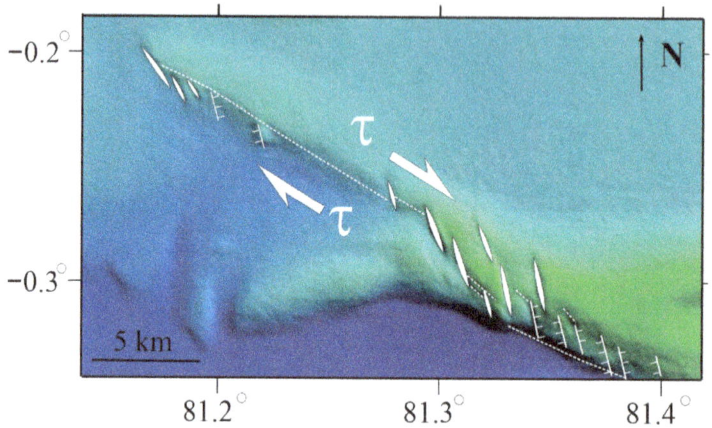

Figure 5. The fragment of the seabed relief map of the Central Indian Ocean Basin (part of the "branch" polygon, Western domain of the Main Anticline). White lenses—tension cracks; white dashed lines—shear fractures; comb—normal faults (ticks directed towards the fault dip); τ—tangential stress.

The second domain is distinguished in the central part of the Main Anticline. Here, the normal faults often form paired tension structures, resulting in the formation of small grabens (pull-apart type basins) bounded by shear fractures parallel to the axis of the Main Anticline (Figure 6). The tension cracks in this part are oriented fairly strictly at an angle of 45° to the strike of the Main Anticline and are usually associated with longitudinal shear fractures, which are morphologically expressed as long narrow furrows or steep scarps. Such a combination of morphostructural elements is typically formed in a dextral simple shear environment.

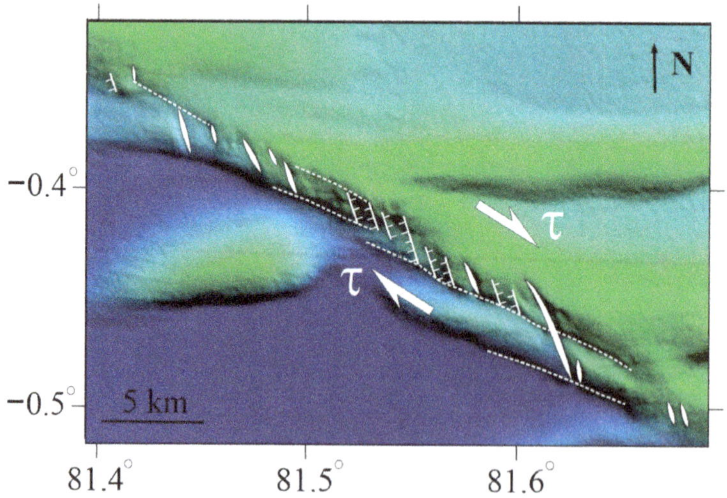

Figure 6. The fragment of the seabed relief map of the Central Indian Ocean Basin (part of the "branch" polygon, Central domain of the Main Anticline). The legend is the same as in Figure 5.

The most complex structure is found at the eastern end of the Main Anticline (Figure 7). The fold hinge rises sharply here, and the fold is truncated by a transverse fault and lacks a pericline. Three sets of narrow, N-S trending gaps are identified north of the transverse normal fault. These do not belong to the previously described fracture sets and compensate for the meridional compression. In the most uplifted part, the tension cracks across the entire anticline, not just its southern limb. In this place, they clearly underwent a counterclockwise rotation. Such a combination of morphostructural elements is typically formed in a simple shear environment. However, such an interpretation encounters a number of contradictions, which are discussed below.

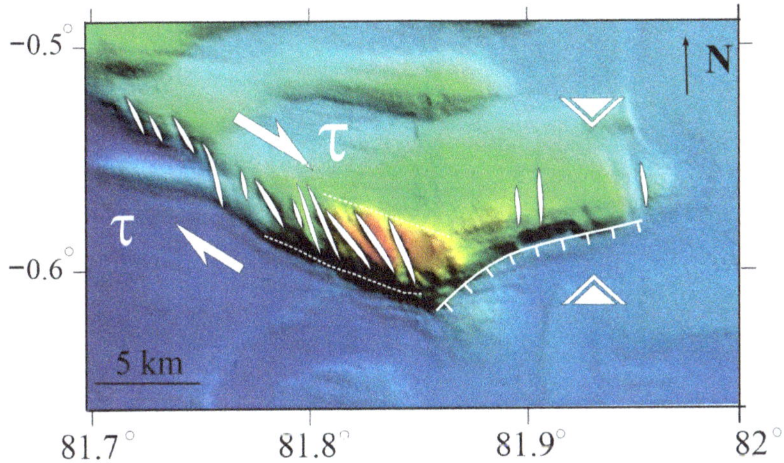

Figure 7. The fragment of the seabed relief map of the Central Indian Ocean Basin (part of the "branch" polygon, Eastern domain of the Main Anticline). White double triangles show the main compression direction. The other legend is the same as in Figure 5.

Two series of sublatitudinal ridges are associated with the central ridge, which from a structural standpoint represent linear and brachyform anticlines. One series is located within the northern limb of the block, while the other is located within the southern limb. All folds have approximately the same morphology—gentle northern limbs and steep southern ones, i.e., the folds have inclined axial planes with southern vergence. The southern limbs of the folds are complicated by faults, expressed as deep narrow depressions in the relief. Despite the similarities described, the northern and southern groups of folds have fundamentally different structures.

The southern group consists of five folds (a1–a5 in Figure 8), associated with the Main Anticline (Figure 8). In addition to the common parameters (inclined axial surfaces, vergence), the folds share another feature. All of them formed on convergent bends of a series of faults that dichotomize from the southern edge of the Main Anticline, dipping northward and forming a "horse-tail" structure. The faults are morphologically expressed as sharp scarps and, less frequently, as broad depressions. Furthermore, each of the folds in the southern group has specific features that distinguish it from the others.

Fold "a1" has a curved axial surface, convex to the north (Figure 8). At the bend, the hinge of the anticline undulates, and a saddle is formed on the fold. The western pericline of the fold is separated from it by a wide strait, most likely formed in a tension zone. The part of fold "a1" adjacent to the Main Anticline is disrupted by a series of tension cracks, similar to those localized within the Main Anticline.

Fold "a2" is also adjacent to the Main Anticline and is bounded to the south by an arcuate fault (Figure 8). Morphologically, the fold represents a gently dipping anticline that virgates westward. The part of fold "a2" adjacent to the Main Anticline is disrupted by a series of tension cracks, similar to those localized within the Main Anticline.

Fold "a3" is very similar in structure to fold "a1" (Figure 8). The axial surface of the anticline is also curved, and the fold is bounded on the south by an arcuate fault. Directly at the bend, the hinge undulates, resulting in a deep saddle on the anticline. The part of fold "a3" adjacent to the Main Anticline is disrupted by a pair of tension cracks, similar to those localized within the Main Anticline itself.

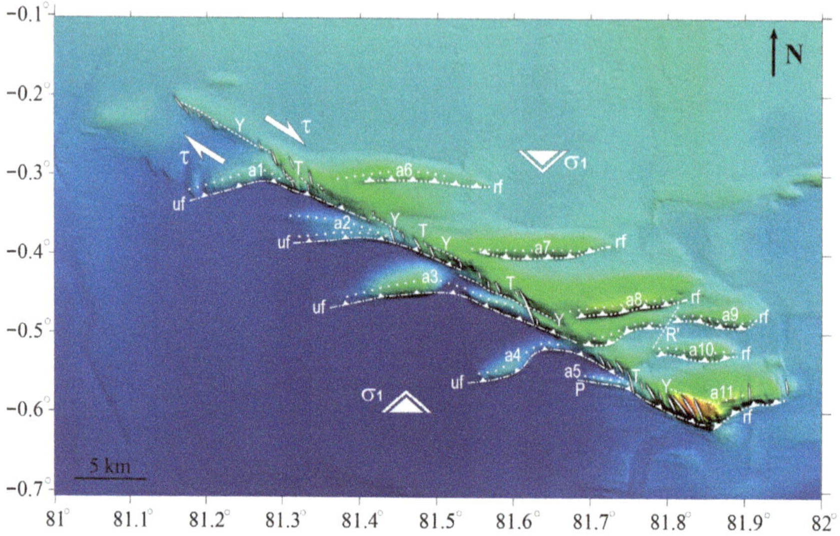

Figure 8. The fragment of the seabed relief map of the Central Indian Ocean Basin, polygon "branch". Dash-dotted lines mark inclined faults (triangular ticks indicate dip direction), while dotted lines indicate fold axes. Y—general Riedel shift, P—secondary Riedel shears, T—extension fractures, R′—antithetic shears, a1–a11—folds, rf—reverse faults, uf—faults with actively sinking foot walls. For other symbols and signs, see Figure 5.

The anticline "a4" is fundamentally different from the others. The axial surface of the fold and the fault bounding it to the south are S-shaped (Figure 8).

The anticline "a5" is the simplest of the folds in the southern group. It is short and very slightly curved, although in the fracture zone, the fold is disrupted by a large tension crack common with the crack on the southern limb of the Main Anticline (Figure 8).

The northern group of folds (a6–a11 on Figure 8) consists of at least seven anticlines, morphologically expressed as relatively wide ridges up to 100 m high. Unlike the folds in the southern group, they are arranged similarly: gently inclined anticlines and gently dipping anticlines, sometimes with weakly curved axial surfaces. The folds differ in the curvature of their axial planes. Fold "a6" has a gentle northward curvature, fold "a7" has a gentle southward curvature, fold "a8" is a straight axial plane, and fold "a9" exhibits a left-lateral offset of its axial plane, likely related to antithetic faults R′.

All folds are bounded on the south by faults, which are expressed as deep valleys and escarpments. A distinctive feature of the fold distribution is that the intervals between folds gradually decrease from west to east, and in the easternmost folds, individual transverse tears of a meridional strike appear.

3.2. Seismic Profiling

Previous CSP surveys revealed a 2.5–3 km thick sedimentary cover in the studied area of the Central Basin [10,11,19]. CSP with a powerful airgun source showed that the sediments and basement are disrupted by pronounced conformable folds and faults, typical of the entire IDIOL. Figure 9 shows a fragment of one of the CSP profiles which was acquired using a less powerful airgun source. It is one of the seismic profiles of the 22nd cruise of the R/V "Professor Shtokman" that traverses the linear tectonic block, which is discussed in this article (the profile's location is shown on Figure 1b). The oceanic basement is not visible on this profile, because the penetration was performed with a low-power pneumatic source of 0.3 L. However, the data clearly demonstrate tectonic disturbances in the upper layer of the sedimentary cover up to 1 km thick. The structural unconformity "A", which marks the main phase of intraplate deformation in the Late Miocene (~8 Ma), is clearly visible. Unconformity "AA," which separates the deformed Late Cretaceous–Early Pliocene complex from the weakly deformed Late Pliocene–Quaternary complex, is also evident. This unconformity is dated to the Early Pliocene (4.0–3.5 Ma) [11].

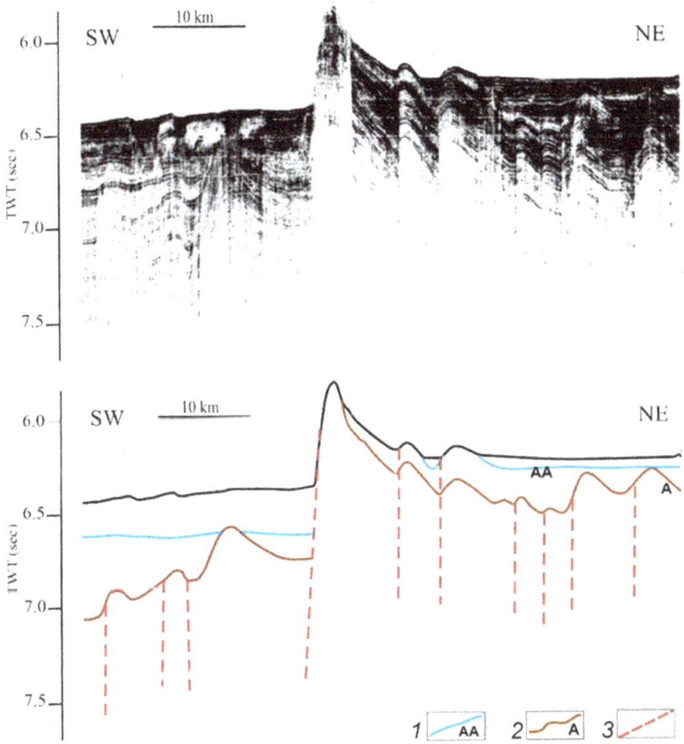

Figure 9. Seismic profile (initial and interpreted) of the 22nd cruise of the R/V "Professor Shtokman"; the line position is shown in Figure 1b [11,19]. 1. AA—Early Pliocene unconformity; 2. A—Late Miocene unconformity corresponding to the main phase of intraplate deformation (~8 Ma); 3—Faults.

The maximum signal penetration depth on the seismic profiles of SO258/2 R/V Sonne during the cruise does not exceed 100 m (Figure 10). Within this depth interval, the wave pattern on the seismic acoustic section is fairly laterally homogeneous. However, the bottommost parts of the section are high-amplitude, while deeper levels are characterized by moderate to low amplitudes. The folded zone is also covered by a sedimentary cover, which is clearly visible on the gentle flanks of the folds. On steeper flanks and slopes of small local morphostructures, the apparent thickness of the layered stratum is significantly reduced.

Within the sedimentary cover on the polygon "branch", three seismic units were identified on the profiles based on the seismostratigraphic criteria (Figure 10) [52]. Seismic unit 1 (SU1) exhibits high-amplitude, continuous reflections forming a layered wave pattern. It is distinguished on almost all seismic sections, especially in the bottom relief depressions, lying in places with a structural unconformity (overlying) on the underlying deposits. The thickness varies from 0 to 6 m. It is very rare in the folded zone of the polygon. Below follows seismic unit 2 (SU2), which is characterized by high-amplitude layered discontinuous horizontal boundaries. The thickness varies from the first meters in the folded parts of the polygon to 35–40 m in the gentle ones. Within SU2, high-amplitude individual horizons are distinguished, which are confidently traced in the gentle parts of the seismic section and gradually pinch out as they approach the folded blocks. However, in general, SU2 is characterized by a highly heterogeneous lateral wave pattern. At the same time, in some places, SU2 is also characterized by an unconformity of the overlying type at the base. Further down is seismic unit 3 (SU3), the base of which cannot be determined due to the insufficient penetrating ability of the profiler. It is distinguished everywhere and is represented by a thin-layered alternation of low- and medium-amplitude reflectors. The visible parts of the cross-sections of the folded zones are also composed of SU3. However, in some areas in the folds, there is an alternation of zones with different degrees of stratification.

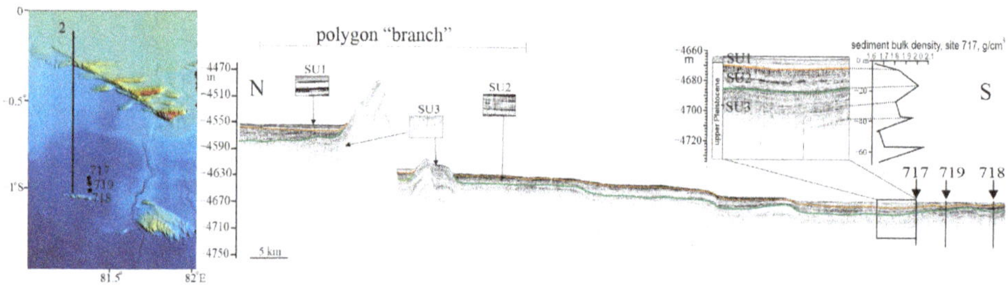

Figure 10. Main seismic units (SU1, SU2, SU3) within the study area and around the oceanic drilling sites ODP 717–719 (location of line 2 is shown in Figure 1b). In the inset, there is the correlation of the seismic section with the sediment density curve of site 717 according to [53]. The boundary between SU1 and SU2 is shown in orange; the boundary between SU2 and SU3 is shown in green; black short narrow—approximate position of site 717 on line 2.

No geological fieldwork was conducted in the area of the polygon. To assign a lithostratigraphic correlation to the identified seismic units and to estimate the approximate age of the folds, we used deep-sea drilling data. Ocean Drilling Program (ODP) sites 717–719 [53] were drilled 70 km south of the studied tectonic block "branch" (Figure 10). They are located on a single conventional sub-meridional line, parallel to the meridional traverses 2 along which comprehensive geophysical studies were conducted during cruise SO 258/2 of the R/V Sonne. The sites are located at the same distance from line 2. The correlation with data from site 717 and the analysis of sections from boreholes 719 and 718 allowed us to provide an approximate lithologic-stratigraphic reference for the selected seismic units [53]. The well sections consist mainly of silty turbidites with a thin interbed of calcareous clays. SU1 represents an alternation of muds and pelagic clays of the Holocene-upper parts of the Upper Pleistocene. SU2 is composed of micaceous silty turbidites with layers of calcareous clays of the Upper Pleistocene. According to well data, the sediments of this part of the section are coarser-grained, which has affected the increase in density and decrease in porosity and has possibly caused a sharp change in the wave pattern. On seismic profiles obtained during drilling, the part of the section identified by us as SU2 and SU3 continues up to 150 m [53]. Deeper lies a complex of turbidites that is exposed in the crests of anticline folds. The age of these turbidites is also Late Pleistocene [53].

The correlation of bathymetric and seismic survey data revealed that in folded zones, within the hinges and limbs of anticlinal folds, sediments with a visible thickness of the first tens of meters outcrop (Figure 10). These sediments exhibit a clear acoustic stratification with extended parallel reflecting boundaries and are likely composed of syn-deformational turbidites of SU3.

The studied linear tectonic block is currently seismically active, with the nearest earthquake epicenters located 20 km to the northeast and 100 km to the southwest, having an average magnitude of 5 (Figure 4a).

4. Discussion

The block "branch" is located in the area of a major reorganization of the spreading system in the Indian Ocean during the Late Cretaceous period, approximately 100 million years ago [21,32,34]. It separates oceanic crust areas of the Central Indian Ocean Basin formed at spreading ridges with different orientations. It is reflected in the different orientations of linear magnetic anomalies in these regions: NE-SW in the north of the Bay of Bengal and E-W in the basin south of the equator. The different spreading directions are also evident in the orientations of paleo-transform faults trending NW-SE and meridionally, respectively. It can be assumed that the block "branch" is located in a suture zone dividing lithospheric blocks of different ages. Similar structures can form during the development of a new spreading ridge propagating into older oceanic lithosphere [54,55].

It is important to note the fundamental difference between the new structural–tectonic results presented in this paper and those previously derived from a spars grid of seismic profiles. These articles discuss the structural aspects of the area of the Intraplate Deformation of the Indian Ocean Lithosphere (IDIOL) in the Central Indian Ocean Basin showing east–west trending Riedel shears coinciding with reverse faults formed by the reactivation of ancient spreading faults [14,56]. These ideas are reiterated in a recent article by M. Desa and M. Ramana (2021), where an interpretation of a seismic profile crossing the "branch" tectonic block is provided. The reverse faults identified by these authors are also considered to have formed during the reactivation of normal spreading faults [56]. However, the described major NW-SE shear in the "branch" block does not fit into this model, as it formed later in already existing spreading crust during plate kinematics reorganization. The east–west trending right-lateral en echelon compression zones (anticlines, reverse faults) coincide with the fault trends described by Bull and Scrutton [56] but have a different origin.

Thus, the structure of the tectonic block «branch» combines several standard structural elements: (1) a main linear anticline with elements of oblique shear fractures; (2) dextral en echelon compression structures (inclined folds, reverse faults), and (3) left-stepping en echelon tensile structures (tension fractures, pull-aparts) (Figure 8). This paragenesis may correspond to a mechanical setting of simple shear, allowing for its possible interpretation within the Riedel model [57]. In this interpretation, the Major Anticline corresponds to the Riedel general right-lateral shear zone (Y in Figure 8). From the southeast to the northwest, the faults of the major shear zone bifurcate to the left (southward) at an angle of 5–15°. These synthetic faults are likely secondary Riedel shears (P). Extension fractures within the Major Anticline represent Riedel "T" fractures, which, according to this model, should be oriented at an angle of 45° to the general shear zone (Figure 8). However, this paragenesis does not strictly fit into the Riedel model, meaning its formation environment does not fully correspond to the mechanical setting of simple shear for several reasons. First, the simple shear mechanical environment assumes a constant volume of the deforming block, and compression deformations across the main shear are not accounted for in this model. In addition, despite the apparent external similarity in the structures of the northern and southern limbs of the main shear, they are, in fact, organized differently. The northern limb structure is interpreted as a dextral en echelon fold system with southward vergence, where anticlines are closely associated with the shear itself. The folds in the southern limb are not directly associated with the main shear but form compensatory structures at convergent bends of P-shears, which develop from bifurcating faults of the main shear zone (Figure 8). Morphologically, the folds of the southern limb are in many ways similar to those of the northern limb (inclined, sharply asymmetrical, with rounded hinges, bounded to the south by inclined faults), but they have less distinctness and amplitude. The main difference between the paragenesis of the northern and southern limbs lies in the identical (southern) vergence of the folds in both limbs. This does not correspond to the standard Riedel model, in which the vergence opposite limbs should be opposite. The most likely reason for this situation is the difference in the rheological properties of the rocks that make up the limbs of the shear zone. The northern limb appears to be lighter, while the southern one is heavier. As a result, the faults that bound en echelon folds to the south in the different limbs have different kinematics. In the northern limb, they formed as reverse faults ("rf" in Figure 8). Within the northern limb, they developed as reverse faults ("rf" in Figure 8) with an active hanging wall. In the southern limb, these faults formed through the underthrusting of the active footwall beneath the hanging wall ("uf" in Figure 8). These faults represent a rough kinematic analog of a subducting plate.

It should also be kept in mind that, with the general sub-meridional compression, the northern limb of the shear zone is most likely overthrust onto the southern one, that is, the general shear is not vertical but dips to the north–north–east and represents not just a shear but a reverse fault (Figure 11). Therefore, the Main Anticline has a sharply asymmetrical structure. Its southern limb is not only broken by a series of detachments; it is probably also steeper.

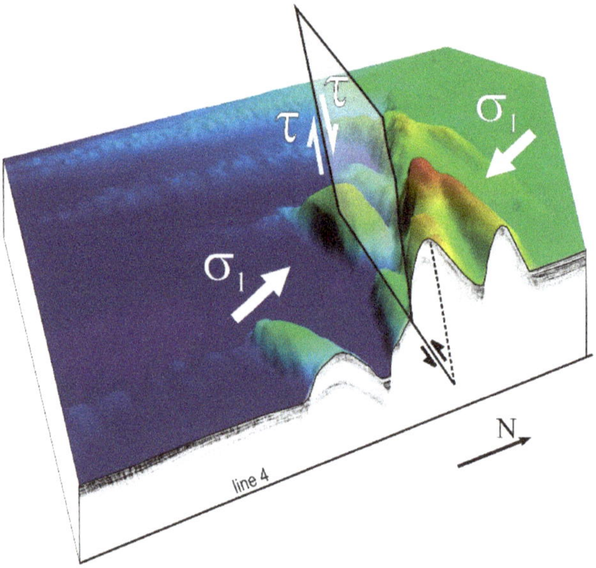

Figure 11. A 3D model of the studied area with a cross-section along line 4 (location of line 4 is shown in Figure 1b) The plane marks the approximate position of the main shear zone. σ1—compressive stress; τ—shear stress; black arrows indicate a direction of the vertical displacement of blocks.

It is important to note that, in the Riedel shear model, the angle between the direction of compression and the general shear should be 45°, but in reality, it is even smaller when considering the angle of internal friction. In our case, the angle between the direction of compression and the general shear is significantly greater than 45°. Moreover, the direction of compression (meridional) is fixed by specific compressional structures, in our case—latitudinal anticlines, as well as strictly meridional tension fractures at the eastern end of the zone. This situation is probably explained by the fact that this shear formed on a previously existing boundary of two oceanic crust blocks, rather than being newly formed as a result of compression. Finally, the main shear zone of the block "branch" is expressed as a compressional structure (an anticline), allowing the overall block formation setting to be interpreted as transpression. This aligns with the definition by Sanderson and Marchini [58], who modeled transpression as a deformation involving shear accompanied by shortening (compression) across the fault plane and vertical extension along the plane, essentially combining pure shear and simple shear mechanisms.

Thus, the structure of the isolated block "branch" can be modeled as a transpression-related en echelon structural pattern.

5. Conclusions

For the first time, a detailed bathymetric survey of the seabed using multibeam was carried out in a local area of intraplate deformation.

In the studied area, a tectonically isolated block (branch) was identified, whose analysis allows it to be considered as a structural paragenesis of multi-scaled structural elements (folds and various types of faults).

Within the block, three structural domains were identified and mapped, characterized by structural elements such as left-stepping en echelon short fault arrays, tension cracks, left-stepping en echelon arrays, and others. The morphostructure of the domains is described in detail. Based on a detailed morphostructural analysis, a series of maps and diagrams were created to illustrate the modern morphology and evolution of the identified structures

The structure of the studied tectonic block combines several standard structural elements: (1) a major linear anticline with shear fracture elements, (2) dextral en echelon compression structures (inclined folds and reverse faults), and (3) left-stepping en echelon tensile structures (tension fractures and pull-aparts). This paragenesis largely corresponds to a mechanical setting of simple shear, suggesting its possible interpretation within the Riedel model.

Additionally, structural analysis revealed that submeridional compression in this region was compensated not by the formation of large sub-latitudinal compression zones but by the development of conjugated local oblique shear zones, which overall correspond to the mechanical setting of dextral transpression, i.e., a combination of simple and pure shear. This led to the formation of distinctive structural parageneses, in which en echelon compression and extension zones, complicated by Riedel shears, are combined.

Author Contributions: Conceptualization, V.V.Y., O.V.L. and A.V.T.; methodology, A.V.T., O.V.L. and V.V.Y.; software, V.V.Y., A.V.T., A.D.R.G. and I.A.V.; validation, V.V.Y., O.V.L. and A.V.T.; formal analysis, V.V.Y. and O.V.L.; investigation, O.V.L., A.V.T., V.V.Y., I.A.V., Y.G.M. and A.D.R.G.; resources, V.V.Y.; data curation, O.V.L. and V.V.Y.; writing—original draft preparation, O.V.L., A.V.T., V.V.Y., Y.G.M., I.A.V. and Y.G.M.; writing—review and editing, V.V.Y. and O.V.L.; visualization, Y.G.M., A.V.T. and A.D.R.G.; supervision, V.V.Y. and O.V.L.; project administration, O.V.L. and V.V.Y.; funding acquisition, V.V.Y. All authors have read and agreed to the published version of the manuscript.

Funding: This work was performed within the framework of project No. FMWE-2024-0019 (O. Levchenko, Yu. Marinova, I. Veklich). The APC was funded by Instituto Potosino de Investigación Científica y Tecnológica, A.C. and by grant CONAHCYT project A1-S-29604 to Vsevolod Yutsis.

Institutional Review Board Statement: Not applicable.

Informed Consent Statement: Not applicable.

Data Availability Statement: The gravity datasets for this study can be found in the [GFZ Data Services]. EIGEN-6C4 The latest combined global gravity field model including GOCE data up to degree and order 2190 of GFZ Potsdam and GRGS Toulouse [https://doi.org/10.5880/icgem.2015.1]. Magnetic data are available in the National Centers for Environmental Information [https://www.ncei.noaa.gov (accessed on 10 February 2024)]. Multibeam raw data files are available through Alfred Wegener Institute, Helmholtz Centre for Polar and Marine Research, Bremerhaven, PANGAEA, [https://doi.org/10.1594/PANGAEA.881319].

Acknowledgments: We would like to express our sincere gratitude to the Chief Scientist of Cruise SO258/2, RV Sonne, Wolfram Geissler, for providing bathymetric data acquired by a multibeam echosounder and seismic data acquired by a Parasound profiler. We would also like to thank the entire cruise crew and scientific team for their invaluable assistance in data acquisition. Special thanks go to A.B. Kirmasov for his expert advice on structural analysis. V. Yutsis expresses his appreciation to the Instituto Potosino de Investigación Científica y Tecnológica (IPICYT) for administrative and financial support. The authors are grateful to the unknown reviewers for their valuable comments and suggestions, which greatly improved the manuscript.

Conflicts of Interest: The authors declare no conflicts of interest.

References

1. Bergman, E.A.; Solomon, S.C. Earthquake source mechanisms from body-waveform inversion and intraplate tectonics in the northern Indian Ocean. *Phys. Earth Planet. Inter.* **1985**, *40*, 1–23. [CrossRef]
2. Weissel, J.K.; Anderson, R.N.; Geller, C.A. Deformation of the Indo-Australian plate. *Nature* **1980**, *287*, 284–291. [CrossRef]
3. Geller, C.A.; Weissel, J.K.; Anderson, R.N. Heat transfer and intraplate deformation in the central Indian Ocean. *J. Geophys. Res.* **1983**, *88*, 5560–5570. [CrossRef]
4. Levchenko, O.V.; Merklin, L.R.; Neprochnov, Y.P. The folded features in the Central Indian Ocean Basin. *Geotectonica* **1985**, *1*, 15–23.
5. Yutsis, V.; Levchenko, O.; Ivanenko, A.; Veklich, I.; Turko, N.; Marinova, Y. New Insights into the Seamount Structure of the Northern Part of the Ninetyeast Ridge (Indian Ocean) through the Integrated Analysis of Geophysical Data. *J. Mar. Sci. Eng.* **2023**, *11*, 924. [CrossRef]
6. Hengesh, J.V.; Whitney, B.B. Transcurrent reactivation of Australia's western passive margin: An example of intraplate deformation from the central Indo-Australian plate. *Tectonics* **2016**, *35*, 1066–1089. [CrossRef]
7. Cianfarra, P.; Locatelli, M.; Capponi, G.; Crispini, L.; Rossi, C.; Salvini, F.; Federico, L. Multiplereactivations of the Rennick Graben-Fault system (northern VictoriaLand, Antarctica): New evidencefrom paleostress analysis. *Tectonics* **2022**, *41*, e2021TC007124. [CrossRef]
8. Cianfarra, P.; Pinheiro, M.R.; Villela, F.N.J.; Salvini, F. Intraplate Strike-Slip Corridor within South America (NE Border of the Paraná Basin) Unveiled by Structural Analysis of Faults and Fracture Swarms. *Geosciences* **2022**, *12*, 101. [CrossRef]
9. Pinardi, S.; Cox, R.T.; Leslie, D.; Hasan, M.R.; Adhikari, N.; van Arsdale, R.B. Topographic ridges express late Quaternary faulting peripheral to the New Madrid seismic zone, intraplate USA: Their tectonic implications. *Geomorphology* **2025**, *468*, 109502. [CrossRef]
10. Levchenko, O.V.; Milanovsky, V.E. Intraplate deformation in the center of the Indian Ocean from detailed tectonic mapping data. *Dokl. Earth Sci.* **1999**, *365*, 792–797.
11. Verzhbitsky, V.E.; Levchenko, O.V. Detailed Structure of Intraplate Deformation Zone in the Central Basin of the Indian Ocean (As deduced from three continuous profiling sites). *Geotectonics* **2002**, *36*, 498–513.
12. Wiens, D.A.; Stein, S.; Demets, C.; Gordon, R.G.; Stein, C. Plate tectonic models for Indian Ocean "intraplate" deformation. *Tectonophysics* **1986**, *132*, 37–48. [CrossRef]
13. Neprochnov, Y.P.; Levchenko, O.V.; Merklin, L.R.; Sedov, V.V. The structure and tectonics of the intraplate deformation area in the Indian Ocean. *Tectonophysics* **1988**, *156*, 89–106. [CrossRef]
14. Bull, J.M.; Scrutton, R.A. Fault reactivation in the central Indian Ocean and the rheology of oceanic lithosphere. *Nature* **1990**, *344*, 855–858. [CrossRef]
15. Chamot-Rooke, N.; Jestin, F.; de Voogd, B.; Phedre working group. Intraplate shortening in the Central Indian Ocean determined from a 2100-km-long north-south deep seismic reflection profile. *Geology* **1993**, *21*, 1043–1046. [CrossRef]
16. Levchenko, O.V.; Evsyukov, Y.D.; Subrahmanyam, C.; Vital, G.S.; Drolia, R.K. Detailed survey of the ocean bottom structure in the Central Indian Ocean Intense Deformation Zone: Tectonic implications. *Mar. Geol.* **1993**, *115*, 165–171. [CrossRef]
17. Murthy, K.S.R.; Neprochnov, Y.P.; Levchenko, O.V.; Rao, T.C.S.; Milanovsky, V.E.; Lakshminarayana, S. Some new observations on the intra-plate deformation in the Central Indian Ocean Basin (CIOB). *Mar. Geol.* **1993**, *114*, 185–193. [CrossRef]
18. Beekman, F.; Bull, J.M.; Cloetingh, S.; Scrutton, R.A. Crustal fault reactivation facilitating lithospheric folding in the Central Indian Ocean. *Geol. Soc. Spec. Pub.* **1996**, *99*, 251–263. [CrossRef]
19. Neprochnov, Y.P.; Gopala Rao, D.; Subrahmanyam, C.; Murthy, K.S.R. *Intraplate deformation in the Central Indian Ocean Basin*; Geological Society of India Memoir: Bangalore, India, 1998; p. 250.
20. Krishna, K.S.; Bull, J.M.; Scrutton, R.A. Evidence for multiphase folding of the central Indian Ocean lithosphere. *Geology* **2001**, *29*, 715–718. [CrossRef]
21. Desa, M.A.; Ramana, M.V. Integrated analysis of magnetic, gravity and multichannel seismic reflection data along a transect southeast of Sri Lanka, Bay of Bengal: New constraints. *Mar. Geol.* **2021**, *438*, 106543. [CrossRef]
22. Desa, M.; Ramprasad, T.; Kamesh Raju, K.A. An integrated geophysical study east of the southern Chagos Laccadive Ridge Complex, Central Indian Ocean Basin: Implications for the Rodriguez Triple Junction dynamics in the Late Cretaceous. *Mar. Geol.* **2019**, *414*, 47–63. [CrossRef]
23. Yatheesh, V.; Dyment, J.; Bhattacharya, G.C.; Royer, J.Y.; Kamesh Raju, K.A.; Ramprasad, T.; Chaubey, A.K.; Patriat, P.; Srinivas, K.; Choi, Y. Detailed Structure and Plate Reconstructions of the Central Indian Ocean Between 83.0 and 42.5 Ma (Chrons 34 and 20). *J. Geophys. Res. Solid. Earth* **2019**, *124*, 4305–4322. [CrossRef]
24. Yadav, R.; Pandey, D.K.; Ningthoujam, L.S.; Negi, S.S. Quantifying Structural Deformation History in the Central Indian Ocean. *Izv. Phys. Solid. Earth* **2023**, *59*, 1094–1112. [CrossRef]

25. Cochran, J.R. Himalayan uplift, sea level and the record of the Bengal Fan sedimentation at the ODP Leg 116 Sites. In *Proceedings of the Ocean Drilling Program, 116 Scientific Results*; Cochran, J.R., Stow, D.A.V., Eds.; Publications Distribution Center, Ocean Drilling Program, 1000 Discovery Drive: College Station, TX, USA, 1990; Volume 116, pp. 397–414.
26. Ryan, W.B.F.; Carbotte, S.M.; Coplan, J.O.; O'Hara, S.; Melkonian, A.; Arko, R.; Weissel, R.A.; Ferrini, V.; Goodwillie, A.; Nitsche, F.; et al. MultiResolution Topography synthesis. *Geochem. Geophys. Geosyst.* **2009**, *10*, Q03014. [CrossRef]
27. Levchenko, O.V.; Veklich, I.A. Mosaic of deformed tectonic blocks in the central basin of the Indian ocean. *Dokl. Earth Sci.* **2022**, *505*, 465–470. [CrossRef]
28. Krocker, R.; Geissler, W.H.; Dorschel, B. Swath Sonar Bathymetry During R/V Sonne Cruise SO258/2 with Links to Multibeam Raw Data Files. Alfred Wegener Institute, Helmholtz Centre for Polar and Marine Research, Bremerhaven, PANGAEA, 2018. [CrossRef]
29. Dziewonski, A.M.; Chou, T.-A.; Woodhouse, J.H. Determination of earthquake source parameters from waveform data for studies of global and regional seismicity. *J. Geophys. Res.* **1981**, *86*, 2825–2852. [CrossRef]
30. Ekström, G.; Nettles, M.; Dziewonski, A.M. The global CMT project 2004-2010: Centroid-moment tensors for 13,017 earthquakes. *Phys. Earth Planet. Inter.* **2012**, *200–201*, 1–9. [CrossRef]
31. U.S. Geological Survey, Earthquake Lists, Maps, and Statistics. 2020. Available online: https://www.usgs.gov/natural-hazards/earthquake-hazards/lists-maps-and-statistics (accessed on 1 December 2023).
32. Desa, M.A.; Ramana, M.V. Middle Cretaceous geomagnetic field anomalies in the Eastern Indian Ocean and their implication to the tectonic evolution of the Bay of Bengal. *Mar. Geol.* **2016**, *382*, 111–121. [CrossRef]
33. Ningthoujam, L.S.; Pandey, D.K.; Nair, N.; Yadav, R.; Khogenkumar, S.; Negi, S.S.; Kumar, A. Plume-ridge interactions in the Central Indian Ocean Basin: Insights from new wide-angle seismic and potential field modelling. *Tectonophysics* **2022**, *824*, 229222. [CrossRef]
34. Desa, M.A.; Ramana, M.V.; Ramprasad, T. Seafloor spreading magnetic anomalies south off Sri Lanka. *Mar. Geol.* **2006**, *229*, 227–240. [CrossRef]
35. Levchenko, O.V.; Geissler, W.H. Geophysical Investigations in the Eastern Indian Ocean in Cruise SO258/2 of the RV Sonne (Germany). *Oceanology* **2019**, *59*, 467–469. [CrossRef]
36. Sandwell, D.T.; Müller, R.D.; Smith, W.H.F.; Garcia, E.; Francis, R. New Global Marine Gravity Model from CryoSat-2 and Jason-1 Reveals Buried Tectonic Structure. *Science* **2014**, *346*, 65–67. [CrossRef]
37. Tozer, B.; Sandwell, D.T.; Smith, W.H.F.; Olson, C.; Beale, J.R.; Wessel, P. Global Bathymetry and Topography at 15 Arc Sec: SRTM15+. *Earth Space Sci.* **2019**, *6*, 1847–1864. [CrossRef]
38. Foerste, C.; Bruinsma, S.L.; Abrykosov, O.; Lemoine, J.-M.; Marty, J.C.; Flechtner, F.; Balmino, G.; Barthelmes, F.; Biancale, R. EIGEN-6C4 The latest combined global gravity field model including GOCE data up to degree and order 2190 of GFZ Potsdam and GRGS Toulouse. 2014. GFZ Data Services. *ICGEM Int. Cent. Glob. Earth Models* 2014. [CrossRef]
39. Ince, E.S.; Barthelmes, F.; Reißland, S.; Elger, K.; Förste, C.; Flechtner, F.; Schuh, H. ICGEM—15 Years of Successful Collection and Distribution of Global Gravitational Models, Associated Services, and Future Plans. *Earth Syst. Sci. Data* **2019**, *11*, 647–674. [CrossRef]
40. National Centers for Environmental Information, NCEI Database. NGDC. Available online: https://www.ngdc.noaa.gov/mgg/trk/trackline (accessed on 15 December 2023).
41. Salem, A., Williams, S.; Fairhead, D.; Smith, R.; Ravat, D. Interpretation of magnetic using tilt-angle derivatives. *Geophysics* **2008**, *73*, 14JF-Z11. [CrossRef]
42. Stewart, I.C.F.; Miller, D.T. Directional tilt derivatives to enhance structural trends in aeromagnetic grids. *J. Appl. Geophys.* **2018**, *159*, 553–563. [CrossRef]
43. McAdoo, D.C.; Sandwell, D.T. Folding of the oceanic lithosphere. *J. Geophys. Res.* **1985**, *90*, 8563–8569. [CrossRef]
44. Zuber, M.T. Compression of oceanic lithosphere: An analysis of intraplate deformation in the Central Indian Ocean. *J. Geophys. Res.* **1987**, *92*, 4817–4825. [CrossRef]
45. Gerbault, M. At what stress level is the Central Indian Ocean lithosphere buckling? *Earth Planet. Sci. Lett.* **2000**, *178*, 165–181. [CrossRef]
46. Kazmin, V.G.; Levchenko, O.V. Recent deformations of the Indo-oceanic lithosphere. In *Recent Tectonic Activity of the Earth and Seismicity*; Pushcharovsky, Y.M., Ed.; Nauka: Moscow, Russia, 1987. (In Russian)
47. Shemenda, A.I. Modelling of Intraplate Deformations in NE Indian Ocean. *Geotektonika* **1989**, *3*, 37–49. (In Russian)
48. Stein, C.A.; Cloetingh, S.; Wortel, R. SEASAT-derived gravity constraints on stress and deformation in the northeastern Indian Ocean. *Geophys. Res. Lett.* **1989**, *16*, 823–826. [CrossRef]
49. Evsyukov, Y. New data on the structure of uplifts in the equatorial part of the Central Indian Ocean Basin. *Dokl. Akad. Nauk. SSSR* **1991**, *320*, 677–681.
50. Kirmasov, A.B. *Fundamentals of Structural Analysis*; Nauchnyi Mir: Moscow, Russia, 2011; p. 368. (In Russian)
51. Tevelev, A.V. *Structural Geology. Manual*; SIC INFRA-M: Moscow, Russia, 2023; p. 342. (In Russian) [CrossRef]
52. Catuneanu, O. *Principles of Sequence Stratigraphy*, 2nd ed.; Elsevier: Amsterdam, The Netherlands, 2022; p. 375.
53. Cochran, J.R.; Stow, D.A.V.; Amano, K.; Balson, P.S.; Brass, G.; Corrigan, J.; Raman, C.V.; Tiercelin, J.-J.; Townsend, M.R.; Wijayananda, N.P. *Ocean Drilling Program Leg 116 Citations*; Ocean Drilling Program: College Station, TX, USA, 1988; Volume 116, p. 388.

54. Dubinin, E.P.; Ushakov, S.A. *Ocean Riftogenesis;* GEOS: Moscow, Russia, 2001; p. 293. (In Russian)
55. Delescluse, M.; Chamot-Rooke, N. Instantaneous deformation and kinematics of the India-Australia plate. *Geophys. J. Int.* **2007**, *168*, 818–842. [CrossRef]
56. Bull, J.M.; Scrutton, R.A. Seismic reflection images of intraplate deformation, central Indian Ocean, and their tectonic significance. *J. Geol. Soc.* **1992**, *149*, 955–966. [CrossRef]
57. Riedel, W. Zur mechanik geologischer brucherscheinungen. *Cent. Fur Minerol. Geol. Und Paleontol.* **1929**, *vol. 1929B*, 354–368.
58. Sanderson, D.J.; Marchini, W.R.D. Transpression. *J. Struct. Geol.* **1984**, *6*, 449–458. [CrossRef]

Journal of
*Marine Science
and Engineering*

MDPI

Article

Enhanced Continental Weathering Triggered the Anoxia of Seawater and Mass Extinctions During the Late Ordovician

Pan Tang [1], Xiangrong Yang [1,2,*] and Detian Yan [3]

[1] School of Geosciences, Yangtze University, Wuhan 430100, China; tangpan2013@126.com
[2] Hubei Key Laboratory of Marine Geological Resources, China University of Geosciences, Wuhan 430074, China
[3] School of Earth Resources, China University of Geosciences, Wuhan 430074, China
* Correspondence: m15727044911@163.com

Abstract: During the Late Ordovician period, changes in climate and mass extinctions were observed; however, the factors influencing these phenomena have not been fully understood. In order to understand the relationships among redox water conditions, climates, and mass extinctions in the Late Ordovician, this study analyzes the chemical index of alteration (CIA) in shales and $^{87}Sr/^{86}Sr$ in carbonate leachates as proxies of changes in chemical weathering intensity and chemical weathering rate in the Late Ordovician (mainly from Katian to Hirnantian). The results show that an enhanced chemical weathering rate (increased $^{87}Sr/^{86}Sr$ ratios) and decreased chemical weathering intensity (decreased CIA values) characterized the late Katian, which might be attributed to the global orogenesis and enhanced precipitation/runoff under the warming climate (late-Boda warming). This enhanced chemical weathering rate contributed to the CO_2 drawdown in the *P. pacificus* biozone, corresponding to the initiation of cooling and further glaciation. Meanwhile, the enhanced weathering-induced high primary productivity could have contributed to the expansion of anoxic seawater in the Katian, which further caused the Katian extinction. The Hirnantian Glaciation was characterized by high $^{87}Sr/^{86}Sr$ ratios in carbonates and extremely low CIA values in shales, which were likely related to the exposure of continents during low sea level and the glacial grinding of unweathered rocks. This study shows that the highest denudation rate and lowest chemical weathering intensity in the Hirnantian stage might have resulted in enhanced CO_2 release and contributed to the end of glaciation.

Keywords: Late Ordovician; Boda warming; Hirnantian Glaciation; mass extinctions; chemical weathering intensity

1. Introduction

The Late Ordovician witnessed significant geological events involving climate changes, mass extinction, and biological recovery [1–4]. The former includes the warming of the climate (Boda warming event in late Katian) and the cooling of the climate (glaciation in Hirnantian), and the latter generally includes the Great Ordovician Biodiversification Event (GOBE) and Late Ordovician mass extinctions (LOMEs). Extensive studies have shown that Boda warming and Hirnantian Glaciation were related to changes in atmospheric CO_2 and temperature caused by a combination of intense volcanic eruption, increases in continental weathering, and organic carbon burial [5–9]. Two phases of extinction marked the LOMEs in traditional understandings, and new studies have found that the LOMEs happened within the middle–late Katian (named as Katian extinction) [10,11]. A variety of extinction mechanisms have been proposed, mostly involving temperature, volcanism, and euxinia in seawater [12–17].

The chemical weathering of silicate rock entails the dissolution of silicate minerals and the formation of secondary minerals [18]. This process transforms atmospheric CO_2

into dissolved HCO_3^- and releases significant amounts of nutrients (such as phosphorus and iron) into the ocean, thereby substantially influencing the paleoclimate, primary productivity, and oceanic hydrochemistry conditions [19–21]. Factors such as surface temperature, tectonic activity, volcanic eruptions, and land-plant colonization generally impact the rate of chemical weathering [22–24]. It is suggested that the intense tectonic activity during the Late Ordovician not only led to crustal uplift [5] but also facilitated the southward movement of the Appalachian orogenic belt relative to the Laurentian continent [8], both of which enhanced continental weathering. Moreover, the greenhouse climate resulting from significant volcanic activity in the late Katian and a plentiful supply of weathered protoliths further amplified continental weathering [25,26]. In contrast, the global cooling in the Hirnantian was characterized by weak chemical weathering intensity [27].

Silicate weathering mainly includes chemical and physical denudation rates (denudation rate = chemical weathering rate + physical weathering rate), with chemical weathering intensity representing the ratio of the chemical weathering rate to the denudation rate [28]. Generally, a linear array on the logarithmic plot of chemical weathering versus physical erosion rate could be observed under a low erosion rate (transport-limited), in which case chemical weathering intensity is enhanced due to thick soil (completely weathered) sequences [28]. In contrast, silicate weathering shows substantial scatter away from the trend at higher erosion rates (kinetically limited) when chemical weathering rates are higher [28]. Traditionally, weathering-controlled changes in climate mean that an enhanced chemical weathering rate, rather than increased chemical weathering intensity, contributes to the drawdown of atmospheric CO_2 [18–20]. The relationship between climate and weathering is complicated, and several hypotheses have been proposed. For example, a high chemical weathering intensity triggered CO_2 drawdown following a carbon release event caused by volcanism [29], a low chemical weathering intensity and high chemical weathering rate occurred under a warming climate [30,31], and the silicate weathering rate reached a maximum for intermediate denudation rates and further contributed to CO_2 drawdown. In contrast, extremely low denudation rates (supply limited regime) and high denudation rates (kinetically limited regime) can decrease or even reverse CO_2 sequestration [32].

The strontium (Sr) isotopic compositions in seawater are influenced by continental runoff and mantle contributions from hydrothermal systems [33–38]. As a result, seawater $^{87}Sr/^{86}Sr$ ratios could be used as significant indicators for continental weathering trends in geological periods because these $^{87}Sr/^{86}Sr$ ratios could be used to constrain continental weathering fluxes of dissolved Sr [33]. Weathering proxies in mudstones include the chemical alteration index (CIA), plagioclase alteration index (PIA), and chemical weathering index (CIW) [39–41]. All of these weathering proxies should reflect the changes in chemical weathering intensity. In south China, the redox water conditions, hydrodynamic conditions, volcanism, and evolution of biomass have been widely analyzed in previous studies [3,10,16,29]. Consequently, detailed investigations into the weathering trends (chemical weathering rate and intensity) of the Late Ordovician (using weathering indices in shales and $^{87}Sr/^{86}Sr$ in carbonate leachates) are essential to elucidate the critical relationships between continental weathering, paleoclimate, and the LOMEs.

2. Geological Setting

South China was located as a separate continent in the Ordovician (Figure 1A) [42,43], consisting of the Yangtze Block in northwest and the Cathaysia Block in the southeast. The collision of them happened in the early Neoproterozoic (Figure 1B), leading to the compression between the Yangtze Block and Cathaysia Block [44,45]. The graptolite in the Katian includes *Dicellograptus complanatus*, *Dicellograptus complexus*, *Paraothograptus pacificus* biozones, *Metabolograptus extraodinarius*, and *Metabolograptus persculptus* biozones in the Hirnantian and *Akidograptus ascensus*, *Parakidograptus acuminatus*, *Cystograptus vesiculosus*, and *Coronograptus cyphus* biozones in the Rhuddanian (Figure 1C) [46,47].

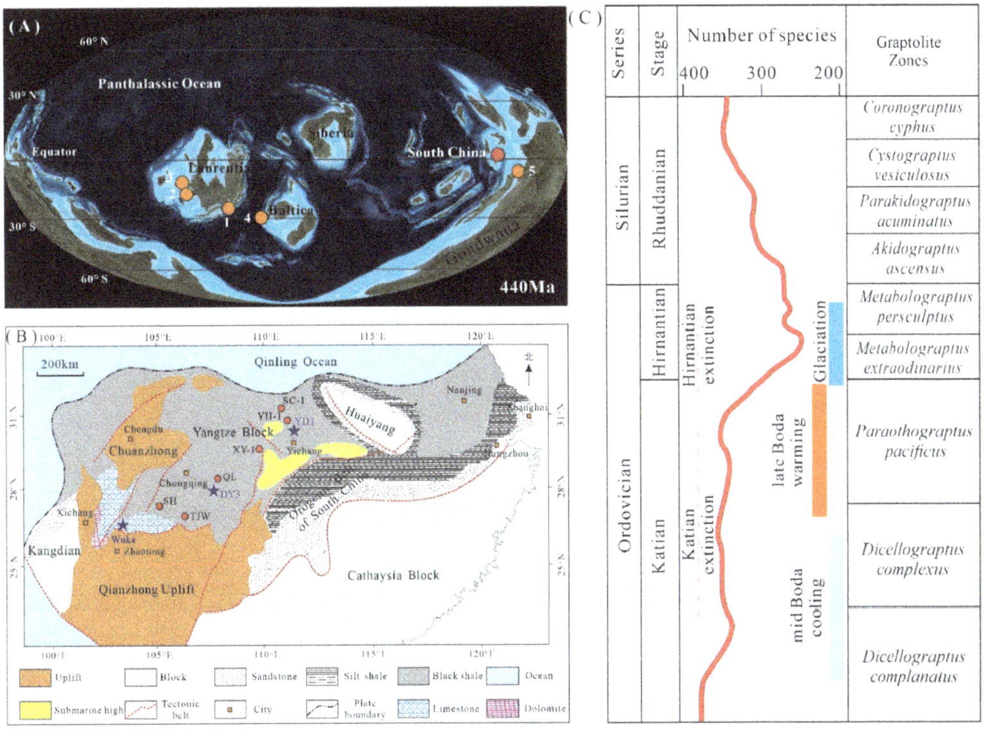

Figure 1. (**A**) Global paleogeography(~440 Ma, modified after http://deeptimemaps.com, accessed on 1 January 2023) during the Late Ordovician and Early Silurian period. 1: Dob's Linn section, Scotland; 2: Vinini Creek section, Nevada; 3: Monitor Range, Nevada; 4: Holy Cross Mountains; 5: Parahio Valley India Himalaya section, India. (**B**) Early Silurian paleogeographic map showing the distribution of the lithofacies of the Yangtze area [43] showing the sites Wuke (WK), Shuanghe (SH), Tianjiawan (TJW), Mouchuangou (MCG), Qiliao (QL) sections and DY3, XY-1, YD1, Yihuang-1 (YH-1), Shenci-1 (SC-1) boreholes. (**C**) Time scale and graptolite biozones are from [46,47]; generic diversity across Late Ordovician to Early Silurian is from [10].

A shallow water section (Wuke section) was located in western Sichuan Province, south China (Figure 1B) [17]. The stratigraphy of the Wuke (WK) section is organized into the Linxiang (LX) Formation, the Tiezufeike (TZFK) Formation, and the Butuo (BT) Formation, in stratigraphic order. The LX Formation (~5 m) is primarily characterized by argillaceous limestones and calcareous shales. The Tiezufeike Formation (~15 m) is characterized predominantly by limestones and dolomitic limestones, with the upper part of Tiezufeike Formation containing a significant presence of shelly fauna, notably the Hirnantian fauna. The Butuo Formation (~>20 m) mainly consists of laminated limestones and calcareous mudstones, alternating with layers of argillaceous siltstones and mudstones.

The two boreholes (YD1 and DY3) analyzed in this study were located on the Upper Yangtze platform (Figure 1B). The boreholes are categorized into the Lingxiang Formation, Wufeng Formation, Guanyinqiao Bed, and Longmaxi Formation, arranged in stratigraphic order. The Wufeng and Longmaxi formations primarily consist of black shale, whereas the Guanyinqiao Bed comprises carbonaceous and muddy limestones containing abundant benthic fauna [16].

Environmental differences were noted between the western margin of the Yangtze Block, characterized by a deep shelf (shale dominated), and the southwestern Yangtze Block, characterized by a shallow shelf (carbonates dominated). In the deep shelf area, a widespread distribution of organic-rich shales of the Wufeng (WF) and Longmaxi (LMX) formations were observed [13,16,17]. The siliceous and argillaceous shales of the Wufeng and Longmaxi formations, dating from the latest Hirnantian to Aeronian, were deposited under anoxic water conditions [13,16,17]. The controlling mechanisms for this included intense upwelling, volcanism, and thermohaline stratification. In the shallow shelf areas, minor carbonate facies were present along the flanks of the Kangdian and Central Guizhou uplifts, where the waters likely became less anoxic and upwelling weakened compared to those in the deep-water shelf [41,43]. Also, the carbonates in the shallow shelf were deposited under more oxic water conditions, leading to less total organic matter in the rocks [17].

3. Methods

In this study, a total of 15 carbonate samples from the Wuke section and 29 shale samples from DY3 and YD1 boreholes were performed for geochemical analyses.

3.1. Major and Trace Elements

The melting method was used for the pretreatment of samples for major element analyses, with the cosolvent being lithium fluoride, lithium metaborate, and lithium tetraborate and the oxidant being ammonium nitrate and lithium bromide. The melting temperature was set to 1050 °C, and the melting time was 15 min. An X-ray fluorescence spectrometer (XRF) was used for the analysis of major elements, with the standard curve using Chinese National Standard Material (GBW07101-14) [48] and a relative standard deviation (RSD) of less than 2% [48]. The analytical precision was better than 10%.

The trace elements were analyzed by an Agilent 7700e ICP-MS at the Wuhan Sample Solution Analytical Technology Co., Ltd., Wuhan, China. First, fresh and fine-grained shale samples were powdered into 0.075 mm-diameter pellets, and these powders were placed in an oven at 105 °C for approximately 12 h. Second, approximately 50 mg of the sample was dissolved in HNO_3 and HF. Third, HNO_3, Milli-Q water, and an internal standard solution were added. Fourth, the final solution was transferred to a polyethylene bottle and diluted to 100 g by adding 2% HNO_3. The analytical precision was better than 5%.

The CIA is calculated as follows: CIA = $Al_2O_3/(Al_2O_3 + CaO^* + Na_2O + K_2O) \times 100$ [49]. In the equation, CaO^* represents the CaO content in the silicate, and thus, it is calculated following the method: CaO^* is initially corrected using P_2O_5 contents ($CaO - 10/3 \times P_2O_5$) [50]. If the calculated value exceeds the Na_2O content, the CaO^* value is set as the Na_2O value; otherwise, the CaO^* value is set as the CaO content. Due to the possible addition of K_2O to weathered samples during K-metasomatism, a correction of CIA values (CIA_{corr}) has been proposed as follows: $CIA_{corr} = Al_2O_3/(Al_2O_3 + CaO^* + Na_2O + K_2O^*) \times 100$ [51], where K_2O^* is calculated as follows: $(K_2O^*) = \{m \times [(Al_2O_3)] + m \times [(CaO^*) + (Na_2O)]\}/(1 - m)$, with $m = (K_2O)/[(Al_2O_3) + (CaO^*) + (Na_2O) + (K_2O)]$.

To eliminate the influence of K_2O, Harnois [39] proposed the CIW proxy ($CIW = Al_2O_3/(Al_2O_3 + CaO^* + Na_2O) \times 100$). Fedo et al. [51] defined the PIA proxy as a necessary weathering intensity, which is calculated as follows: $PIA = (Al_2O_3 - K_2O)/(Al_2O_3 - K_2O + CaO^* + Na_2O) \times 100$. In addition, the compositional maturity of a rock (ICV) is expressed as $ICV = (Fe_2O_3 + K_2O + Na_2O + CaO + MgO + MnO + TiO_2)/Al_2O_3$ [52].

3.2. Elemental Concentrations and $^{87}Sr/^{86}Sr$ Analysis in Carbonates

About 100 mg of carbonate samples were rinsed three times with the Milli-Q water (18.2 MΩ). The carbonate samples were leached twice with 0.05 M acetic acid at room temperature for 24 h. Fractions of the carbonate leachates were retained in 2% HNO_3 for elemental analysis using an Elan Quadrupole ICP-OES. The reference material JLs-1was analyzed [53] and indicates that accuracy and precision were less than ±7% for all elemental concentrations reported here.

Sr isotopic ratios were measured using the Neptune plus MC-ICP-MS at the Wuhan Sample Solution Analytical Technology Co., Ltd., Wuhan, China. The Sr isotope analysis of carbonate leachates involves three steps [54]. Sample digestion: about 50–200 mg of sample power (200 mesh) was placed in an oven at 105 °C for drying and was then dissolved using the HNO_3 and HF. The solution was evaporated to dryness and was dissolved in 1.0 mL of 2.5 M HCl again. Column chemistry: the supernatant solution was loaded into an ion-exchange column packed with AG50W resin. The Sr fraction was eluted using 2.5 M HCl and gently evaporated to dryness prior to mass-spectrometric measurement. Sr isotope analyses were conducted at the Wuhan Sample Solution Analytical Technology Co., Ltd., Hubei, China, using a Neptune Plus MC-ICP-MS. International NIST 987 standard [54] was measured every seven samples analyzed, with the $^{87}Sr/^{86}Sr$ ratio of 0.710242 ± 14 (2SD, n = 345) having been achieved, which is identical within error to their published values 0.710248 ± 12 [55]. Additionally, the BCR-2 (basalt) and RGM-2 (rhyolite) yielded results of 0.705012 ± 22 (2SD, n = 63) and 0.704173 ± 20 (2SD, n = 20) for $^{87}Sr/^{86}Sr$, respectively.

4. Results

4.1. Major Elements

The result of major element concentrations can be found in the Supplementary Files (Tables S1 and S2). Si, Al, and Ca are the primary elements and show high concentrations in the studied shales, while Fe, Mg, Na, K, Ti, and P are the minor elements and show low concentrations (Tables S1 and S2). Generally, Si, Al, Na, and K are linked with silicate minerals, and Ca and Mg are linked with carbonate minerals. The Si percentage varies from 35.7% to 70.0% in DY3 (except for one marlstone with low SiO_2 contents (13.4%) in Guanyinqiao Member) and from 56.0% to 77.7% in YD1, Al varies from 3.2% to 11.5% in DY3 and from 4.7% to 15.7% in YD1, and Ca varies from 1.7% to 26.0% in DY3 and from 0.6% to 15.7% in YD1.

All the CIA, CIA_{corr}, CIW, and PIA profiles in DY3 and YD1 boreholes exhibit decreasing trends from the Katian to Hirnantian (Figure ??), with values ranging from 58.1 to 66.0, from 59.2 to 67.9, from 69.5 to 81.0, and from 62.1 to 75.4 in DY3 and from 60.4 to 68.3, from 61.1 to 70.5, from 69.3 to 83.5, and from 64.0 to 78.4 in YD1. The ICV shows variable values throughout the Katian to Hirnantian rocks, ranging from 0.98 to 14.4 in DY3 and from 0.90 to 4.43 in YD1.

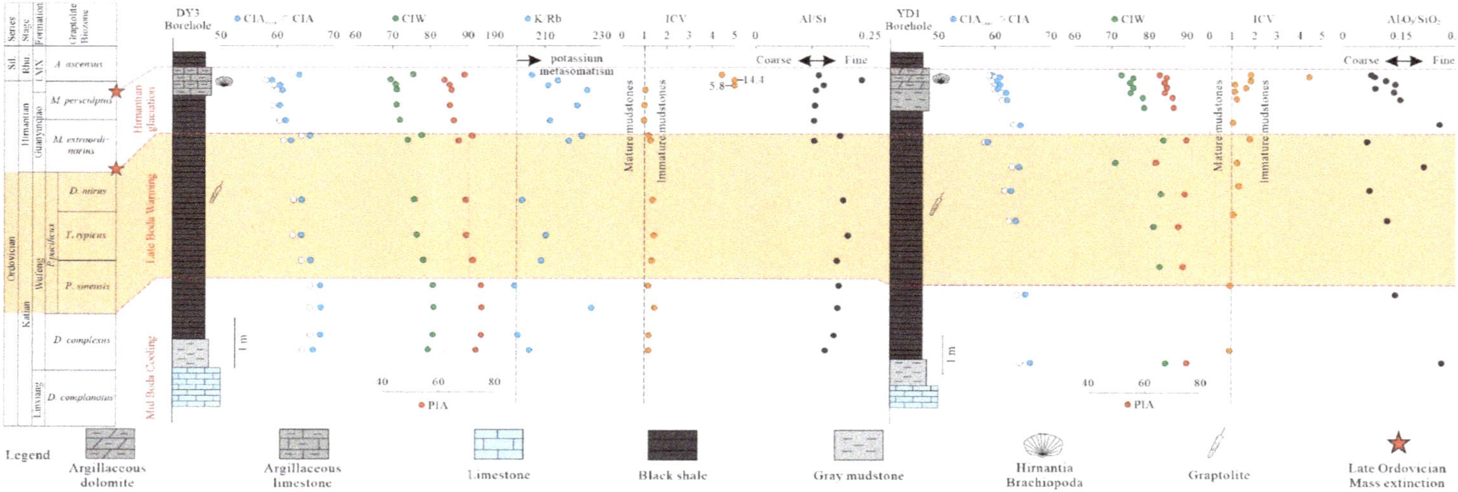

Figure 2. Stratigraphic CIA, CIA_{corr}, CIW, PIA, K/Rb, ICV, and Al/Si of the Wufeng Formation and Guanyinqiao Member in DY3 and YD1 boreholes. The orange area represents the period of late-Boda warming climate, the black dotted lines represents 210 for K/Rb ratio and 1 for ICV [56,57], with high K/Rb ratios indicating minor potassium metasomatism and low ICV values indicating mature mudstones. Sil.: Silurian, Rhu.: Rhuddanian, LMX: Longmaxi Formation.

4.2. Trace Elements

The trace elements in shales of DY3 borehole have been analyzed in this study (Table S1). Sc, Hf, Rb, La, and Th in the Katian show steady contents, varying from 8.40 μg/g to 11.2 μg/g, from 2.51 μg/g to 4.01 μg/g, from 104 μg/g to 148 μg/g, from 31.8 μg/g to 47.6 μg/g, and from 12.5 μg/g to 17.8 μg/g (Table S1). The Sc, Hf, Rb, Th in these rocks show relatively low contents compared to those in PAAS (Sc: 16, Hf: 5, Rb: 160), while the La and Th show more variables when compared with those in PAAS (La: 38, Th:14.6). Sc, Co, Rb, La, Th in the Hirnantian show variable contents, varying from 3.60 μg/g to 9.40 μg/g, from 1.59 μg/g to 4.17 μg/g, from 38.2 μg/g to 107 μg/g, from 15.5 μg/g to 69.2 μg/g, and from 4.85 μg/g to 13.8 μg/g (Table S1). Except for La, all other trace elements show lower contents relative to those in PAAS.

4.3. $^{87}Sr/^{86}Sr$ and Elements in Carbonate Leachates

The results of $^{87}Sr/^{86}Sr$ and elements in carbonate leachates can be found in the Supplementary Files (Table S3). $^{87}Sr/^{86}Sr$ have stratigraphic variations in the WK section, increasing from 0.708492 to 0.709022 in the lower Tiezufeike Formation and decreasing from 0.709022 to 0.708130 in the upper Tiezufeike Formation. In addition, the Sr and Mn in carbonate leachates show variable contents throughout the Tiezufeike Formation, ranging from 301.1 μg/g to 1446.6 μg/g, and ranging from 17.0 μg/g to 183.7 μg/g, respectively. The corresponding Mn/Sr and Sr/Ca ratios vary between 0.03 ppm/% and 0.50 ppm/% and between 29.91 ppm/% and 190.68 ppm/%, respectively.

5. Discussion

5.1. δ^{13} C Records and Climate Events in the Late Ordovician

Carbon isotopes are identified as a useful tool for determining local and global stratigraphic correlations. We compared the $\delta^{13}C$ values in Late Ordovician sediments of the WK section [58] with those in other published studies [59–62]. In south China, the $\delta^{13}C_{org}$ values are high in the *D. complanatus* Biozone (Figure ??; Linxiang Formation), and they shift toward low values within the *D. complexus* Biozones (lower Wufeng Formation in shale-dominated sections and lower Tiezufeike Formation in carbonate-dominated section). The $\delta^{13}C_{org}$ values remained low throughout the lower-middle *P. pacificus* biozone; subsequently, the $\delta^{13}C_{org}$ values began to increase slowly in the upper *P. pacificus* biozone and more rapidly in the lower Hirnantian (*P. persculptus*) zone and peaked in the upper part of the *M. extraordinarius* biozone. This peak in high $\delta^{13}C_{org}$ (~−25‰) is concurrent with the occurrence of the Hirnantia–Dalmanitina fauna [58], predominantly a cold/cool-water fauna found in the upper Tiezufeike Formation in WK section and Guanyinqiao Member in the DY and YD boreholes (Figure ??).

Various climatic events, including the early and late-Boda warming, mid-Boda cooling, and Hirnantian Glaciation, have been delineated through petrological and geochemical evidence, with $\delta^{13}C$ record correlation utilized to constrain the timing of these geological phenomena (Figure ??; [63]). $\delta^{13}C$ values remained stable and high during the late Katian (from *D. complanatus* to *D. complexus*), aligning with the mid-Boda cooling [61]. Subsequently, $\delta^{13}C$ values decrease in the *P. pacificus* biozone, reflecting intense volcanism and late-Boda warming [64]. An increase in $\delta^{13}C$ is noted in the uppermost *P. pacificus* Biozone, equivalent to the upper Tiezufeike Formation in WK, the upper Wufeng Formation in Tianjiawan and Qiliao, and the Vinini Formation in Vinini Creek (Figure ??). A positive $\delta^{13}C$ excursion is observed within the *M. persculptus* and *M. persculptus* biozones, alongside a significant regression and the most extensive Hirnantian glacial event. By the onset of the Silurian, $\delta^{13}C$ values revert to lower levels similar to those before the Hirnantian Glaciation, marking the end of the glaciation and a transition to a warmer climate (Figure ??).

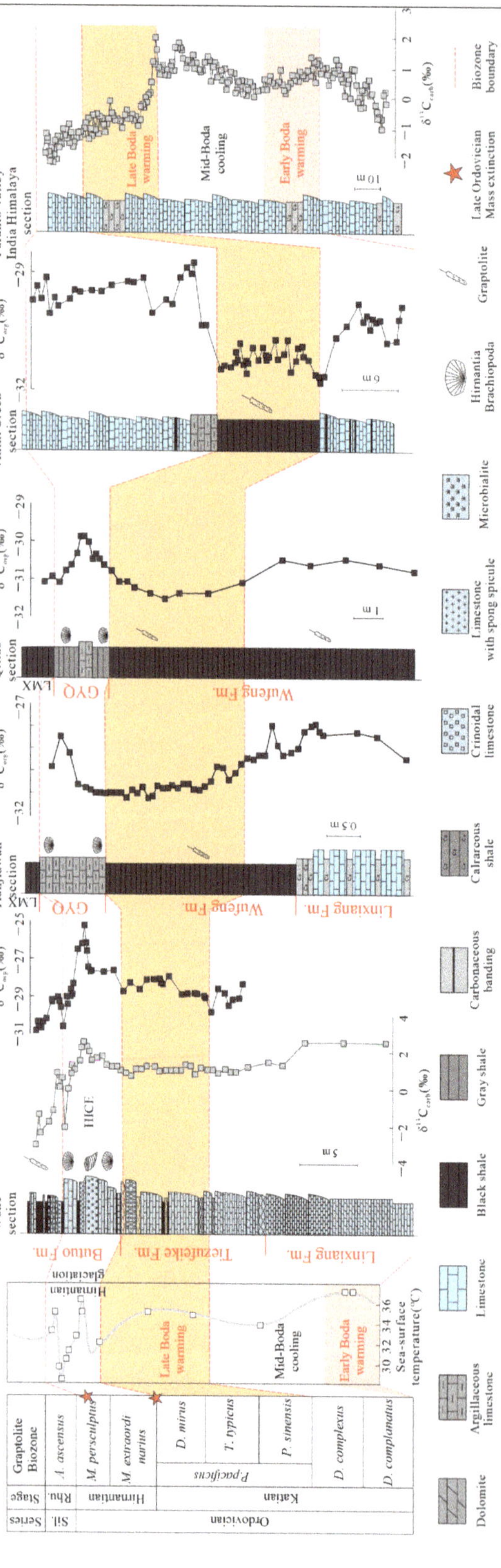

Figure 3. $\delta^{13}C$ profiles of the Late Ordovician–Early Silurian strata in the Wuke, TJW, QL, Vinini Creek, and Parahio Valley India Himalaya [59–61]. The sea surface temperature is based on the clumped oxygen isotope data [62]. The pink and orange areas represent the periods of early and late-Boda warming climates, respectively. Sil.: Silurian, Rhu.: Rhuddanian, GYQ: Guanyinqiao Formation.

5.2. The Reliability of the Weathering Proxies in Shales and Carbonates

5.2.1. Evaluation of Weathering Proxies in Shales

The weathering proxies in shales are widely utilized to evaluate silicate weathering [65]. In this study, CIA, CIA_{corr}, CIW, and PIA show decreasing trends throughout the DY3 and YD1 boreholes (Figure ??; [66,67]) and positive correlations among these proxies could be observed (Figure 4A–C; [68,69]). A thorough evaluation of these proxies is essential because they might be affected by source rocks, hydrodynamic cycles, and post-depositional diagenesis (Figure 4A–F). In this case, this study uses the trace element concentrations and their ratios, Al_2O_3/SiO_2, and K/Rb, to evaluate the influences of the above factors.

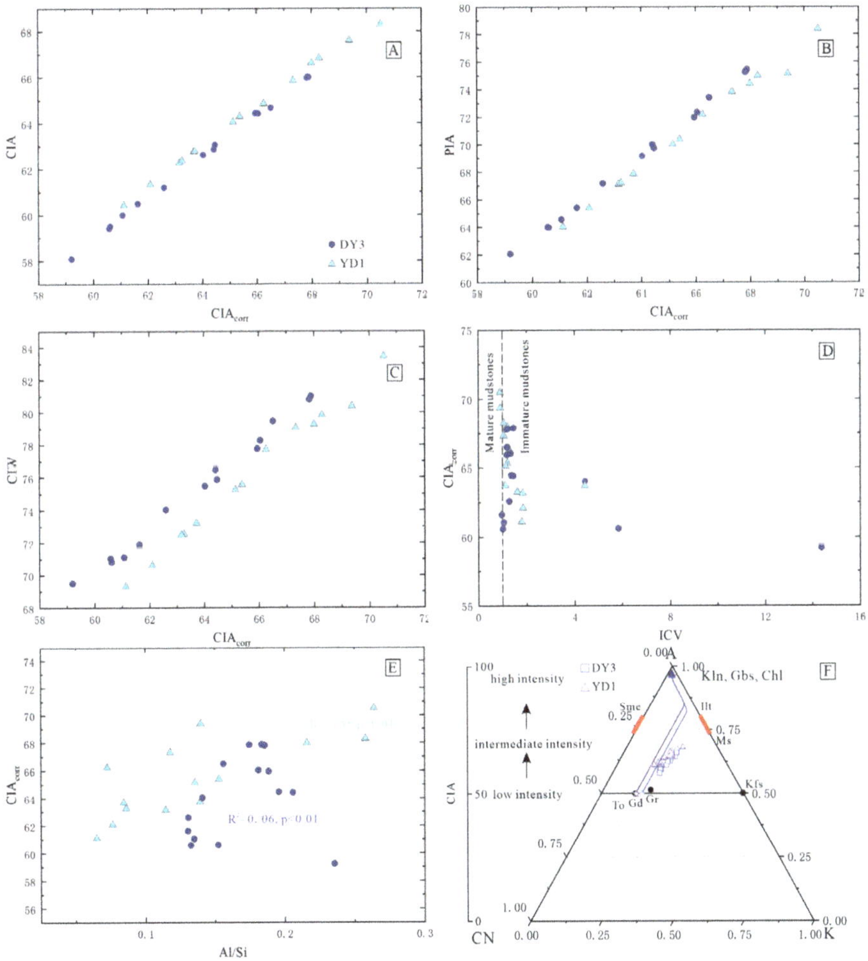

Figure 4. Crossplots of CIA_{corr} versus (**A**) CIA, (**B**) PIA, and (**C**) CIW. (**D**) Plots of analyzed samples of DY3 and YD1 on the A-CN-K diagram [49–51]. A: Al_2O_3, CN: $CaO^* + Na_2O$, K: K_2O, To: tonalite, Gd: granodiorite, Gr: granite, Kln: kaolinite, Gbs: gibbsite, Chl: chlorite, Ilt: Illite, Ms: muscovite, Kfs: K-feldspar. The arrows represent the weathering trends for YD1 and DY3 rocks. Crossplots of CIA_{corr} versus (**E**) ICV, (**F**) Al/Si. The black dotted line in (**D**) represents 1 for ICV [67], with low ICV values indicating mature mudstones.

Traditionally, Zr, Ti, Al, and their ratios are used to identify the lithology of parental rocks; however, Zr is strongly controlled by the presence of zircons and hence the sedimentary rocks, as zircons tend to be enriched in the silt-size fraction. In addition, Ti abundances in shales generally co-vary with Al, because both elements are immobile during chemical weathering and associated with secondary clay minerals [68]. So, the granulometry of shales is likely the dominant parameter controlling Al_2O_3/TiO_2 ratios in shales. Instead, we choose elements such as Hf, La, Th, and Sc in clastic rocks as effective indicators for determining the types of source rocks and tectonic settings [69]. The discrimination diagrams indicate that all the samples are near the regions of felsic to intermediate rocks (Figure 5A,B; [70–72]).

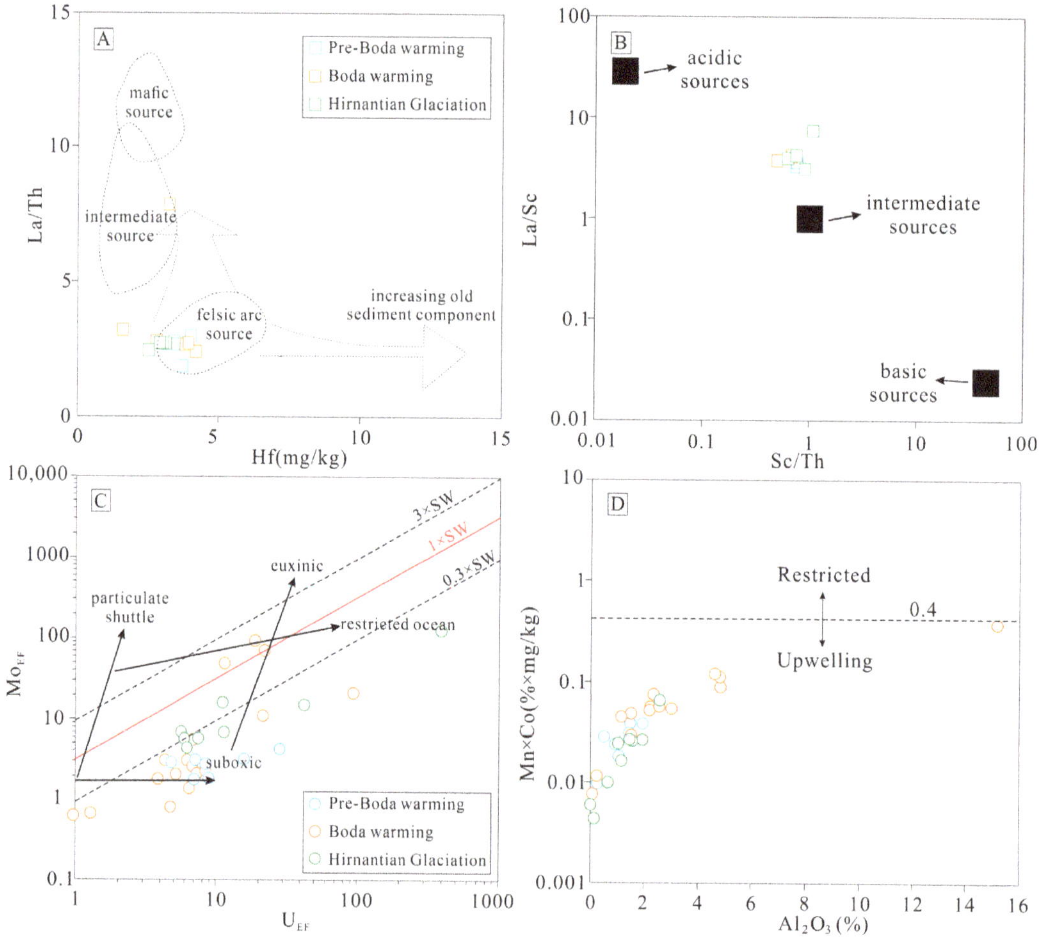

Figure 5. Crossplots of (**A**) Hf versus La/Th, (**B**) Sc/Th versus La/Sc, and (**C**) Mo_{EF} versus U_{EF}; the SW means the ratio of Mo_{EF}/U_{EF} in modern seawater, the solid lines reflect the changes in redox water conditions, restricted and upwelling settings [70,71]. (**D**) Mn × Co versus Al_2O_3; the dotted line represents the boundary value (0.4) between restricted and upwelling settings [72]. All the major and trace elements used above are analyzed for whole rocks from [58].

Additionally, A-CN-K diagrams are utilized to determine the types of source rocks. The deviation of our samples from the ideal weathering trend of source rocks (parallel to the A-CN axis) in the Al_2O_3- ($CaO^* + Na_2O$)$-$ K_2O (A-CN-K) ternary diagram (Figure 4F).

The corrected CIA values (CIA_{corr}) are distributed along the ideal weathering trends of source rocks, and the intersection between the weathering trend line and feldspar join represents the source rock. In this study, the ternary diagrams show that the samples from two boreholes converge towards a point representing the source rocks (granodiorite or granite; Figure 4F). These comprehensive observations strongly support the conclusion that Late Ordovician shales derived from a consistent protolith source. Therefore, the observed variations in weathering indicators throughout the Katian and Hirnantian periods are unlikely to be attributable to changes in the types of source rocks.

Adding K_2O during diagenesis can affect the accuracy of the CIA values because the samples are characterized by less weathering than their actual state. The K/Rb ratio can be used to assess the degree of K mobility due to metasomatism, with the rocks having lower K/Rb ratios (>220) in potassium metasomatism [66]. We found that our samples are characterized by slightly high K/Rb ratios (200–227), indicating minor influences of diagenesis (Figure ??). The CIA_{corr} values are calculated to avoid the influence of potassium metasomatism. In our study, a substantial positive correlation between CIA_{corr} and CIA indicates that the correction of the CIA preserves the temporal weathering trend (Figure 4A). Furthermore, the variations in weathering proxies might be related to multiple sedimentary cycles before burial, thus influencing the actual weathering experience of the protolith in source areas. Generally, the ICV is a reliable indicator for distinguishing mature and immature mudstones, with mature mudstones experiencing several weathering cycles and exhibiting low ICVs (<1). In contrast, immature mudstones exhibit high ICVs > 1 [67]. In our study, all the shales have high ICVs > 1, with just three samples showing slightly low values (0.9, 0.92, 0.98), suggesting that they experienced only one cycle of weathering (Figure 4D). The Al/Si ratio is a grain size proxy, with fine particles having high Al_2O_3/SiO_2 ratios and coarse components having low Al/Si ratios. There is moderate correlation between CIA_{corr} and Al_2O_3/SiO_2 in YD1, which could not be observed in DY3 (Figure 4E). We believe that there is a negligible influence of grain size on the variations in weathering proxies for the DY3. In YD1, the coarse part is characterized by low chemical weathering intensity, likely enhanced erosion of unweathered rocks with high size.

5.2.2. Evaluation of Carbonate $^{87}Sr/^{86}Sr$

The Sr isotopic compositions in carbonates are susceptible to early diagenetic transformations and late diagenetic fluid reactions [73]; therefore, evaluating the reliability of carbonate $^{87}Sr/^{86}Sr$ before using it as the primary seawater record is important [74]. The seawater Sr is incorporated into carbonate minerals in the synsedimentary stage; however, the Sr concentrations in carbonate minerals could be modified during diagenesis [75]. Generally, 300 ppm Sr appears to be the lower threshold value, in which case the bulk carbonate could record the primary $^{87}Sr/^{86}Sr$ ratios [37]. All of the studied samples have high Sr contents (Figure ??) and there is no obvious relationship between Sr and $^{87}Sr/^{86}Sr$ (Figure ??), suggesting that primary seawater $^{87}Sr/^{86}Sr$ is potentially preserved in bulk carbonate.

Mn tends to be incorporated into carbonate minerals when bulk carbonate reacts with nonmarine diagenetic fluids, leading to higher Mn concentrations and radiogenic $^{87}Sr/^{86}Sr$ ratios in bulk carbonate [74]. As a result, Mn/Sr ratios are often used to discriminate diagenetic alterations, with unaltered bulk carbonates characterized by low Mn/Sr ratios [74]. In the WK section, the bulk carbonates have low ratios (Mn/Sr < 1; Figure ??), conforming to the geochemical criteria for Sr isotopes [74,75]. Moreover, bulk carbonate diagenesis should be characterized by low $\delta^{18}O$ values and more radiogenic $^{87}Sr/^{86}Sr$ ratios than unaltered bulk carbonate, as diagenesis results in high temperatures. However, there are no correlations between Mn/Sr (Figure 7A), Sr/Ca (Figure 7B), or $\delta^{18}O$ (Figure 7C) and $^{87}Sr/^{86}Sr$, indicating that our samples were not altered by diagenesis.

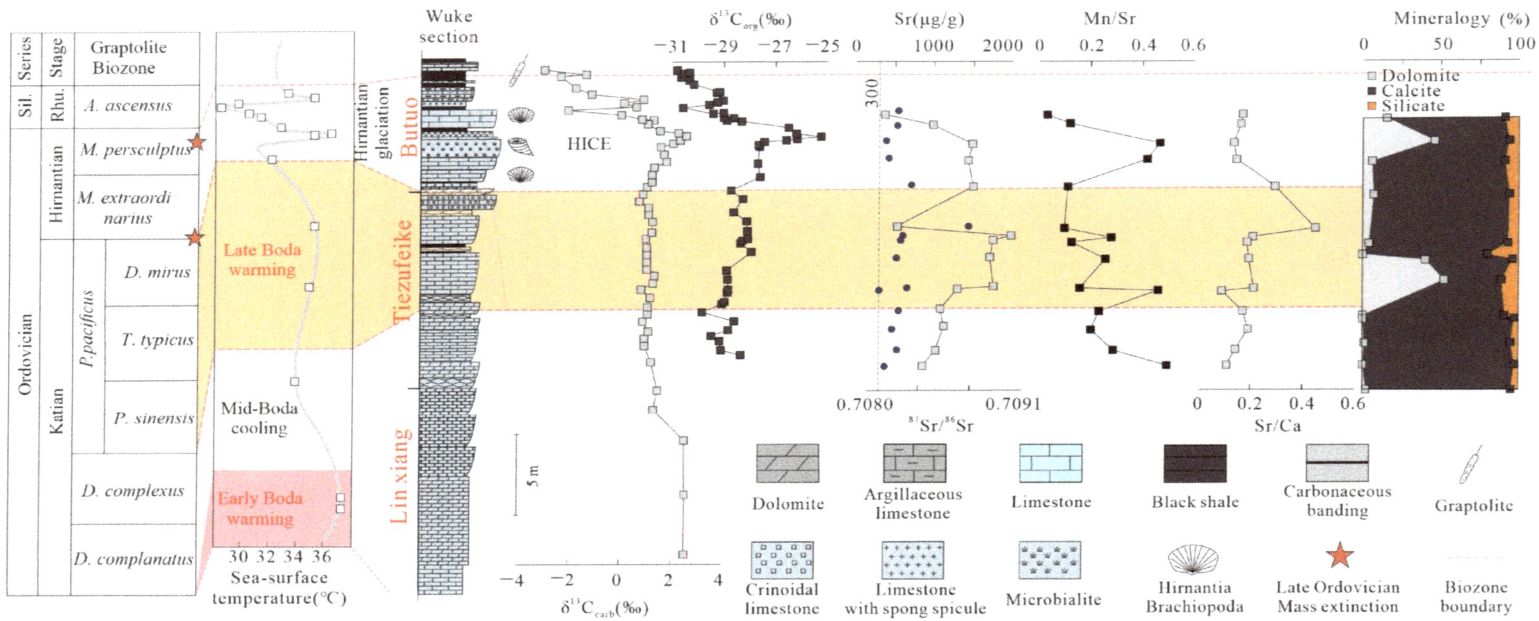

Figure 6. Stratigraphic $\delta^{13}C_{org}$, $\delta^{13}C_{carb}$, $^{87}Sr/^{86}Sr$, Mn/Sr, Sr/Ca, and mineralogy of the Linxiang and Tiezufeike formations in WK section. The sea surface temperature is based on the clumped oxygen isotope data [62]. The pink and orange areas represent the periods of early and late-Boda warming climates, respectively. Sil.: Silurian, Rhu.: Rhuddanian.

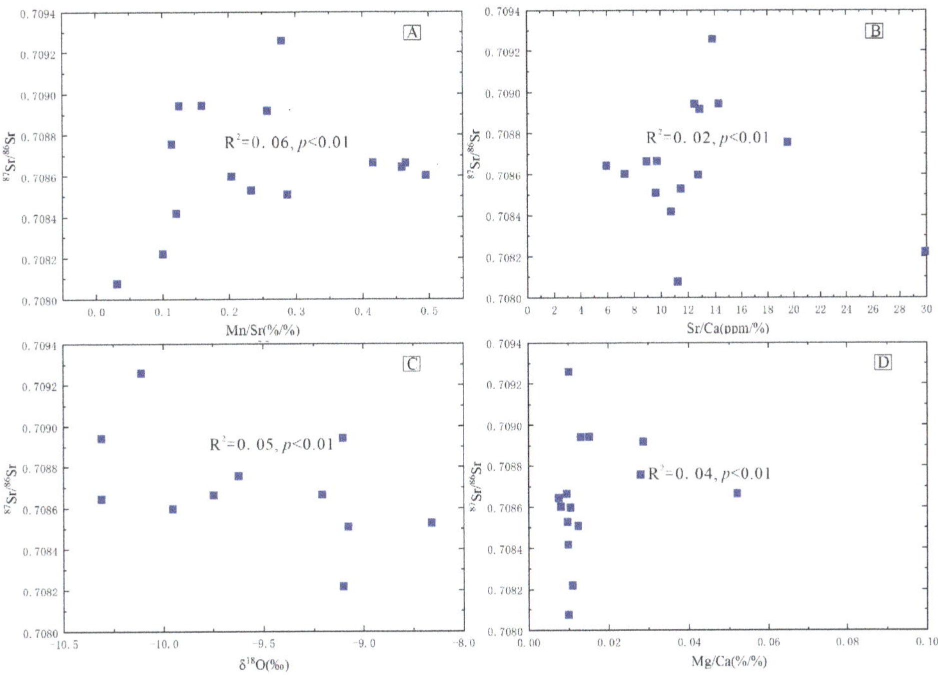

Figure 7. Crossplots of $^{87}Sr/^{86}Sr$ versus (**A**) Mn/Sr, (**B**) Sr/Ca, (**C**) $\delta^{18}O$, and (**D**) Mg/Ca.

Dolomitization is a possible process that can alter the geochemical composition (e.g., Mg/Ca, Mn/Sr, and $^{87}Sr/^{86}Sr$ ratios) of limestones [76]. However, the peak $^{87}Sr/^{86}Sr$ ratios do not correspond to the high dolomite contents in the WK section (Figure ??), and there is no correlation between Mg/Ca and $^{87}Sr/^{86}Sr$ (Figure 7D), suggesting that the observed variation in $^{87}Sr/^{86}Sr$ is not controlled by primary minerals or dolomitization during diagenesis.

5.3. Fluctuations in Weathering Proxies in the Late Ordovician

5.3.1. Fluctuations in CIA_{corr} Values in the Late Ordovician

Weathering proxies, including the CIA, CIA_{corr}, PIA, and CIW, have been used to analyze weathering trends throughout geological history. Previous studies have evaluated the limits of the chemical weathering intensity; however, there are some differences in the threshold values among strong, moderate, and weak [13,45,46]. In this study, the degree of chemical weathering intensity could be divided into intense (~75–100), intermediate (~65–75), and weak (~50–65) chemical weathering intensity [45,46]. It is broadly acknowledged that regions experiencing moderate to high chemical weathering intensity are distinguished by elevated CIA and CIA_{corr} values. In contrast, low CIA and CIA_{corr} values are indicative of areas with reduced chemical weathering intensity [13,16].

Within the Yangtze Block, CIA and CIA_{corr} values demonstrate a progressive decline from the Linxiang Formation to the Wufeng Formation [77,78], despite some variance in absolute values across different boreholes and sections (Figure ??). Similarly, in Scotland, CIA_{corr} and CIA values also exhibit a marked decrease, with the lower part of the Hartfell shales characterized by higher CIA values and the upper part by lower CIA values (Figure ??; [27]). Consequently, the observed weathering trends based on CIA and CIA_{corr} in the late Katian are deemed reliable, reflecting consistency across global boreholes and sections (Figure ??). This suggests a diminishing trend in chemical weathering intensity during the late Katian period (from *D. complanatus* to *P. pacificus*). Of course, the climatic zonality could influence the chemical weathering intensity, with high latitude areas generally showing low temperatures and low chemical weathering intensity and low latitude areas generally showing high temperatures and high chemical weathering intensity. In this case, the climatic zonality is another possibility influencing the differences in the CIA values between the China and Scotland sections, although this requires more exploration in the future.

In south China, CIA and CIA_{corr} exhibit lower values in the Hirnantian rocks (CIA average values for the DY and YD boreholes: 60 and 63; CIA_{corr} average values for the DY and YD boreholes: 61 and 64) than in the late Katian rocks (CIA average values for the DY and YD boreholes: 64 and 66; CIA_{corr} average values for the DY and YD boreholes: 69 and 67) (Table 1; Figure ??). This result is consistent with previous studies, suggesting that the rocks deposited during the Hirnantian Glaciation experienced weak chemical weathering intensity [16]. However, increases in CIA and CIA_{corr} could be observed in Scotland, in contrast to what has been observed in south China. This study attributes these differences in CIA trends to the different weathering regimes under the same climatic conditions. For example, the continental weathering in south China was characterized by the glacial grinding of unweathered rocks with low CIA values; however, a more incongruent weathering regime likely characterized the Scotland rocks based on the high lithium isotopes [27].

Table 1. The average values of CIA and CIA_{corr} in the rocks of SD1, DY3, Shuanghe, Qiliao, XY-1, YD1 in south China.

Borehole/Section — CIA or CIA_{corr} Value	SD1	DY3	Shuanghe	Qiliao	XY-1	YD1
Average CIA value in Hirnantian	63	60	/	59	59	63
Average CIA_{corr} value in Hirnantian	69	61	59	/	60	64
Average CIA value in late-Boda warming	66	64	/	66	62	66
Average CIA_{corr} value late-Boda warming	72	69	65	/	64	67

5.3.2. Fluctuations in Seawater $^{87}Sr/^{86}Sr$ in the Late Ordovician

Several seawater $^{87}Sr/^{86}Sr$ records from the Late Ordovician have been reported [79]. The overall published datasets from the mid–late Katian have shown relatively stable and lower seawater $^{87}Sr/^{86}Sr$ values (~0.7080), which are attributed to the enhanced weathering of fresh volcanic rocks [80]. However, all previous studies focus more on the Upper Ordovician carbonate successions, especially from the Tremadocian to Sandbian Stages [80]. In fact, there is still the lack of a detailed and high-resolution seawater $^{87}Sr/^{86}Sr$ curve from Katian to Hirnantian, which could provide important constraints on the changes in climates.

The overall $^{87}Sr/^{86}Sr$ ratios in Late Ordovician section from south China are higher than those in the southwestern Ontario borehole [81] and the Copenhagen Canyon section (Figure ??). However, the initial $^{87}Sr/^{86}Sr$ ratios in the Wuke section are similar to the published conodont apatite $^{87}Sr/^{86}Sr$ ratios in the late Sandbian and early Katian from the south China sections. It seems that the anomaly of high $^{87}Sr/^{86}Sr$ ratios in the late Katian might be caused by a restricted ocean, as is observed in the late Ediacaran West Gondwana basins [82] and the latest Ediacaran Sichuan Basins [83]. This condition of the restricted ocean in the Late Ordovician has also been observed by others, who suggest that moderate restricted conditions characterize the Yangtze shelf sea and likely influence the degree of water mass restriction [84,85]. However, we have re-analyzed the major and trace elements in the WK section to identify if our samples were deposited under a restricted basin [58]. The enrichments of Mo and U in sediments have widely been used to reconstruct the redox state of water columns, as well as the changes in open (seawater) versus enclosed (lake/lagoon) conditions [86]. The results show that a transition from suboxic to sulfidic water conditions could be observed from the early to late Katian, which is followed by suboxic water conditions in the Hirnantian (Figure 5C). Therefore, the majority of samples in our studied section were deposited in open marine settings, instead of restricted ocean. In addition, Co and Mn contents in marine sediments were effective proxies for distinguishing the restricted and upwelling settings due to the significant differences in the supply of these metals to the water column [72]. We found that the majority of samples have stable low Co×Mn values (<0.1), indicating open marine settings or upwelling settings (Figure 5D). As a result, we would like to highlight that the changes in carbonate $^{87}Sr/^{86}Sr$ ratios in the Wuke section still record the global trend of seawater $^{87}Sr/^{86}Sr$ curve. However, the reasons for the overall high values will be considered in future works.

Several similarities are found in the Late Ordovician $^{87}Sr/^{86}Sr$ curves between south China and other areas. First, both the $^{87}Sr/^{86}Sr$ curves in south China and southwestern Ontario show low Sr isotopes which is consistent with the overall ratios globally, likely resulting from enhanced continental basalt weathering, a sustainable increase in oceanic hydrothermal Sr flux, or a combination of both [87]. Second, the $^{87}Sr/^{86}Sr$ ratios become much more positive in the late Katian, reaching high Sr isotope ratios not attained until the Silurian found in previous studies [80]. Third, the prolonged high $^{87}Sr/^{86}Sr$ ratios in the Hirnantian have been observed in south China, southwestern Ontario, and the Copenhagen Canyon (Figure **??**). These radiogenic $^{87}Sr/^{86}Sr$ ratios in Glaciation were caused by the enhanced mechanical erosion driven by sea-level fall or the preferential weathering of biotite in cooling climates [79]. Therefore, we still believe that our Sr isotope record has global-scale implications for the changes in silicate weathering, hydrothermal activity, and the weathering of fresh volcanic rocks.

The seawater $^{87}Sr/^{86}Sr$ is controlled by both river influx from continents and hydrothermal influx [36]. A major drop in marine $^{87}Sr/^{86}Sr$ records across the middle–late Ordovician boundary (Darriwilian-Sandbian) was attributed to enhanced continental basalt weathering [88], subduction-related volcanism, or a sustainable increase in oceanic hydrothermal Sr flux [87]. The initial low $^{87}Sr/^{86}Sr$ ratios in the lower most Tiezufeike Formation are consistent with the hypotheses about basalt weathering and hydrothermal input; therefore, the influence of these factors should be taken into account. First, A-CN-K diagrams are utilized to determine the types of source rocks, with the results showing that all the samples are sourced from intermediate-felsic igneous rocks, as a result, the effect of changes in the types of source rocks on the $^{87}Sr/^{86}Sr$ ratios is minor. Second, the contribution of hydrothermal activity to the deposition of Late Ordovician shales in south China is mainly based on the positive Eu anomaly (Eu/Eu*), because a reduction of Eu^{3+} to Eu^{2+} would occur under the extreme reduction in hydrothermal fluids [89,90]. The results show that weak hydrothermal activity (occasionally slightly positive Eu anomaly) exists on the northern and southwestern margins of the Upper Yangtze Platform, situated in an extensional tectonic background [85]; thus, the possibility of decreasing hydrothermal Sr flux could be excluded. In this case, we just retain the spreading rate or the hydrothermal Sr-input as a steady condition when considering that the spreading rate or the hydrothermal Sr-input is unknown at present. Anyway, we would like conduct more modeling work about the influences of spreading rate or the global hydrothermal Sr-input on the Late Ordovician Sr isotopes in seawater. Third, sea-level-fall-induced restricted seawater conditions are unlikely to lead to the higher $^{87}Sr/^{86}Sr$ ratios because the geochemical characteristics in bulk rocks above have identified open ocean settings. Overall, the increase in $^{87}Sr/^{86}Sr$ ratios has been attributed to the increasing continental crust erosion during the Taconic orogeny or the sea-level drop related to the onset of Hirnantian Glaciation.

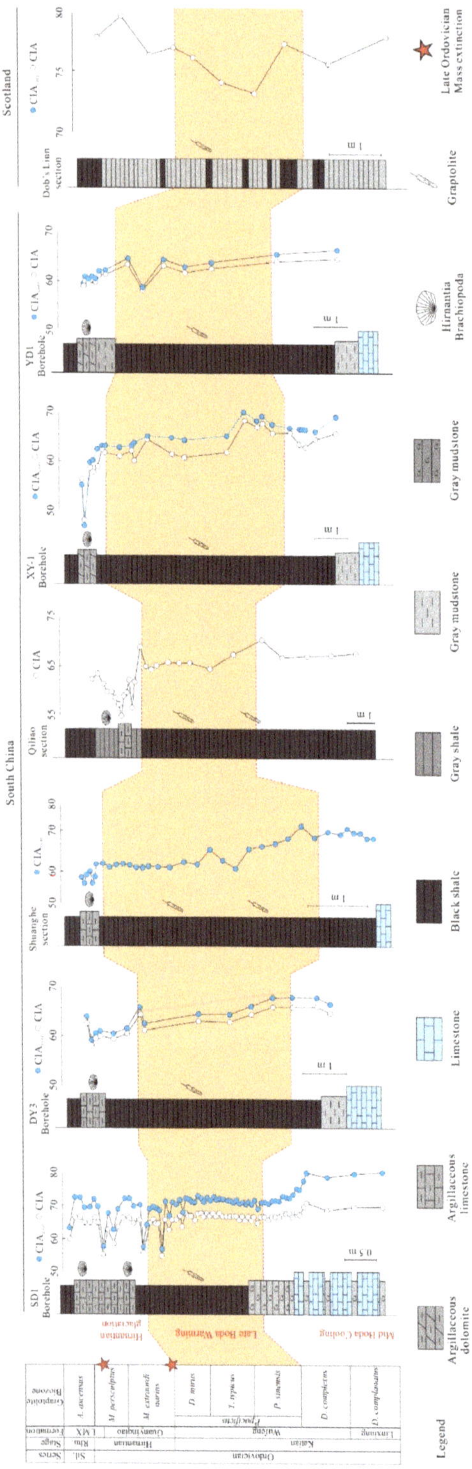

Figure 8. Correlation of CIA and CIA_{corr} in the SD1, DY3 (this study), SH, QL, XY-1, YD1(this study), and Dob's Linn [16,59]. The orange area represents the period of late-Boda warming climates. Sil.: Silurian, Rhu.: Rhuddanian, LMX: Longmaxi.

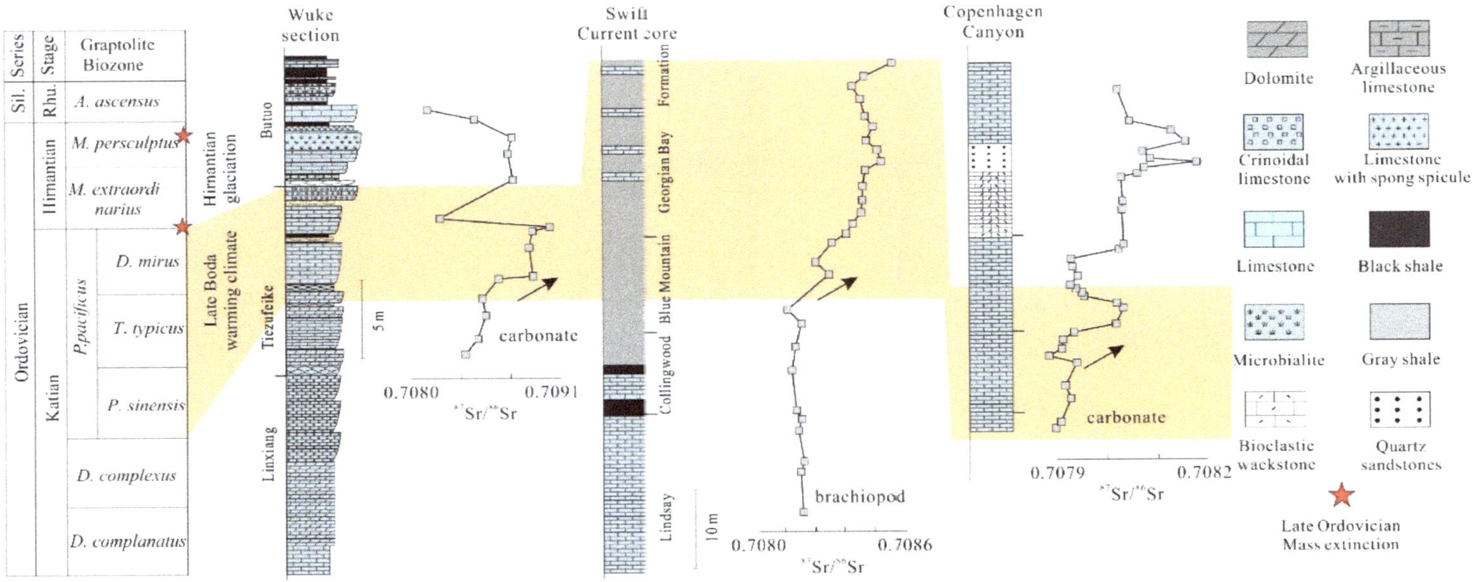

Figure 9. Stratigraphic $^{87}Sr/^{86}Sr$ of the Late Crdovician sediments in WK (this study), Swift Current core, Copenhagen Canyon sections [79,81]. The orange area represents the period of late-Boda warming climates. Sil.: Silurian, Rhu.: Rhuddanian, LMX: Longmaxi.

5.4. The Changes in Continental Weathering in the Late Ordovician

Traditional studies have argued that continental weathering is controlled by plant coverage, temperature, runoff, and tectonic processes [91]. First, given that the first non-vascular land plants were only just evolving and colonizing the continents in the mid–late Ordovician, it is probable that the weathering pattern and corresponding clay types were different and less abundant [92]. For example, illites are thought to dominate prior to terrestrialization by plants [92]. In fact, the clay minerals of the Late Ordovician rocks in both south China and Scotland were dominated by illite [27,89]. In addition, the occurrence of non-vascular land plants would be suggested to induce the enhanced formation of clay minerals caused by a decreased denudation rate and increased chemical weathering intensity, leading to the preservation of cations in clays and decreased consumption of atmosphere CO_2 [27]. However, both the changed types of clay and increased chemical weathering intensity were inconsistent with the observation in our geochemical records. As a result, the effect of missing plant coverage on the records of the weathering trends in this study might be minor.

Generally, the elevated temperature could enhance the feldspar dissolution, weathering rate, chemical weathering intensity (high CIA values), and further CO_2 consumption [4]. Hydrologic regulation (precipitation and runoff) has also widely been proposed to control the silicate weathering, with increased runoff causing a high weathering rate in low-relief areas [91]. Some studies further suggested that the runoff-induced high denudation rate would contribute to the high chemical weathering and low chemical weathering intensity [31]. Tectonic uplift can enhance the chemical weathering rate by accelerating the rate of denudation [22]. In this case, high erosion and a rapid supply of fresh minerals would result in a short time to reach weathering equilibrium, so the weathering rate, solute concentrations, and fluxes are high [91].

The weathering flux to the oceans should be maximized when thermodynamic equilibrium between the dissolving and precipitating minerals is approached. Furthermore, peak weathering and CO_2 drawdown occur with moderate erosion rates. In contrast, low and high erosion rates would decrease or reverse CO_2 sequestration. In a simplified model, the low denudation rate would result in a supply limited regime (with thick soils isolating the bedrock from climatic conditions), while the high denudation rate would cause the kinetically limited regime (with rocks being eroded before undergoing significant hydrolysis) [32].

A certain proportion of sedimentary rocks in paleo-catchments will necessarily exaggerate the numerical estimate of the CIA. Therefore, the proposed weathering trends using CIA values of Late Ordovician shales need be checked in the future. Anyway, the declining CIA values from the *D. complanatus* to *P. pacificus* biozones might suggest diminishing chemical weathering intensity (Figure **??**). However, oxygen isotope evidence indicates that a mid-Boda cooling climate happened within the *D. complanatus* and *D. complexus* biozones and a late-Boda warming climate happened in the *P. pacificus* biozone (Figure 10) [93]. Moreover, the CIA trend is contrary to the observation that increasingly radiogenic $^{87}Sr/^{86}Sr$ occur within the *P. pacificus* biozone and indicate an augmented weathering flux (Figure 10). Therefore, this study believes that the decrease in CIA values and chemical weathering intensity in late Katian should not just be controlled by temperature regulation. Instead, tectonic and hydrologic regulations likely account for the changes in continental weathering. During the mid-Boda cooling interval, the chemical weathering and denudation rates were much lower due to the low temperatures and precipitation/runoff. In this interval, the low denudation rate would result in more clay formation and thick soils isolating the bedrock from weathering, which is the supply limited regime (high W/D and chemical weathering intensity; Figure 11).

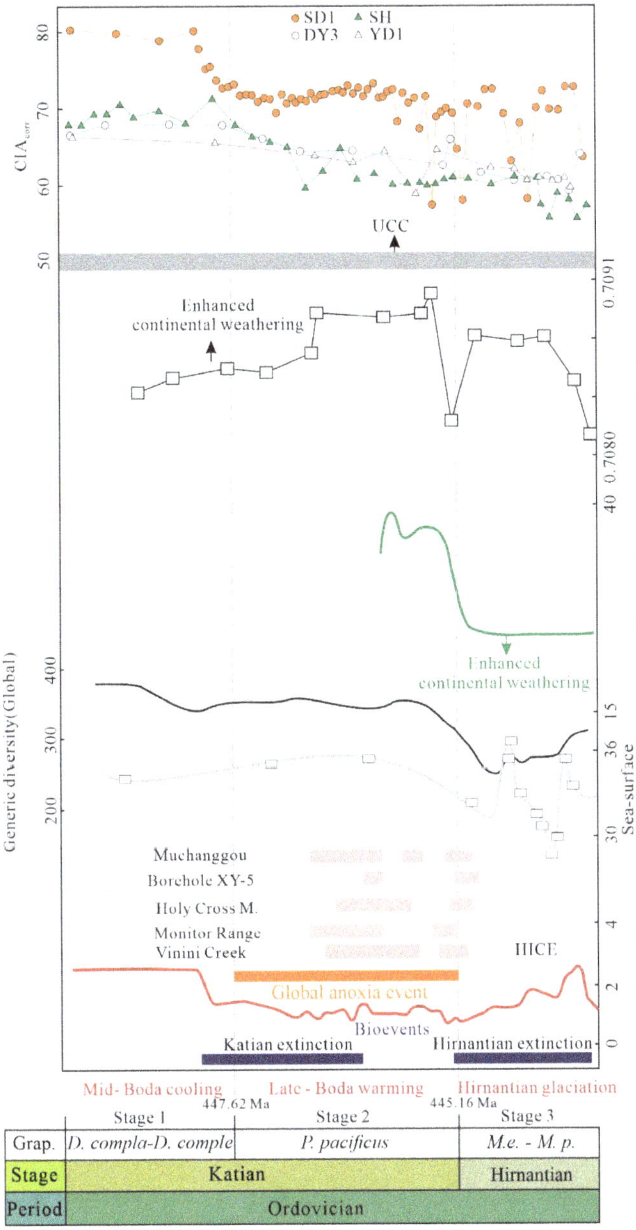

Figure 10. Trends in CIA_{corr}, $^{87}Sr/^{88}Sr$ (this study), and $\delta^{34}S$ in carbonate-associated sulfate (CAS). Generic diversity, Katian, and Hirnantian extinctions are based on [10]. Global anoxia event is based on [17]. The gray brand represents the CIA values for UCC (upper continental crust). Grap.: Graptolite Biozone, *D. compla.-D.comple.*: *Dicellograptus complanatus-Dicellograptus complexus*, *M.e.-M. p.*: *Metabolograptus extraodinarius-Metabolograptus persculptus*.

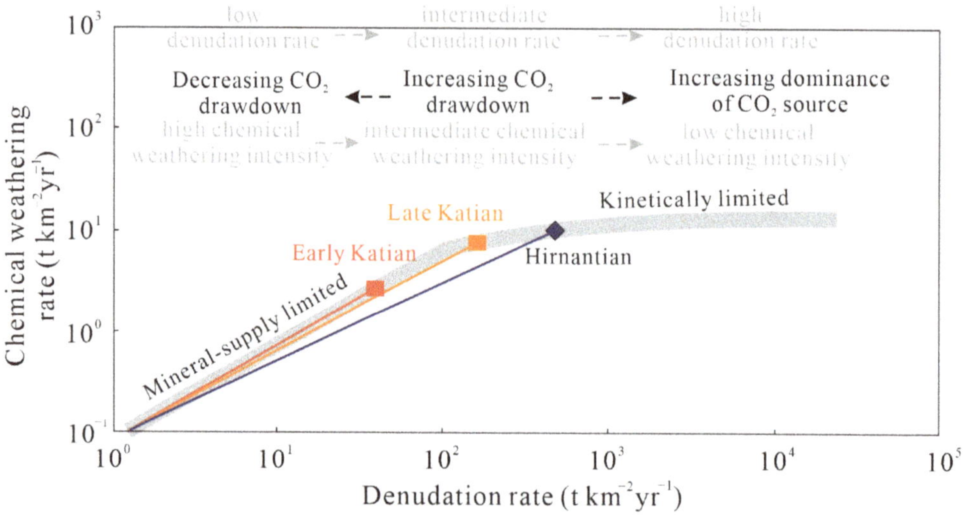

Figure 11. Relationship between denudation rate and chemical weathering rate [28,32]. The gray line shows a linear relation under mineral-supply limited regimes and a nonlinear relation in kinetically limited regimes. Peak weathering rate and CO_2 drawdown occurs with moderate erosion rate; in contrast, low and high erosion rates would decrease or even reverses CO_2 sequestration. The red square represents the early Kaitan, with low denudation rate and high chemical weathering intensity. The orange square represents the late Kaitan, with intermediate denudation rate and intermediate chemical weathering intensity. The blue square represents the Hirnantian with high denudation rate and low chemical weathering intensity.

A warming climate and orogenesis likely contributed to reducing chemical weathering intensity within the *P. pacificus* biozone. The warming climate is caused by global volcanism and the injection of CO_2 into the atmosphere (Figure 10; [94]), with the high temperatures in this interval leading to an increase in runoff of 6.8 to 2.3%/°C between high- and low-latitude rivers, respectively [95]. In this case, the rapid erosion of rocks and an exceptionally increased denudation rate would occur due to increased precipitation/runoff, as extensively reported in Early Silurian warming climates [17]. The orogeny in the late Katian has been reported globally, such as the Kwangsian Orogeny in south China, the Caledonian Orogen in Scotland, or the Taconic orogeny in the Appalachian margin of Laurentia, ensuring the continual exposure of fresh material [96–98], which tends to weather more rapidly. The intermediate denudation rate and enough fresh rocks would result in moderate erosion and intermediate weathering intensity (Figure 11).

Glaciation occurred within the *M. extraordinarius* and *M. persculptus* zones [93]. The weathering trends in Hirnantian Glaciation are subject to debate, with several perspectives reported. First, some studies propose that low temperatures and glacial ice cover could lower the chemical weathering rate, leading to a low chemical weathering intensity (low W/D ratio) [16]. Second, others suggest that glaciation enhances the denudation rate while reducing the chemical weathering intensity (low W/D ratio) [99]. However, the Yangtze Platform was located in subtropical regions during the Ordovician and hence was not subject to glacial weathering. Third, a cooling climate prolongs the continental residence time of water, allowing more clay formation, thereby reducing global weathering flux and promoting more incongruent weathering (intermediate W/D ratio) [27]. Given the high $^{87}Sr/^{86}Sr$ ratios and extremely low CIA values in Hirnantian Glaciation, this study suggests that the Hirnantian Glaciation was characterized by increased chemical weathering and denudation rates but low chemical weathering intensity (low W/D ratio). The amplified clastic fluxes from continents and increased continental material transported into the ocean are corroborated by low εNd(t) values and high TiO_2/Al_2O_3, Zr/Al_2O_3 ratios [100]. Increased erosion during low sea level in glaciation, the glacial grinding of unweathered rocks, or a combination of them, caused the kinetically limited regime, which contributed to the increased denudation rate. Furthermore, the high seawater $^{87}Sr/^{86}Sr$ ratios and low sulfate–sulfur isotopes could support the increased chemical weathering. The high $^{87}Sr/^{86}Sr$ ratios might be driven by glacial grinding and abrasion that produced fine-grained Rb-rich glacial till with minerals such as biotite [101], as well as the preferential weathering of biotite with radiogenic $^{87}Sr/^{86}Sr$ in glaciation [101]. A large negative sulfur excursion ($\delta^{34}S$ in carbonate-associated sulfate) occurred throughout the Hirnantian Glaciation, and increasing continental weathering (especially sulfides) was required based on the geochemical box modeling [102]. Overall, the increased denudation rate exceeded the increased chemical weathering rate, leading to the lowest chemical weathering intensity in glaciation (Figure 11).

5.5. Implications for the Changes in Global Climates and LOMEs

The Late Ordovician and Early Silurian transition is characterized by an expansion of anoxia in the deeper oceans which resulted in the deposition of organic-rich shales which later became oil- and gas-source rocks [16,41,103]. Both the orbital forcing and paleoenvironmental changes have been explained for the deposition of those source rocks [9,104]. In the late Katian, a warming event (Boda event) was restricted to the upper part of the Ka4 time slice [105] or the late Katian *P. pacificus* biozone [63]. Previous findings imply that volcanism-induced paleoenvironmental repercussions in late Katian had long-term effects on the global system rather than short-term impacts on paleoenvironments and paleoclimates [106]. Therefore, volcanism likely contributed to the increase in pCO_2, negative $\delta^{13}C$ excursion, and the late-Boda warming event [26]. In the *P. pacificus* interval, higher continental runoff is expected to have occurred, due to warming and a humid climate caused by intense volcanism. Furthermore, high runoff accelerates chemical weathering and denudation. The enhanced chemical weathering rate increased the consumption of atmospheric CO_2 and the nutrient influx into the ocean, leading to high primary productivity in surface waters [43,45]. The enhanced chemical weathering rate and weathering-induced high primary productivity contributed to the enrichment of organic matter, lower pCO_2, and lower global temperatures. Therefore, this study suggests that the weathering rate reaches maximum values for intermediate denudation rates and weathering intensity (Figure 11), which further contributes to CO_2 drawdown and cooling. Therefore, this study suggests that if intense volcanism can push the climate system out of balance (Boda warming), silicate weathering controlled by temperature and runoff would help to allow climatic recovery, even triggering glaciation [107]. We found that high chemical weathering and denudation rates, and the lowest chemical weathering intensity characterize the Hirnantian Glaciation. The high weathering rate seems to violate the fact that low temperatures and glacial cover characterized the Hirnantian. However, an increased chemical weathering rate in glaciations has also been proposed, with the factors being attributed to the enhanced erosion of silicate, and weathering of sulfides and carbonates [108]. In fact, the enhanced carbonate and sulfide weathering in the Hirnantian is supported by the high carbon isotopes and low sulfate-sulfur isotopes [102]. In this case, an increase in sulfide and carbonate weathering leads to a possible source of CO_2 followed by a longer sink of CO_2. This hypothesis is consistent with the observation of CO_2 release with the highest weathering and denudation rates (Figure 11). Overall, the highest denudation rate and lowest chemical weathering intensity in Hirnantian should result in the CO_2 release and termination of glaciation.

Traditional views suggested that the LOMEs were marked by two phases in the Late Ordovician (Figure 10). This study considered that the first phase corresponded to the start of glaciation and was within the boundary between *Paraothograptus pacificus* and *Metabolograptus extraodinarius* graptolite zones, and the second phase corresponded to the termination of glaciation and happened in the latest Hirnantian and earliest Rhuddanian periods (Figure 10; [4]). However, new studies found that the Late Ordovician extinctions happened within the middle–late Katian (named the Katian extinction) and also suggested a rapid global expansion of OMZs throughout the late Katian, based on trace elemental concentrations and Tl and U isotopes (Figure 10; [102]). This study suggests that weathering-induced high primary productivity and seawater anoxia likely played significant roles in controlling the LOMEs, given the coupling of high weathering rates and expanded anoxic seawater in the late Katian (Figures 10 and 12). Previous studies suggested that oxygen level rather than temperature was the main factor controlling the Late Ordovician extinctions for as long as 35 million years [109]. The Late Ordovician crisis includes three rapid declines in species richness, with the first decline occurring in the late Katian, the second decline occurring in the earliest Hirnantian, and the third decline occurring in late Hirnantian (Figure 10; [110]). It is believed that weathering-induced seawater anoxia could explain the crisis in the late Katian, while the development of more intense euxinia and volcanism throughout the water column likely drove the second and third extinction in the Hirnantian (Figure 10; [14]).

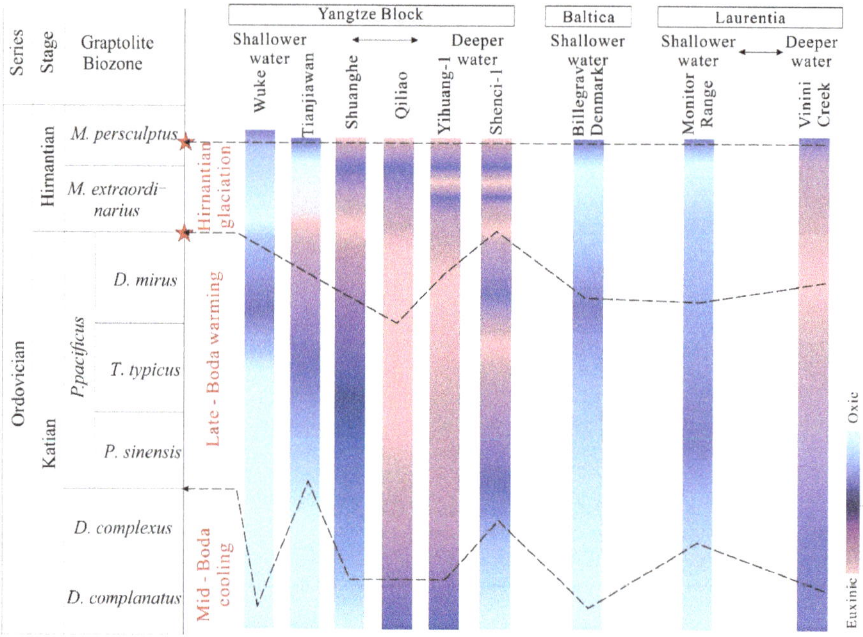

Figure 12. Scheme illustrating redox dynamics during the Ordovician and Silurian transition; the data are from [14,16,59].

6. Conclusions

The weathering trends during the Late Ordovician (late Katian-Hirnantian) have been evaluated by the chemical index of alteration (CIA) in shales and $^{87}Sr/^{86}Sr$ in carbonate leachates. The intervals within the *D. complanatus* and *D. complexus* biozones were marked by a low weathering rate (low $^{87}Sr/^{86}Sr$ ratios), and by high chemical weathering intensity (high CIA values). Conversely, the intervals within the *P. pacificus* biozone showed a high weathering rate (high $^{87}Sr/^{86}Sr$ ratios) and intermediate chemical weathering intensity (intermediate CIA values). This increase in weathering rate and reduction in chemical weathering intensity should be attributed to global orogenesis and enhanced precipitation/runoff under the warming climate. The Hirnantian Glaciation was characterized by high $^{87}Sr/^{86}Sr$ ratios and extremely low CIA values, which might be related to the exposure of continents during low sea level and the glacial grinding of unweathered rocks, with high $^{87}Sr/^{86}Sr$ ratios and low CIA values.

The changes in weathering had significant influences on the climatic changes and mass extinctions in the Late Ordovician. The intermediate weathering rate and intensity contributed to the CO_2 drawdown in the *P. pacificus* biozone, corresponding to the initiation of cooling and further glaciation. Meanwhile, the enhanced weathering-induced high primary productivity resulted in the expansion of anoxic seawater in the Katian, which further caused Katian extinction. In the Hirnantian Glaciation, the highest denudation rate and lowest chemical weathering intensity likely resulted in enhanced CO_2 release and contributed to the termination of glaciation.

Supplementary Materials: The following supporting information can be downloaded at: https://www.mdpi.com/article/10.3390/jmse12122237/s1, Table S1: Major and trace element concentrations in the DY3 borehore; Table S2: Major elements concentrations in the YD1 borehore; Table S3: Elemental concentrations and $^{87}Sr/^{86}Sr$ analysis in the Wuke section.

Author Contributions: Methodology, D.Y.; Formal analysis, X.Y.; Writing—original draft, P.T. All authors have read and agreed to the published version of the manuscript.

Funding: The project was supported by the Open Funds for Hubei Key Laboratory of Marine Geological Resources, China University of Geosciences, No. MGR202407, the National Natural Science Foundation of China (42402123, 41690131), and Beijing Nova Program (Z211100002121136).

Institutional Review Board Statement: Not applicable.

Informed Consent Statement: Not applicable.

Data Availability Statement: The data presented in this study are available on request from the corresponding author.

Conflicts of Interest: The authors declare no conflicts of interest.

References

1. Melchin, M.J.; Mitchell, C.E.; Holmden, C.; Štorch, P. Environmental changes in the Late Ordovician-early Silurian: Review and new insights from black shales and nitrogen isotopes. *Geol. Soc. Am. Bull.* **2013**, *125*, 1635–1670. [CrossRef]
2. Harper, D.A.T.; Hammarlund, E.U.; Rasmussen, C.M.Ø. End Ordovician extinctions: A coincidence of causes. *Gondwana Res.* **2014**, *25*, 1294–1307. [CrossRef]
3. Jin, C.S.; Liao, Z.W.; Lash, G.G. High-frequency redox variation across the Ordovician–Silurian transition, South China. *Palaeogeogr. Palaeoclimatol. Palaeoecol.* **2021**, *566*, 110218. [CrossRef]
4. Deng, K.; Yang, S.Y.; Guo, Y.L. A global temperature control of silicate weathering intensity. *Nat. Commun.* **2022**, *13*, 1781. [CrossRef] [PubMed]
5. Kump, L.R.; Arthur, M.A.; Patzkowsky, M.E.; Gibbs, M.T.; Pinkus, D.S.; Sheehan, P.M. A weathering hypothesis for glaciation at high atmospheric pCO_2 during the Late Ordovician Interpreting carbon—Isotope excursions: Carbonates and organic matter. *Palaeogeogr. Palaeoclimatol. Palaeoecol.* **1999**, *16152*, 18173–19887.
6. Finlay, A.J.; Selby, D.; Gröcke, D.R. Tracking the Hirnantian glaciation using Os isotopes. *Earth Planet. Sci. Lett.* **2010**, *293*, 339–348. [CrossRef]
7. Lenton, T.; Michael, C.; Martin, J.; Pires, N.; Dolan, L. First plants cooled the Ordovician. *Nat. Geosci.* **2012**, *5*, 86–89. [CrossRef]
8. Swanson-Hysell, N.L.; Macdonald, F.A. Tropical weathering of the Taconic orogeny as a driver for Ordovician cooling. *Geology* **2017**, *45*, 719–722. [CrossRef]
9. Sproson, A.D.; von Strandmann, P.A.E.; Selby, D.; Jarochowska, E.; Frýda, J.; Hladil, J.; Loydell, D.K.; Slavík, L.; Calner, M.; Maier, G.; et al. Osmium and lithium isotope evidence for weathering feedbacks linked to orbitally paced organic carbon burial and Silurian glaciations. *Earth Planet. Sci. Lett.* **2022**, *577*, 117260. [CrossRef]
10. Deng, Y.; Fan, J.X.; Zhang, S.H.; Fang, X.; Chen, Z.Y.; Shi, T.K.; Wang, H.W.; Wang, X.B.; Yang, J.; Hou, X.D. Timing and patterns of the Great Ordovician Biodiversification Event and Late Ordovician mass extinction: Perspectives from South China. *Earth Sci. Rev.* **2021**, *220*, 103743. [CrossRef]
11. Harper, D.A.T. Late Ordovician Mass Extinction: Earth, fire and ice. *Natl. Sci. Rev.* **2024**, *11*, nwad319. [CrossRef] [PubMed]
12. Sheehan, P.M. The Late Ordovician mass extinction. *Annu. Rev. Earth Planet. Sci.* **2001**, *29*, 331–364. [CrossRef]
13. Yan, D.T.; Chen, D.Z.; Wang, Q.C.; Wang, J.G. Large-scale climatic fluctuations in the latest Ordovician on the Yangtze block, South China. *Geology* **2010**, *38*, 599–602. [CrossRef]

14. Hammarlund, E.U.; Dahl, T.W.; Harper, D.A.T.; Bond, D.P.G.; Nielsen, A.T.; Bjerrum, C.J.; Schovsbo, N.H.; Schönlaub, H.P.; Zalasiewicz, J.A.; Canfield, D.E. A sulfidic driver for the end-Ordovician mass extinction. *Earth Planet. Sci. Lett.* **2012**, *331*, 128–139. [CrossRef]
15. Jones, D.; Martini, A.; Fike, D. A volcanic trigger for the Late Ordovician mass extinction? Mercury data from south China and Laurentia. *Geology* **2017**, *45*, 631–634. [CrossRef]
16. Zou, C.N.; Qiu, Z.; Poulton, S.W.; Dong, D.Z.; Wang, H.Y.; Chen, D.Z.; Lu, B.; Shi, Z.S.; Tao, H.F. Ocean euxinia and climate change "double whammy" drove the Late Ordovician mass extinction. *Geology* **2018**, *46*, 535–538. [CrossRef]
17. Liu, M.; Chen, D.Z.; Jiang, L.; Stockey, R.G.; Aseal, D.; Zhang, B.; Liu, K.; Yang, X.R.; Yan, D.T.; Planavsky, N.J. Oceanic anoxia and extinction in the latest Ordovician. *Earth Planet. Sci. Lett.* **2022**, *588*, 117553. [CrossRef]
18. Wang, P.; Du, Y.S.; Yu, W.C.; Algeo, T.J.; Zhou, Q.; Xu, Y.; Qi, L.; Yuan, L.J.; Pan, W. The chemical index of alteration (CIA) as a proxy for climate change during glacial-interglacial transitions in Earth history. *Earth Sci. Rev.* **2020**, *201*, 103032. [CrossRef]
19. Walker, J.C.G.; Hays, P.B.; Kasting, J.F. A negative feedback mechanism for the long-term stabilization of earth's surface temperature. *Geophys. Res. Lett.* **1981**, *86*, 9776–9782. [CrossRef]
20. Berner, R.A. Weathering, plants, and the long-term carbon cycle. *Geochim. Et Cosmochim. Acta* **1992**, *56*, 3225–3231. [CrossRef]
21. Föllmi, K. The phosphorus cycle, phosphogenesis and marine phosphate-rich deposits. *Earth Sci. Rev.* **1996**, *40*, 55–124. [CrossRef]
22. Raymo, M.E.; Ruddiman, W.F. Tectonic forcing of late Cenozoic climate. *Nature* **1992**, *359*, 117–122. [CrossRef]
23. Caves, J.K.; Ibarra, D.E.; Von Blanckenburg, F. Neogene cooling driven by land surface reactivity rather than increased weathering fluxes. *Nature* **2019**, *571*, 99–102. [CrossRef] [PubMed]
24. Wang, Y.Y.; Liang, K.; Xiao, Y.L.; Chen, B.; Shan, E.H.; Yang, T.Y.; Zhang, M.; Sun, H.; Gu, H.O.; Tong, F.T.; et al. Carbonate lithium isotope systematics indicate cooling triggered mass extinction during the Frasnian-Famennian transition. *Glob. Planet. Chang.* **2023**, *230*, 104284. [CrossRef]
25. Buggisch, W.; Joachimski, M.M.; Lehnert, O.; Bergstrom, M.; Repetski, E.; Webers, F. Did intense volcanism trigger the first Late Ordovician icehouse? *Geology* **2010**, *38*, 327–330. [CrossRef]
26. Lefebvre, V.; Servais, T.; Francois, L.; Averbuch, O. Did a Katian large igneous province trigger the Late Ordovician glaciation? A hypothesis tested with a carbon cycle model. *Palaeogeogr. Palaeoclimatol. Palaeoecol.* **2010**, *296*, 310–319. [CrossRef]
27. Pogge Von Strandmann, P.A.E.; Desrochers, A.; Murphy, M.J.; Finlay, A.J.; Selby, D.; Lenton, T.M. Global climate stabilisation by chemical weathering during the Hirnantian glaciation. *Geochem. Perspect* **2017**, *3*, 230–237. [CrossRef]
28. West, A.J.; Galy, A.; Bickle, M. Tectonic and climatic controls on silicate weathering. *Earth Planet. Sci. Lett.* **2005**, *235*, 211–228. [CrossRef]
29. Shen, J.; Yin, R.S.; Zhang, S.; Algeo, T.J.; Bottjer, D.J.; Yu, J.X.; Xu, G.Z.; Penman, D.; Wang, Y.D.; Li, L.Q.; et al. Intensified continental chemical weathering and carbon-cycle perturbations linked to volcanism during the Triassic–Jurassic transition. *Nat. Commun.* **2022**, *13*, 299. [CrossRef]
30. Yang, J.H.; Cawood, P.A.; Condon, D.J.; Liu, J.Z.; Deng, X.S.; Wang, J.F.; Du, Y.S.; Yuan, D.X. Anomalous weathering trends indicate accelerated erosion of tropical basaltic landscapes during the Permo-Triassic warming. *Earth Planet. Sci. Lett.* **2022**, *577*, 117256. [CrossRef]
31. Pogge von Strandmann, P.A.E.; Jones, M.T.; West, A.J.; Murphy, M.J.; Stokke, E.W.; Tarbuck, G.; Wilson, D.J.; Pearce, C.R.; Schmidt, D.N. Lithium isotope evidence for enhanced weathering and erosion during the Paleocene-Eocene Thermal Maximum. *Sci. Adv.* **2021**, *7*, eabh4224. [CrossRef] [PubMed]
32. Bufe, A.; Rugenstein, J.K.C.; Hovius, N. CO_2 drawdown from weathering is maximized at moderate erosion rates. *Science* **2024**, *383*, 1075–1080. [CrossRef] [PubMed]
33. Veizer, J. Strontium isotopes in seawater through time. *Annu. Rev. Earth Planet. Sci.* **1989**, *17*, 141–167. [CrossRef]
34. Dennison, R.E.; Koepnick, R.B.; Burke, W.H.; Hetherington, E.A. Construction of the Cambrian and Ordovician seawater $^{87}Sr/^{86}Sr$ curve. *Chem. Geol.* **1998**, *152*, 325–340. [CrossRef]
35. McArthur, J.M.; Howarth, R.J.; Bailey, T.R. Strontium isotope stratigraphy: LOWESS version 3: Best fit to the marine isotope curve for 0–509 Ma and accompanying look-up table for deriving numerical age. *J. Geol.* **2001**, *109*, 155–170. [CrossRef]
36. Saltzman, M.R.; Edwards, C.T.; Leslie, S.A.; Dwyer, G.S.; Bauer, J.A.; Repetski, J.; Harris, A.; Bergström, S.M. Calibration of a conodont apatite-based Ordovician $^{87}Sr/^{86}Sr$ curve to biostratigraphy and geochronology: Implications for stratigraphic resolution. *Geol. Soc. Am. Bull.* **2014**, *126*, 1551–1568. [CrossRef]
37. Edwards, C.T.; Saltzma, M.R.; Leslie, S.A.; Bergstrm, S.M.; Sedlacek, A.R.C.; Howard, A.; Bauer, J.A.; Sweet, W.C.; Young, S.A. Strontium isotope ($^{87}Sr/^{86}Sr$) stratigraphy of Ordovician bulk carbonate: Implications for preservation of primary seawater values. *Geol. Soc. Am. Bull.* **2015**, *127*, 1275–1289. [CrossRef]
38. Wang, W.Q.; Garbelli, C.; Zheng, Q.F.; Chen, J.; Liu, X.C.; Wang, W.; Shen, S.Z. Permian $^{87}Sr/^{86}Sr$ chemostratigraphy from carbonate sequences in South China. *Palaeogeogr. Palaeoclimatol. Palaeoecol.* **2018**, *500*, 84–94. [CrossRef]
39. Harnois, L. The CIW index: A new chemical index of weathering. *Sediment. Geol.* **1988**, *55*, 319–322. [CrossRef]
40. Price, J.R.; Velbel, M.A. Chemical weathering indices applied to weathering profiles developed on heterogeneous felsic metamorphic parent rocks. *Chem. Geol.* **2003**, *202*, 397–416. [CrossRef]
41. Chen, W.Z.; Tian, J.C.; Lin, X.B.; Liang, Q.S.; Wang, X.; Yi, D.X.; Li, Y.Y. Climate fluctuations during the Ordovician-Silurian transition period in South China: Implications for paleoenvironmental evolution and organic matter enrichment. *Palaeogeogr. Palaeoclimatol. Palaeoecol.* **2023**, *613*, 111411. [CrossRef]

42. Zhang, G.; Guo, A.; Wang, Y.J.; Li, S.D.; Dong, Y.P.; Liu, S.F.; He, D.F.; Cheng, S.Y.; Lu, R.K.; Yao, A.P. Tectonics of South China continent and its implications. *Sci. China Earth Sci.* **2013**, *56*, 1804–1828. [CrossRef]
43. Lu, Y.B.; Hao, F.; Yan, D.T.; Lu, Y.C. Volcanism-induced late Boda warming in the Late Ordovician: Evidence from the Upper Yangtze Platform, South China. *Palaeogeogr. Palaeoclimatol. Palaeoecol.* **2021**, *578*, 110579. [CrossRef]
44. Chen, X. Influencece of the Late Ordovician glaciation on the basin configuration of the Yangtze Platform in China. *Lethaia* **1984**, *17*, 51–59.
45. Chen, C.; Mu, C.; Zhou, K.K.; Liang, W.; Ge, X.Y.; Wang, X.P.; Wang, Q.Y.; Zheng, B.S. The geochemical characteristics and factors controlling the organic matter accumulation of the Late Ordovician-Early Silurian black shale in the Upper Yangtze Basin, South China. *Mar. Pet. Geol.* **2016**, *76*, 159–175. [CrossRef]
46. Chen, X.; Rong, J.Y.; Fan, J.X.; Zhan, R.B.; Mitchell, C.E.; Harper, D.A.T.; Melchin, M.J.; Peng, P.A.; Finney, S.C.; Wang, X.F. The Global boundary Stratotype Section and Point (GSSP) for the base of the Hirnantian Stage (the uppermost of the Ordovician System). *Epis. J. Int. Geosci.* **2006**, *29*, 183–196. [CrossRef]
47. Gradstein, F.M.; Ogg, J.G.; Schmitz, M.; Ogg, G. (Eds.) *The Geologic Time Scale 2012*; Elsevier: Amsterdam, The Netherlands, 2012; Volume 2.
48. Ge, X.; Chen, D.; Zhang, G.; Huang, T.; Liu, M.; El-Shafeiy, M. Marine redox evolution and organic accumulation in an intrashelf basin, NE Sichuan Basin during the Late Permian. *Mar. Pet. Geol.* **2022**, *140*, 105633. [CrossRef]
49. Nesbitt, H.W.; Young, G.M. Early Proterozoic climates and plate motions inferred from major element chemistry of lutites. *Nature* **1982**, *299*, 715–717. [CrossRef]
50. McLennan, S.M. Weathering and global denudation. *J. Geol.* **1993**, *101*, 295–303. [CrossRef]
51. Fedo, C.M.; Nesbitt, H.W.; Young, G.M. Unraveling the effects of potassium metasomatism in sedimentary rocks and paleosols, with implications for paleoweathering conditions and provenance. *Geology* **1995**, *23*, 921–924. [CrossRef]
52. Cox, R.; Low, D.R.; Cullers, R.L. The influence of sediment recycling and basement composition on evolution of mudrock chemistry in the southwestern United States. *Geochim. Cosmochim. Acta* **1995**, *59*, 2919–2940. [CrossRef]
53. Yang, X.; Yan, D.; Chen, D.; Liu, M.; She, X.; Zhang, J.; Wei, X.; Lu, Z. Spatial variation of carbon isotopic compositions of carbonate and organic matter from the Late Ordovician sedimentary succession in the Yangtze Platform, South China: Implications for sea—Level eustacy and shoaling of marine chemocline. *J. Asian Earth Sci.* **2020**, *202*, 104540. [CrossRef]
54. Xue, S.; Wang, Q.; Deng, J.; Wang, Y.; Peng, T. Mechanism of organic matter assimilation and its role in sulfde saturation of oxidized magmatic ore-forming system: Insights from C-S-Sr-Nd isotopes of the Tulaergen deposit in NW China. *Miner. Depos.* **2022**, *57*, 1123–1141. [CrossRef]
55. Zhang, W.; Hu, Z.C. Estimation of isotopic reference values for pure materials and geological reference materials. *At. Spectrosc.* **2020**, *41*, 93–102. [CrossRef]
56. Van de Kamp, P.C. Potassium distribution and metasomatism in pelites and schists: How and When, relation tpost-depositionalal events. *J. Sediment Res.* **2016**, *86*, 683–711. [CrossRef]
57. Cullers, R.L.; Podkovyrov, V.N. Geochemistry of the Mesoproterozoic Lakhanda shales in southeastern Yakutia, Russia: Implications for mineralogical and provenance control and recycling. *Precambrian Res.* **2020**, *104*, 77–93. [CrossRef]
58. Yang, X.R.; Yan, D.T.; Chen, D.Z.; Liu, M.; She, X.H.; Zhang, B.; Zhang, L.W.; Zhang, J.F. Spatiotemporal variations of sedimentary carbon and nitrogen isotopic compositions in the Yangtze Shelf Sea across the Ordovician-Silurian boundary. *Palaeogeogr. Palaeoclimatol. Palaeoecol.* **2021**, *567*, 110257. [CrossRef]
59. Chen, C. *Research on Paleoceanography, Paleoclimate and Formation Mechanism of Source Rock During Geologic Transition Period from Late Ordovician to Early Silurian in Southern Sichuan Province—Northern Guizhou Province, South China*; China University of Geosciences: Wuhan, China, 2018.
60. Ahm, A.S.; Bjerrum, C.J.; Hammarlund, E.U. Disentangling the record of diagenesis, local redox conditions, and global seawater chemistry during the latest Ordovician glaciation. *Earth Planet. Sci. Lett.* **2017**, *459*, 145–156. [CrossRef]
61. Myrow, P.M.; Fike, D.A.; Malmskog, E.; Leslie, S.A.; Zhang, T.R.; Singh, B.P.; Chaubey, R.S.; Prasad, S.K. Ordovician–Silurian boundary strata of the Indian Himalaya: Record of the latest Ordovician Boda event. *Geol. Soc. Am. Bull.* **2019**, *131*, 881–898. [CrossRef]
62. Finnegan, S.; Bergmann, K.; Eiler, J.M.; Jones, D.S.; Fike, D.A.; Eisenman, I.; Hughes, N.C.; Tripati, A.K.; Fischer, W.W. The magnitude and duration of Late Ordovician-Early Silurian glaciation. *Science* **2011**, *331*, 903–906. [CrossRef]
63. Armstrong, H.A.; Baldini, J.U.L.; Challands, T.J.; Grocke, D.R.; Owen, A.W. Response of the Intertropical Convergence Zone to Southern Hemisphere cooling during Upper Ordovician glaciation. *Palaeogeogr. Palaeoclimatol. Palaeoecol.* **2009**, *284*, 227–236. [CrossRef]
64. Wang, Y.; Tan, J.Q.; Wang, W.H. The influence of Late Ordovician volcanism on the marine environment based on high-resolution mercury data from South China. *Geol. Soc. Am. Bull.* **2023**, *135*, 787–798. [CrossRef]
65. Basu, H.; Dandele, P.S.; Kumar, K.R.; Achar, K.K.; Umamaheswar, K. Geochemistry of black shales from the Mesoproterozoic Srisailam Formation, Cuddapah basin, India: Implications for provenance, palaeoweathering, tectonics, and timing of Columbia breakup. *Chem. Erde.* **2017**, *77*, 596–613. [CrossRef]
66. Bayon, G.; Bindeman, I.N.N.; Trinquier, A.; Retallack, G.J.; Bekker, A. Long-term evolution of terrestrial weathering and its link to Earth's oxygenation. *Earth Planet. Sci. Lett.* **2022**, *584*, 117490. [CrossRef]

67. Perri, F.; Cavalcante, F.; Martín-Martín, M.; Sánchez-Navas, A.; Alcalá, F.J. Cenozoic tectono-sedimentary evolution of the external rif chain (Morocco) derived from mineralogical and geochemical analysis of mudrocks. *Mar. Pet. Geol.* **2024**, *170*, 107124. [CrossRef]
68. Bayon, G.; Toucanne, S.; Skonieczny, C.; Andre, L.; Bermell, S.; Cheron, S.; Dennielou, B.; Etoubleau, J.; Freslon, N.; Gauchery, T.; et al. Rare earth elements and neodymium isotopes in world river sediments revisited. *Geochim. Cosmochim. Acta* **2015**, *170*, 17–38. [CrossRef]
69. Bhatia, M.R.; Crook, K.A. Trace element characteristics of graywackes and tectonic setting discrimination of sedimentary basins. *Contrib. Mineral. Petrol* **1986**, *92*, 181–193. [CrossRef]
70. Algeo, T.J.; Tribovillard, N. Environmental analysis of paleoceanographic systems based on molybdenum–uranium covariation. *Chem. Geol.* **2009**, *268*, 211–225. [CrossRef]
71. Tribovillard, N.; Algeo, T.J.; Baudin, F.; Riboulleau, A. Analysis of marine environmental conditions based on molybdenum–uranium covariation—Applications to Mesozoic paleoceanography. *Chem. Geol.* **2012**, *324–325*, 46–58. [CrossRef]
72. Sweere, T.; van den Boorn, S.; Dickson, A.J.; Reichart, G.-J. Definition of new trace metal proxies for the controls on organic matter enrichment in marine sediments based on Mn, Co, Mo and Cd concentrations. *Chem. Geol.* **2016**, *441*, 235–245. [CrossRef]
73. Sawaki, Y.; Ohno, T.; Tahata, M.; Komiya, T.; Hirata, T.; Maruyama, S.; Windley, B.F.; Han, J.; Shu, D.G.; Li, Y. The Ediacaran radiogenic Sr isotope excursion in the Doushantuo Formation in the Three Gorges area, South China. *Precambrian Res.* **2010**, *176*, 46–64. [CrossRef]
74. Jacobsen, S.B.; Kaufman, A.J. The Sr, C and O isotopic evolution of Neoproterozoic seawater. *Chem. Geol.* **1999**, *161*, 37–57. [CrossRef]
75. Banner, J.L.; Hanson, G.N. Calculation of simultaneous isotopic and trace element variations during water-rock interaction with applications to carbonate diagenesis. *Geochim. Cosmochim. Acta* **1990**, *54*, 3123–3137. [CrossRef]
76. Liu, C.; Wang, Z.; Raub, T.D.; Macdonald, F.A.; Evans, D.A.D. Neoproterozoic cap dolostone deposition in stratified glacial meltwater plume. *Earth Planet. Sci. Lett.* **2014**, *404*, 22–32. [CrossRef]
77. He, L.; Wang, Y.P.; Chen, D.F. Geochemical features of sedimentary environment and paleoclimate during Late Ordovician to Early Silurian in southern Sichuan Basin. *Geochemica* **2019**, *48*, 555–566.
78. Zhang, H. Analysis of Paleoclimate and Provenance Characteristics of Black Shale in Wufeng-Longmaxi Formation in Xianfeng Area, Western Hubei Province. *Coal Geol. Chin.* **2022**, *34*, 22–33, (In Chinese with English Abstract).
79. Hu, D.P.; Zhang, X.L.; Zhou, L.; Finney, S.C.; Liu, Y.S.; Shen, D.; Shen, M.G.; Huang, W.; Shen, Y.N. $^{87}Sr/^{86}Sr$ evidence from the epeiric Martin Ridge Basin for enhanced carbonate weathering during the Hirnantian. *Sci. Rep.* **2017**, *7*, 11348. [CrossRef]
80. Young, S.A.; Saltzman, M.R.; Foland, K.A.; Linder, J.S.; Kump, L.R. A major drop in seawater $^{87}Sr/^{86}Sr$ during the Middle Ordovician (Darriwilian): Links to volcanism and climate. *Geology* **2009**, *37*, 951–954. [CrossRef]
81. Hannigan, R.; Brookfield, M.E.; Basu, A.R. A detailed $^{87}Sr/^{86}Sr$ isotopic curve for the mid-Cincinnatian (Upper Katian-Lower Hirnantian, Upper Ordovician), NE North American Shelf (Ontario, Canada) transition to the Hirnantian glaciation. *Chem. Geol.* **2010**, *277*, 336–344. [CrossRef]
82. Guacaneme, C.; Babinski, M.; Bedoya-Rueda, C.; Paula-Santos, G.M.; Caetano-Filho, S.; Kuchenbecker, M.; Reis, H.L.S.; Trindade, R.I.F. Tectonically-induced strontium isotope changes in ancient restricted seas: The case of the Ediacaran-Cambrian Bambuí foreland basin system, east Brazil. *Gondwana Res.* **2021**, *93*, 275–290. [CrossRef]
83. Zhang, X.; Zhou, G.; Zhang, P.; He, Y.; Wei, Z.; Wang, G.; Zhang, T.; He, W.; Ma, H.; Zhu, C.; et al. Strontium isotope and element constraints on the paleoenvironment of the latest Ediacaran in the Sichuan Basin, southeastern Tibetan Plateau. *Front. Earth Sci.* **2022**, *10*, 865709. [CrossRef]
84. Lu, Y.B.; Jiang, S.; Lu, Y.C.; Xu, S.; Shu, Y.; Wang, Y.X. Productivity or preservation? The factors controlling the organic matter accumulation in the late Katian through Hirnantian Wufeng organic-rich shale, South China. *Mar. Pet. Geol.* **2019**, *109*, 22–35. [CrossRef]
85. Qiu, Z.; Li, Y.F.; Xiong, W.; Fan, T.L.; Zhao, Q.; Zhang, Q.; Wang, Y.M.; Liu, W.; Liang, F.; Zhang, J.Q.; et al. Revisiting paleoenvironmental changes on the Upper Yangtze Blockduring the Ordovician-Silurian transition: New insights from elemental geochemistry. *Sediment. Geol.* **2023**, *450*, 106377. [CrossRef]
86. Abdi, Z.; Rimmer, S.M.; Rowe, H.D.; Nordeng, S. Assessment of bi-metal paleo-redox proxies in the Bakken Formation black shales, Williston Basin, North Dakota, USA. *Chem. Geol.* **2024**, *670*, 122439. [CrossRef]
87. Avila, T.D.; Saltzman, M.R.; Adiatma, Y.D.; Joachimski, M.M.; Griffith, E.M.; Olesik, J.W. Role of seafloor production versus continental basalt weathering in Middle to Late Ordovician seawater $^{87}Sr/^{86}Sr$ and climate. *Earth Planet. Sci. Lett.* **2022**, *593*, 117641. [CrossRef]
88. Nardin, E.; Godderis, Y.; Donnadieu, Y.; Le Hir, G.; Blakey, R.C.; Puceat, E.; Aretz, M. Modeling the early Paleozoic long-term climatic trend. *Geol. Soc. Am. Bull.* **2011**, *123*, 1181–1192. [CrossRef]
89. Ge, X.Y.; Mou, C.L.; Wang, C.S.; Men, X.; Chen, C.; Hou, Q. Mineralogical and geochemical characteristics of K-bentonites from the Late Ordovician to the Early Silurian in South China and their geological significance. *Geol. J.* **2019**, *54*, 514–528. [CrossRef]
90. Olivarez, A.M.; Owen, R.M. The europium anomaly of seawater—Implications for fluvial versus hydrothermal REE inputs to the oceans. *Chem. Geol.* **1991**, *92*, 317–328.
91. Maher, K.; Chamberlain, C.P. Hydrologic regulation of chemical weathering and the geologic carbon cycle. *Science* **2014**, *343*, 1502–1504. [CrossRef]

92. Hazen, R.M.; Sverjensky, D.A.; Azzolini, D.; Bish, D.L.; Elmore, S.C.; Hinnov, L.; Milliken, R.E. Clay mineral evolution. *Am. Mineral.* **2013**, *98*, 2007–2029. [CrossRef]
93. Trotter, J.A.; Williams, I.S.; Barnes, C.R.; Männik, P.; Simpson, A. New conodont $\delta^{18}O$ records of Silurian climate change: Implications for environmental and biological events. *Palaeogeogr. Palaeoclimatol. Palaeoecol.* **2016**, *443*, 34–48. [CrossRef]
94. Hu, D.P.; Li, M.H.; Chen, J.B.; Luo, Q.Y.; Grasby, S.E.; Zhang, T.G.; Yuan, S.L.; Xu, Y.L.; Finney, S.C.; Sun, L.L.; et al. Major volcanic eruptions linked to the Late Ordovician mass extinction: Evidence from mercury enrichment and Hg isotopes. *Glob. Planet. Chang.* **2021**, *196*, 103374. [CrossRef]
95. Chai, Y.F.; Berghuijs, W.R.; Naudts, K.; Janssen, T.A.J.; Yao, Y.; Dolman, H. Using precipitation sensitivity to temperature to adjust projected global runoff. *Environ. Res. Lett.* **2021**, *16*, 124032. [CrossRef]
96. Van Staal, C.; Barr, S. Lithospheric architecture and tectonic evolution of the Canadian Appalachians and associated Atlantic margin. In *Tectonic Styles in Canada: The Lithoprobe Perspective*; Percival, J.A., Cook, F.A., Clowes, R.M., Eds.; Special Paper; Geological Association of Canada: St. John's, NL, Canada, 2012; Volume 49, pp. 41–96.
97. Xu, Y.; Cawood, P.; Du, Y. Intraplate orogenesis in response to Gondwana assembly: Kwangsian orogeny, South China. *Am. J. Sci.* **2016**, *316*, 329–362. [CrossRef]
98. Longman, J.; Mills, B.J.W.; Manners, H.R.; Gernon, T.M.; Palmer, M.R. Late Ordovician climate change and extinctions driven by elevated volcanic nutrient supply. *Nat. Geosci.* **2021**, *14*, 924–929. [CrossRef]
99. Schachtman, N.S.; Roering, J.J.; Marshall, J.A.; Gavin, D.; Granger, D.E. The interplay between physical and chemical erosion over glacial-interglacial cycles. *Geology* **2019**, *47*, 613–616. [CrossRef]
100. Holmden, C.; Mitchell, C.E.; Laporte, D.F. Nd isotope records of late Ordovician sea-level change—Implications for glaciation frequency and global stratigraphic correlation. *Palaeogeogr. Palaeoclimatol. Palaeoecol.* **2013**, *386*, 131–144. [CrossRef]
101. Blum, J.D.; Erel, Y. A silicate weathering mechanism linking increases in marine $^{87}Sr/^{86}Sr$ with global glaciation. *Nature* **1995**, *373*, 415–418. [CrossRef]
102. Kozik, N.P.; Gill, B.C.; Owens, J.D.; Lyons, T.W.; Young, S.A. Geochemical records reveal protracted and differential marine redox change associated with Late Ordovician climate and mass extinctions. *AGU Adv.* **2022**, *3*, e2021AV000563. [CrossRef]
103. Farouk, S.; Lofty, N.M.; Qteishat, A.; Ahmad, F.; Shehata, A.M.; Al-Kahtany, K.; Hsu, C.S. Source and thermal maturity assessment of the Paleozoic-Mesozoic organic matter in the Risha gas field, Jordan. *Fuel* **2023**, *335*, 126998. [CrossRef]
104. Farouk, S.; Abdeldaim, A.; Qteishat, A.; Ahmad, F.; Mohammed, I.; Al-Kahtany, K.; Powell, J.H. Orbital forcing and paleoenvironmental changes across the upper Ordovician glaciation-lower Silurian hot shale in the Risha gas field, northeast Jordan. *Mar. Pet. Geol.* **2024**, *168*, 107040. [CrossRef]
105. Bergström, S.M.; Chen, X.; Gutiérrez-Marco, J.C.; Dronov, A. The new chronostratigraphic classifi cation of the Ordovician System and its relations to major regional series and stages and to $\delta^{13}C$ chemostratigraphy. *Lethaia* **2009**, *42*, 97–107. [CrossRef]
106. Shen, J.; Chen, J.B., Algeo, T.J.; Feng, Q.L.; Yu, J.X.; Xu, Y.G.; Xu, G.Z.; Lei, Y.; Planavsky, N.J.; Xie, S.C. Mercury fluxes record regional volcanism in the South China craton prior to the end-Permian mass extinction. *Geology* **2021**, *49*, 452–456. [CrossRef]
107. Conwell, C.T.; Saltzman, M.R.; Edwards, C.T.; Griffith, E.M.; Adiatma, Y.D. Nd isotopic evidence for enhanced mafic weathering leading to Ordovician cooling. *Geology* **2022**, *50*, 886–890. [CrossRef]
108. Torres, M.A.; Moosdorf, N.; Hartmann, J.; Adkins, J.F.; West, A.J. Glacial weathering, sulfide oxidation, and global carbon cycle feedbacks. *Proc. Natl. Acad. Sci. USA* **2019**, *114*, 8716–8721. [CrossRef]
109. Rasmussen, C.M.; Kröger, B.; Nielsen, M.L.; Colmenar, J. Cascading trend of Early Paleozoic marine radiations paused by Late Ordovician extinctions. *Proc. Natl. Acad. Sci. USA* **2019**, *116*, 7207–7213. [CrossRef]
110. Rong, J.Y.; Harper, D.A.T.; Huang, B.; Li, R.Y.; Zhang, X.L.; Chen, D. The latest Ordovician Hirnantian brachiopod faunas: New global insights. *Earth Sci. Rev* **2020**, *208*, 103280. [CrossRef]

Journal of
Marine Science and Engineering

MDPI

Article

Regional Geological Data on the Volturno Basin Filling and Its Relationship to the Massico Structure (Southern Tyrrhenian Sea, Italy)

Gemma Aiello

Istituto di Scienze Marine (ISMAR), Consiglio Nazionale delle Ricerche (CNR), Sezione Secondaria di Napoli, 80133 Naples, Italy; gemma.aiello@cnr.it

Abstract: We built a regional geological section founded upon the assessment of a seismic line in the Volturno basin, which is situated on the northern Campania continental shelf of the Tyrrhenian margin of Southern Italy. This section has been integrated with multichannel seismic data of Zone E (ViDEPI project) to highlight its relationships with the Massico structure. In the Volturno basin, there are four Pleistocene to Holocene units, recognized based on seismic analysis lie above deep seismo-stratigraphic units, related to Campania Latium carbonate platform and The Frosinone Flysch. Onshore and offshore seismic data, calibrated with lithostratigraphic correlation, have displayed the seismo-stratigraphic framework, including both sedimentary and volcanic seismo-stratigraphic units. Of these, the lavas associated with the Northern Campania Volcanic Zone's Villa Literno volcano are associated with seismic unit 2a. Seismo-stratigraphic data has shown the offshore prolongation of the Massico structure, as involved by normal faults and flower structures. The whole-data interpretation suggests that the tectonic activity acted in correspondence to normal faults, which have controlled half-graben and interposed structural highs, fitting to the regional geological setting of the continental margin.

Keywords: Volturno basin; seismic stratigraphy; Villa Literno volcanic complex; southern Tyrrhenian continental margin

1. Introduction

The half graben offshore basins of Southern Italy represent an important research topic in basin studies on the Campania continental margin and have been investigated in several previous studies [1–5]. Nonetheless, none of these studies have described in detail the stratigraphic relationships existing between the basin filling of the Volturno, one of the most important basins of the Campania continental margin, and the Massico structure. This study analyzes these relationships in detail, putting them in the regional stratigraphic context of the Campania–Latium continental margin, so contributing to the basin studies in Southern Italy.

A half-graben, in contrast to a full graben, is a geological structure surrounded by a fault on one side, whereas a full graben is surrounded by a series of parallel faults. Geological studies on half-graben are numerous and have considered the filling models and their relationships on the extensional basin development [6], their sedimentary models [7], the importance of low-angle normal faults in basin development [8], and the tectonic and bathymetric control factors on the stratigraphic sequences [9].

The half graben offshore basins of Southern Italy have been previously studied considering both the Naples Bay half-graben [3,10,11] and the Salerno Valley [1]. In Naples

Bay, the Acerra–Dohrn canyon fault, a NE-SW-trending regional fault, had a significant stratigraphic control on the seismo-stratigraphic units, whose bathymetric and stratigraphic expression has been recently highlighted [12]. On the Campania–Latium margin, progressive changes in rifting directions have been suggested [4], following the key concepts on the Tyrrhenian basin development, including the continental break-up, the rifting phase and the expansion of oceanic lithosphere [13,14]. In the Tyrrhenian Sea, the rifting developed during classical subdivision of the basin filling in pre-rift, syn-rift, and post-rift stratigraphic sequences. Sartori et al. [14] have examined the very asymmetric stratigraphic relationships between the rifted margins of Sardinia and Campania, which were brought about by a low angle crustal detachment fault that dips eastward. According to this regional structure, the upper plate is outlined by the Campania margin and the one below it by the Sardinia margin [14].

Based on earlier research on the geology of the Tyrrhenian Sea, a regional geological section of the Volturno basin [2,3,15–17] has been constructed in this work using a deep multichannel profile taken during a seismic survey of the Southern Tyrrhenian Sea's extensional regions [2,3,15–17]. The aim is to discuss the seismo-stratigraphic and regional geological data of the basin and its tectonic structure based on seismic data calibrated with the onshore seismic and well data in the Northern Campania Volcanic Zone (NCVZ) [18]. The stratigraphic relationships of the Volturno basin with the adjacent Massico structure have been analyzed in this paper based on the seismic data of the ViDEPI project [19], in the framework of the structural setting of the northern Campania continental margin, which is characterized by half-graben and structural highs. This is in contrast to previous papers on the same area that focused more on the seismo-stratigraphic reconstruction of the Volturno fan delta and on the recognition of a geometry that is characteristic of a fan complex, with a NE-SW-trending fault [2].

2. Geological Setting

The Gulf of Gaeta is located on the northern Campania continental margin, which is part of a large extensional Plio–Pleistocene basin associated with normal and strike-slip faults linked to the evolution of the eastern Tyrrhenian Sea margin (Figure 1). The extensional tectonics have been accompanied by an intense volcanic activity, among them originating from the volcanic area of Phlegraean Fields. On this margin, Plio–Quaternary basin (peri-Tyrrhenian basins) [20], including the Terracina and Gaeta basins, the Capri basin, the Naples Bay, the Salerno Valley, and the Sapri and Paola basins, are located. Below the Plio–Quaternary basin filling, the stratigraphy of the margin shows the seawards extension of the inner tectonic units of the Southern Apennines [21–23] (Figure 1).

The neotectonic phases that influence the tectonic uplift of the Southern Apennines commenced in the Early Pliocene and persisted through various extensional phases until the Middle to Late Pleistocene. The Quaternary marine and continental sedimentation in the Campania coastal plains, which is significantly affected by the tectonic uplift of the Southern Apennines, is characterized by considerable thickness and is frequently composed of cycles of transgressive and regressive facies [2,3,24–27].

The Campania margin's half-graben and the intervening structural highs were formed as a result of two extensional phases that occurred during the Pleistocene. The initial phase took place in the Early Pleistocene and was characterized by NW-SE normal faults, which influenced the development of coastal depressions such as the Campania, Sele, and Volturno Plains. The subsequent phase, occurring at the transition between the Early and Middle Pleistocene, was associated with an NE-SW fault system and played a significant role in the formation of Naples Bay, leading to the development of asymmetrical structures typical of half-graben formations throughout the margin [2,3,5,28,29].

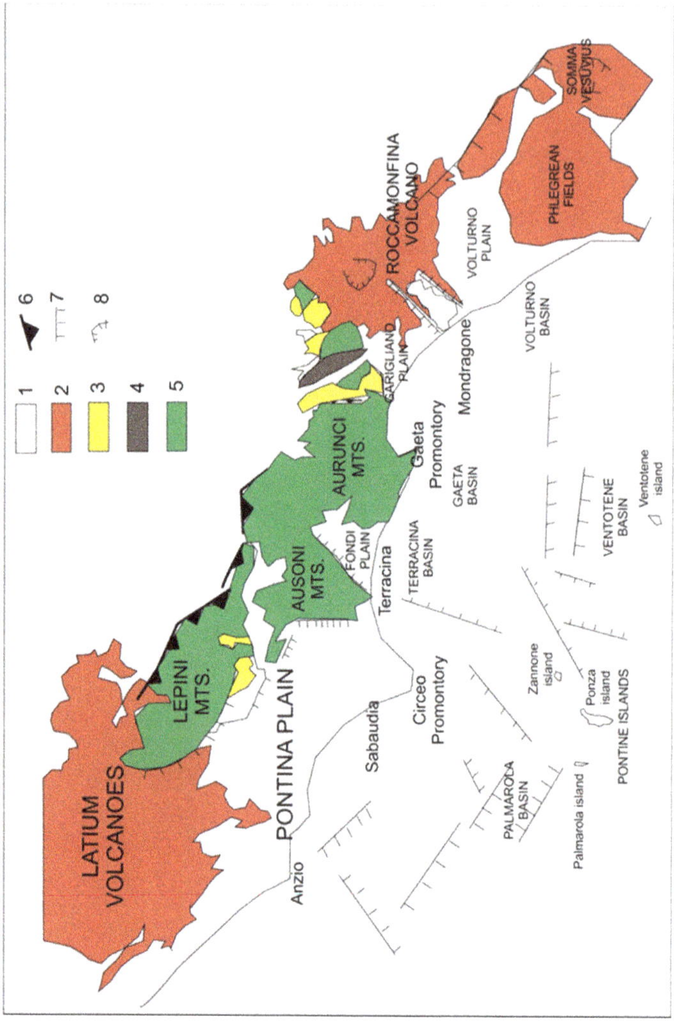

Figure 1. Geological map of the Campania Plain and tectonic lineaments occurring to a regional scale. Key. 1: Continental and marine deposits (Holocene–Late Pleistocene); 2: Volcanic deposits of the Campania–Latium region (Quaternary); 3: Arenaceous-clayey turbidites (Flysch di Frosinone *Auct.*; Tortonian–Messinian); 4: Shallow water organogenic limestones (Calcari a Briozoi e Litotamni; Langhian–Serravallian); 5: Shallow and deep water limestones (Latium–Abruzzi Unit; Jurassic–Cretaceous); 6: Thrust fronts; 7: Normal faults; 8: Caldera rims.

The structural setting of the Campania–Latium continental margin is characterized by structural highs and interposed basins occurring to a regional scale [2,3,5,30,31] (Figure 2). Its tectonic setting is characterized by regional normal faults, from NE-SW to NNE-SSW trending. The Massico Mount constitutes a structural elevation situated along normal faults that trend from northeast to southwest, effectively distinguishing this horst from the coastal depressions of the Garigliano and Volturno plains [32–34]. The down throwing of the coastal structures has been identified based on onshore and offshore seismic interpretation [2,3,32,33].

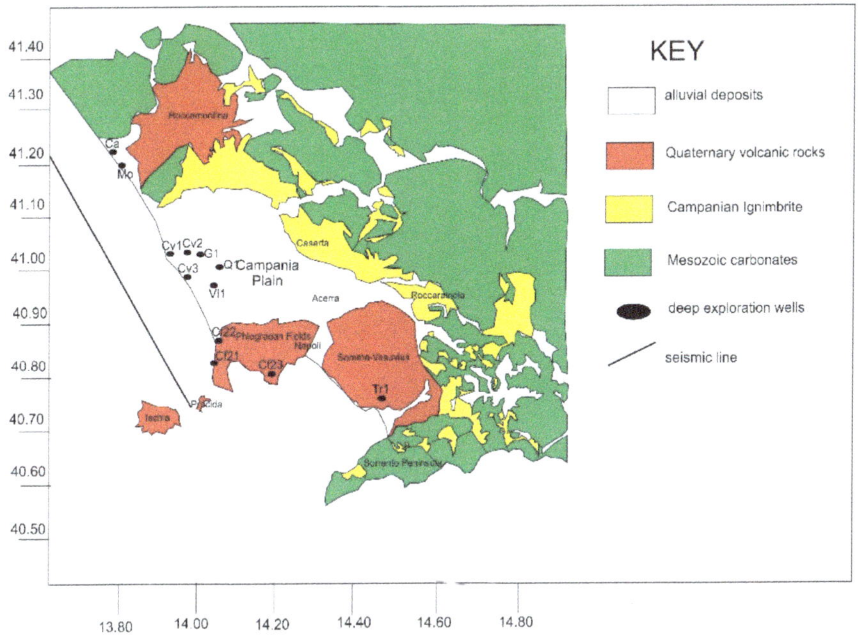

Figure 2. Sketch geological map of the Campania Plain, reporting the location of the seismic profile and exploration wells.

Since the Late Miocene, the Campania Plain has been significantly influenced by extensive volcanic activity. This activity, particularly along the western edge, is closely associated with NE-SW normal faults [18,35]. The volcanic deposits in this region can be categorized into two distinct cycles: the older cycle, which spans from the Miocene to the Pleistocene, consists of calcalkaline, andesitic, and basaltic lavas, represented by the Parete and Villa Literno volcanic complexes. The younger cycle is characterized by alkaline-potassic deposits and is linked to the Plio–Pleistocene extensional tectonics of the Roman co-magmatic province, which includes notable features such as Vesuvius and Campi Flegrei. Significant layers of lavas and pyroclastic materials, dating back approximately 2 million years, have been identified at the Villa Literno 2 and Parete 2 wells [2,3].

In the Northern Campania Volcanic Zone (NCVZ, the stratigraphic data obtained from 600 boreholes have allowed to construct geological sections, showing the stratigraphic architecture of the ignimbrite deposits (Figure 3 [18]. The lowermost unit is made up of marine sediments (Tyrrhenian Layer M1), whose top has been dated back at 126 ky B.P. (Figure 3) [36]. A volcanic sequence overlies the marine deposits and is constituted of massive ignimbrite deposits and lava flows. This unit is overlain by another succession of marine deposits, dated back at 55–50 ky (Tyrrhenian unit M2) [36], which is surrounded by the Campanian Ignimbrite (CI) deposits. The NYT caldera is the preeminent volcano-tectonic lineament of the intermediate Campania Volcanic Zone (CVZ; Figure 3) [18].

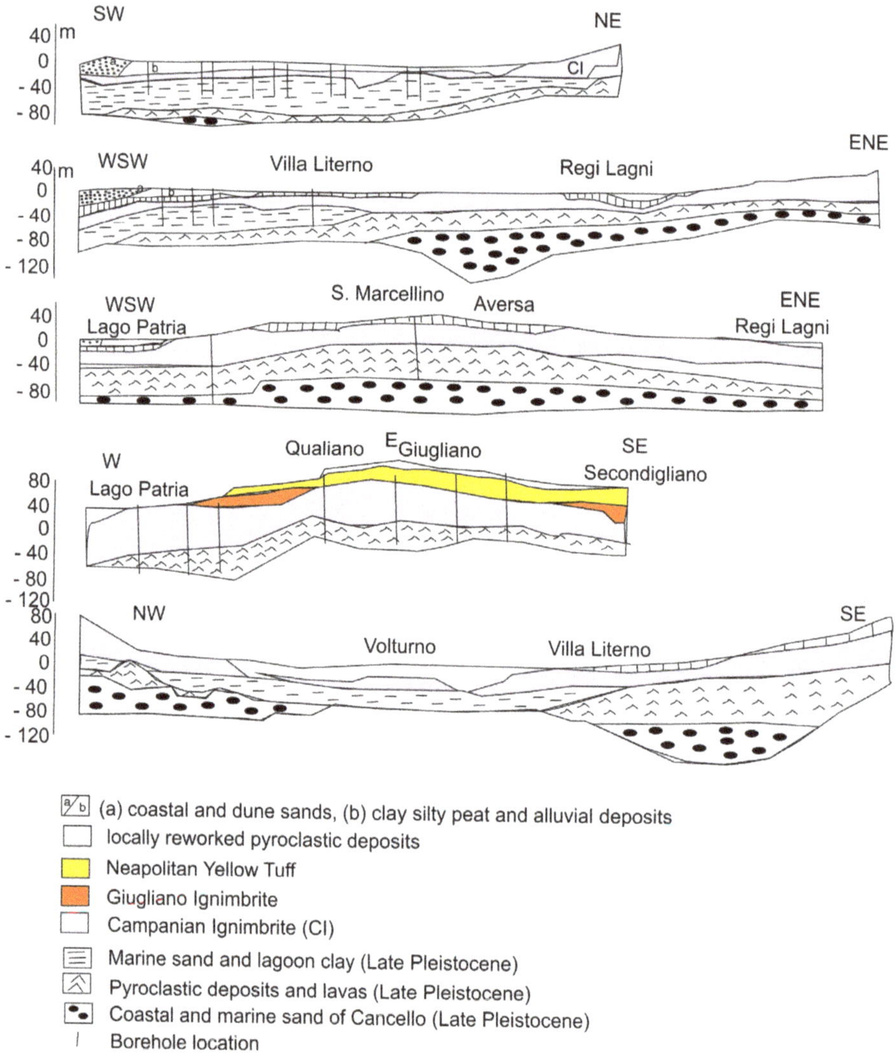

Figure 3. Geological sections of the CVZ (Campania Volcanic Zone; modified after Rolandi et al. [18]).

3. Materials and Methods

The application of a multichannel profile within the Volturno basin facilitated the development of an interpreted geological section that illustrates the stratigraphic connections among the seismo-stratigraphic units, which were previously delineated using seismo-stratigraphic criteria. The seismic profile reveals the geological framework and subsurface structure of the area under investigation, emphasizing the stratigraphic relationships between the acoustic basement and the sedimentary infill of the basin. The multichannel seismic profiles of the ViDEPI project (Table 1) [19], recorded from the Western Geophysical Company (Missouri City, TX, USA), and the well lithostratigraphic data have integrated the seismic dataset. Following Italy's Minister of Industry's pledge to conduct a regional survey of the entire Eastern Tyrrhenian continental shelf, the Western Co. seismic grid was purchased in 1968 [19]. An Aquapulse energy source, 24-trace and 1600-m-long streamer,

2 ms sample rate, 68 m shot and group interval, 10–80 Hz filter, and a recorded length of 4 s were used to obtain these seismic data.

Table 1. Characteristics of the seismic profiles of the Zone E (ViDEPI Project) [19].

Seismic Grid	Seismic Profile	Location
Zone E	E194	Volturno river mouth
Zone E	E196	Cuma offshore
Zone E	E198	Cuma offshore
Zone E	E200	Monte di Procida promontory
Zone E	E101SR, E109	Campania–Latium continental margin

Seismo-stratigraphic techniques and methods have been employed for the geologic interpretation of seismic data and related stratigraphic analysis. The analysis of seismo-stratigraphy relies on recognizing regional unconformities, which in turn allows for the identification of depositional sequences. This process aids in reconstructing the original geometry of sedimentary bodies and their associated environments, as well as facilitating chrono-stratigraphic correlation. High-resolution seismic techniques and sequence stratigraphy have been thoroughly detailed as effective methods for analyzing seismic profiles and these approaches have been applied in the geological interpretation of such profiles [37–43].

Data processing was carried out using specialized software (Promax 2D; Landmark Ltd. (Esher, UK) and Seismic Unix SU44R28). These methods rely on complex mathematical models, enabling us to effectively reduce the impact of multiples (specifically the seabed multiples) and to achieve accurate velocity assessments in generating stacked sections for the geological interpretation. A table was constructed to display the acquisition parameters of the multi-channel seismic survey (Table 2).

Table 2. Acquisition parameters of the multichannel seismic survey.

Seismic Source	n. 2 Guns GI Gun SI/Sodera (210 c.i.)
Length of the seismogram	5 s
Sampling interval	1 ms
Source distance	25 m
Hydrophone distance	12.5 m

The first step involved checking the quality of the data and setting up the field geometry. The purpose of editing the seismic traces was to uncover any seismic traces and spikes within the traces that could cause issues with the Fast Fourier Transform (FFT). A high-level muting was able to remove the seismic signal after the initial seismic arrivals in the seismic traces. Automatic Gain Control (AGC) enabled seismic traces to be normalized. The data processing focused on decreasing the random interference in the seismic data and enhancing the clarity of the seismic waves using deconvolution and spiking techniques.

A velocity analysis was carried out to correct the move-out of the CDP (Common Depth Point) groups, calculating the velocity of various seismic reflectors and generating a final stacked seismic section. Post-stack deconvolution was carried out to eliminate the multiple arrivals. Following deconvolution, a bandpass filter was utilized to boost the seismic signal present along the seismic section.

4. Results

4.1. Seismic Interpretation

The seismic interpretation was carried out using seismo-stratigraphic methods, which enabled the identification of various seismic units (Figures 4 and 5). The lithology of the seismic units was correlated with onshore seismic sections and well lithostratigraphic data.

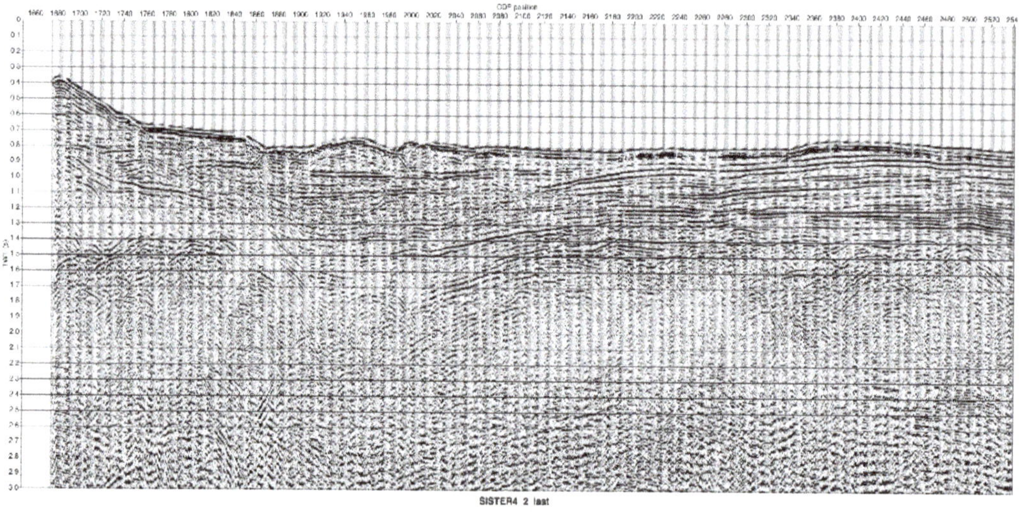

Figure 4. Stacked seismic section SISTER4_2.

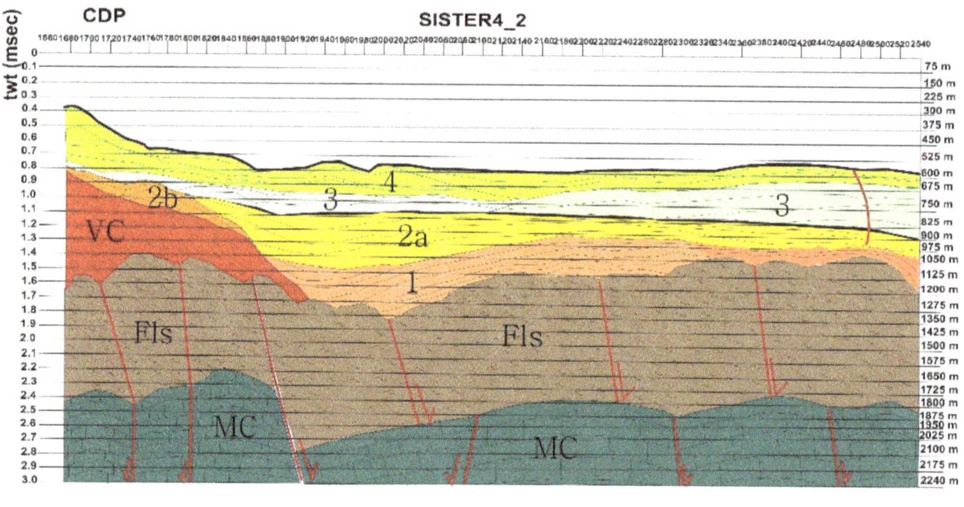

Figure 5. Geological interpretation of the seismic section SISTER4_2. Key. MC: Seismo-stratigraphic unit of the Meso–Cenozoic carbonates. Fls: Seismo-stratigraphic unit of Miocene flysch deposits (Frosinone Flysch). VC: Seismo-stratigraphic unit of the Villa Literno volcanic complex. 1: oldest seismo-stratigraphic unit of the Volturno basin; 2b: relict prograding wedge (Early Middle Pleistocene); 2a: second seismo-stratigraphic unit of the basin filling; 3: seismo-stratigraphic unit composed of alternating sands and clays of a deltaic environment, Pleistocene in age; 4: seismo-stratigraphic unit of coastal clays (Pleistocene).

The lower seismic layer is distinguished by intermittent reflectors with varying amplitudes. The seismic unit's top is found at depths ranging from 1650 m to 2025 m. Some normal faults lower this unit, with dips measuring in the range of tens of meters (Figure 5). The Meso–Cenozoic carbonate unit, identified as MC in Figure 5, is genetically related with the Campania-Lucania carbonate platform, outcropping in the adjacent sectors of the Southern Apennines [21–23].

The seismic layer in question is characterized by a seismic facies that exhibits acoustic transparency and is marked by a limited number of dispersed seismic reflectors. The upper boundary of this seismic unit is found at depths between 1050 m and 1350 m, with an average thickness of approximately 750 m. The presence of normal faults has resulted in a downward displacement of the continuous reflector that defines the top of the seismic unit by several tens of meters. This unit is classified as part of the Frosinone Flysch *Auct.* (FLS in Figure 5) and is associated with the Miocene flysch deposits of the Central Apennines. Its identification has been substantiated through both outcrop investigations and prior seismic surveys conducted in the Volturno offshore region [2,30].

The FLS unit is covered by two distinct seismo-stratigraphic units characterized by facies heterogeneity, one of which is volcanic and the other sedimentary. The volcanic unit, referred to as VC (Figure 5), is wedge-shaped and has been identified as the volcanic deposits associated with the Villa Literno volcanic complex [2,24,25]. This VC unit typically exhibits a thickness of approximately 600 m. Among the oldest volcanic deposits found in the Campania Plain are basalts and andesites, including basaltic lavas that date back 2 million years. These volcanic materials were collected in the Parete-Villa Literno region during geothermal drilling activities [24,25]. The initial unit of the Volturno basin fill (1 in Figure 5) corresponds to the sedimentary unit associated with the VC volcanic complex, with its upper boundary situated at depths between 1050 m and 1125 m.

The VC and 1 seismo-stratigraphic units are overlain by two distinct seismo-stratigraphic units (2a and 2b in Figure 5). The seismo-stratigraphic unit 2b was interpreted as an ancient prograding wedge, down lapping on the volcanic deposits of the VC unit. Moving towards the basin, it passes into the seismo-stratigraphic unit 2a, which is the second unit of the basin filling. The seismo-stratigraphic unit 3 (Figures 4 and 5) is marked by high amplitude like nearly parallel seismic reflectors and is probably composed of alternating sands and clays of a deltaic environment, Pleistocene in age [33]. The seismo-stratigraphic unit 4 (Figures 4 and 5) is characterized by parallel to almost parallel, discontinuous to continuous seismic reflections and has been interpreted as clays deposited in a Pleistocene coastal environment [33].

Some significant seismic profiles pertaining to the ViDEPI project [19] have been analyzed (Figures 6–8). The seismic profile E101SR has shown seismo-stratigraphic units and fault trends down throwing the Meso–Cenozoic carbonates (Figure 6). The fault intersections have shown the offshore prolongation of the Massico Mt. structure. A horst has been identified, involving the seismo-stratigraphic unit correlated with the Meso–Cenozoic carbonate deposits (Figures 6 and 7). Proceeding southwards, an anticline has been identified, interpreted as a positive flower structure (Figure 6). The seismo-stratigraphic 2 is characterized by discontinuous seismic reflectors and has been correlated with the Miocene flysch deposits ("Flysch di Frosinone" *Auct.*), outcropping in the adjacent onshore areas. An Early Pleistocene progradational unit has been recognized (unit 3a in Figure 6), representing the Volturno submerged fan. The seismo-stratigraphic unit 3b has been interpreted as Middle–Late Pleistocene marine and coastal deposits and is characterized by discontinuous and sub-parallel seismic reflectors (Figure 6), whilst the 4 one as Late Pleistocene–Holocene marine and coastal deposits.

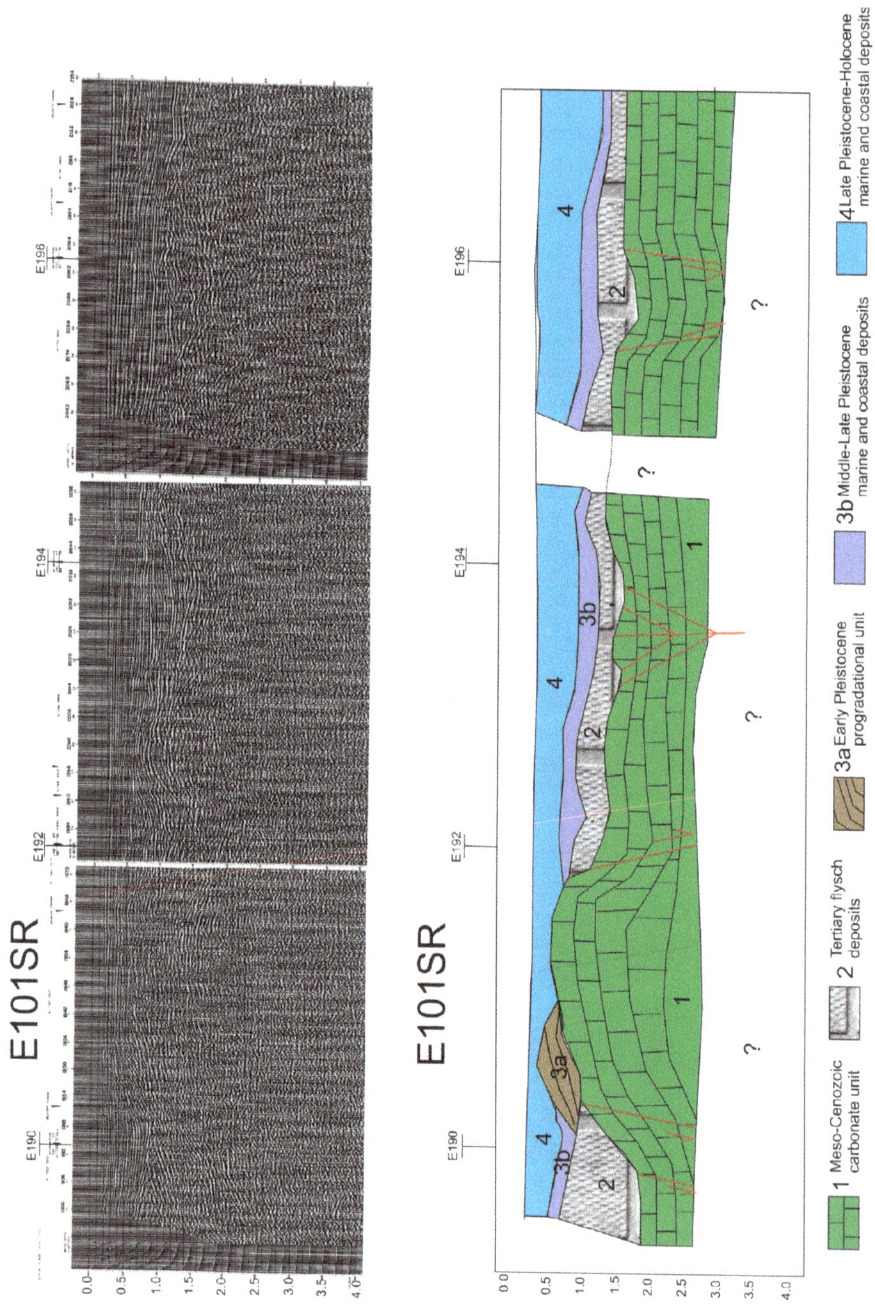

Figure 6. Seismic profile E101SR and corresponding geological interpretation.

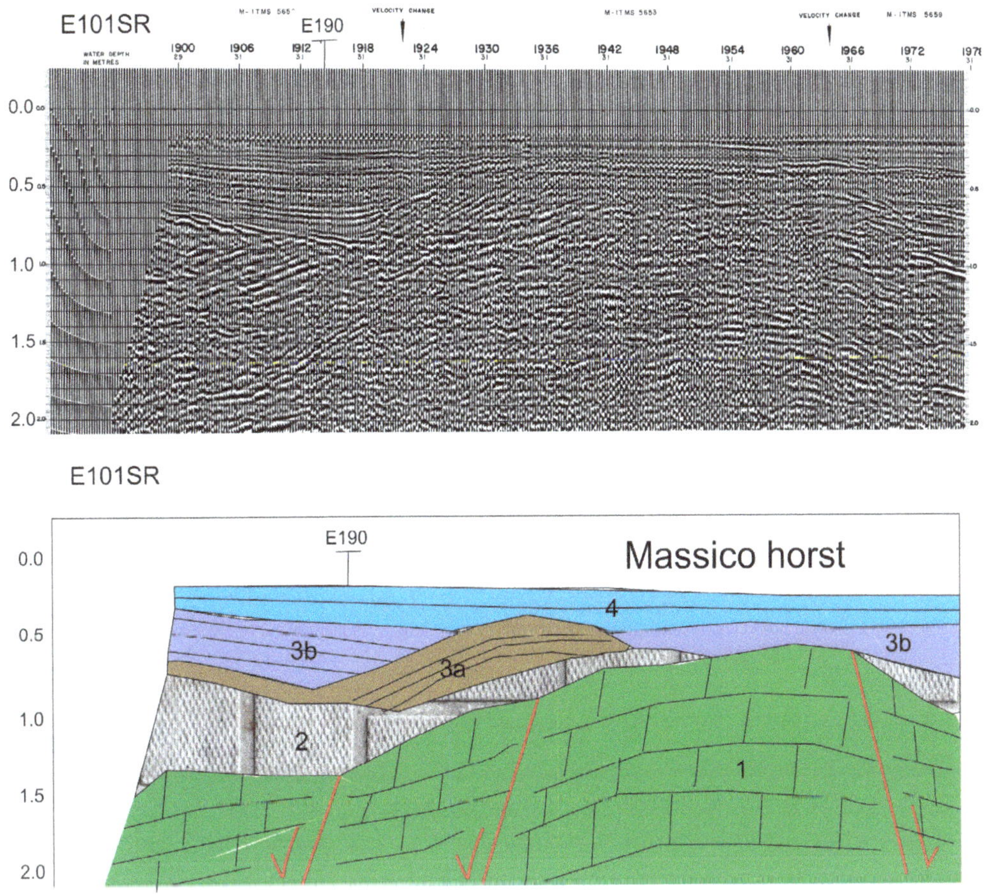

Figure 7. Details of the seismic profile E101SR and corresponding geological interpretation, showing the Massico horst structure.

Detailed seismic sections are provided here, showing the previously mentioned structural location and where faults intersect near the Massico structure (Figures 7 and 8). Figure 7 displays the Massico horst formation, interpreted on detailed seismic profile E101 SR, where the same seismo-stratigraphic units shown in Figure 6 have been identified. Figure 8 highlights the occurrence of flower structures at the top of the Meso–Cenozoic carbonate acoustic basement.

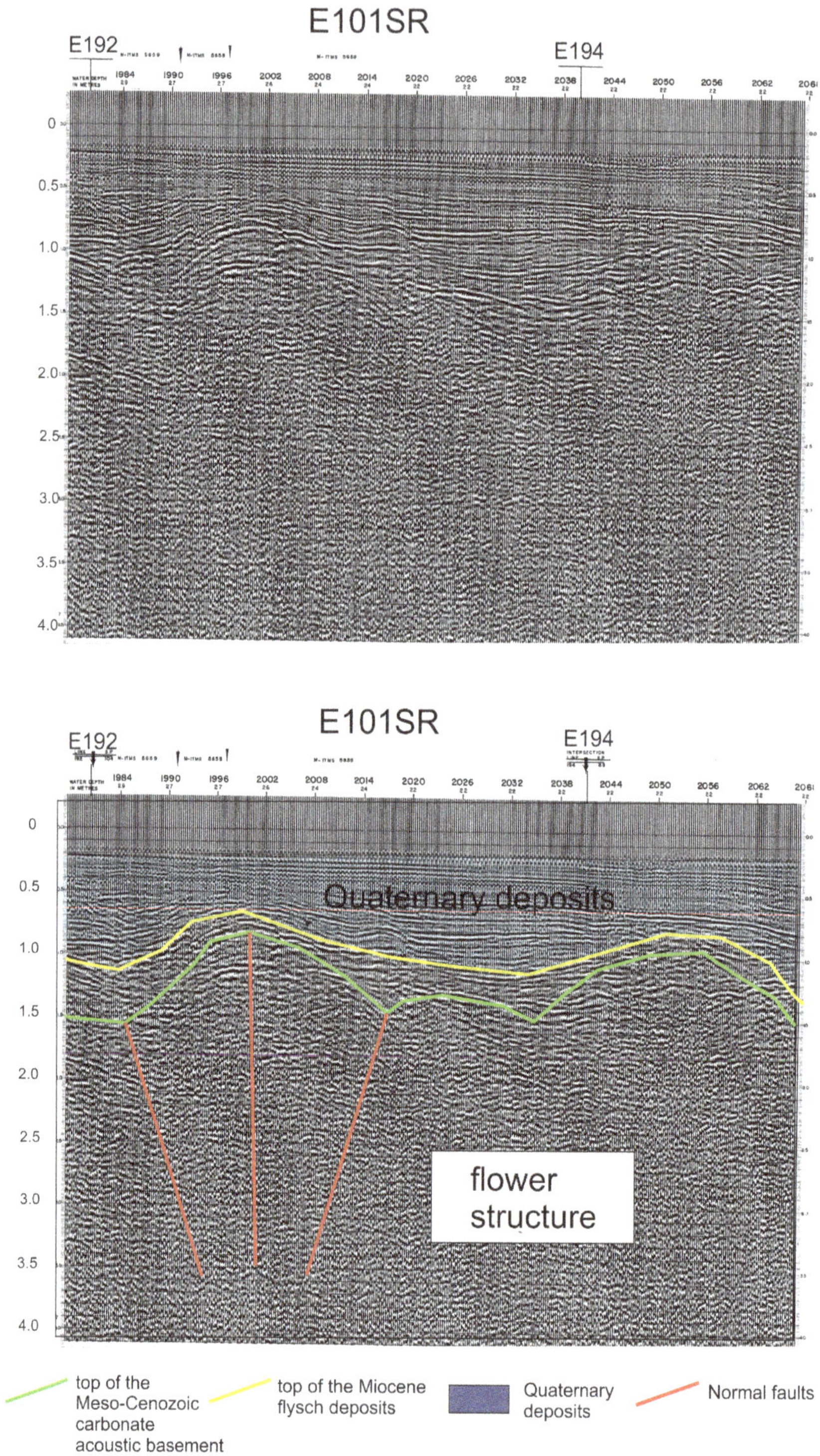

Figure 8. Details of the seismic profile E101SR and corresponding geological interpretation, showing the flower structures involving the top of the Meso–Cenozoic carbonate acoustic basement.

4.2. Stratigraphic Correlation of Well Lithostratigraphic Data

A stratigraphic correlation has been established to compare the rock layers from well data drilled in the Volturno Plain with those from the offshore Phlegrean Fields (Figure 9). This correlation illustrates the distribution and thickness of pyroclastic and alluvial sediments in relation to the lavas, specifically andesites and basalts, associated with the Villa Literno volcanic complex (VC seismo-stratigraphic unit depicted in Figure 5).

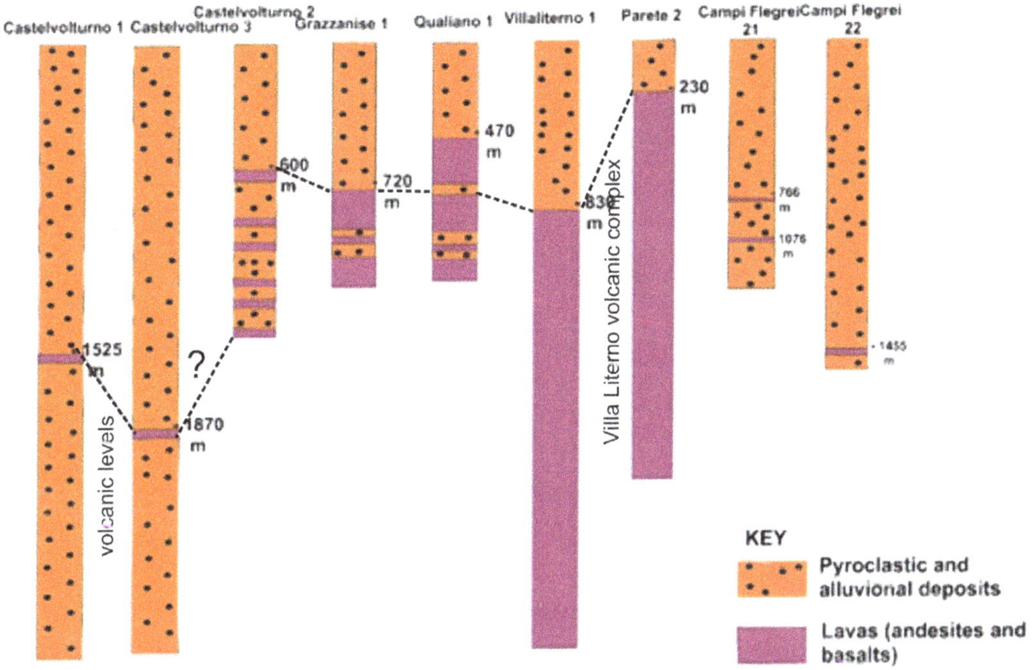

Figure 9. Qualitative stratigraphic correlation between deep exploration wells.

Two primary categories of geological formations have been identified: pyroclastic and alluvial deposits, which constitute the sedimentary material within the Volturno Plain, alongside the volcanic lavas (andesites and basalts) originating from the Villa Literno volcanic complex. The lavas exhibit their minimum thickness at the Castelvolturno 1 and Castelvolturno 3 wells, where these deposits are located at depths of 1525 m and 1870 m, respectively (Figure 9).

A stratigraphic correlation has been conducted for the wells Castelvolturno 2, Grazzanise 1, Qualiano 1, Villa Literno 1, and Parete 2 (see Figure 9). The highest occurrence of volcanic materials was identified at depths of 600 m in Castelvolturno 2, 720 m in Grazzanise 1, 470 m in Qualiano 1, 830 m in Villa Literno 1, and 230 m in Parete 2. Notably, the thickness of the volcanic deposits exhibits an increasing trend when moving from Castelvolturno towards the towns of Parete and Villa Literno (refer to Figure 9).

The Castelvolturno 1 and Castelvolturno 3 wells have revealed sedimentary layers that transition from marine to terrestrial environments, extending to a depth of 3000 m (see Figure 9). Within these sedimentary sequences, two volcanic strata have been identified, located at depths of 1430 and 1450 m in Castelvolturno 1 and between 1800 and 1830 m in Castelvolturno 3.

At the Castelvolturno 2, Grazzanise 1, and Qualiano 1 wells, there is an observed increase in the thickness of the lava layers, which are interbedded with sand and shale deposits. The depths at which these lava formations have been encountered are 475 m, 720 m, and 500 m, respectively (see Figure 9).

At the Villa Literno 1 well, a sequence of 150 m of andesitic tuffs is situated beneath the overlying recent pyroclastic deposits. This is succeeded by 650 m of clastic deposits formed in a marine and transitional setting, along with interbedded layers of tuffs, basalts, and andesites extending from 830 m to 2980 m (Figure 9).

At the Parete 2 well, the initial 300 m consist of recent pyroclastic materials interspersed with clastic deposits. Below this layer, a sequence of alternating basaltic and andesitic lavas extends to the well's total depth of 1800 m.

It is also important to consider in the calibration of the seismic section is the lithostratigraphic data from Mondragone 1 well. These lithostratigraphic data are crucial for the geological interpretation of the deep seismo-stratigraphic units, respectively related to the Miocene flysch deposits and to the Meso–Cenozoic carbonate acoustic basement. The Mondragone 1 well was drilled to a depth of 675 m, penetrating Quaternary deposits consisting primarily of sandy materials with volcanic components and conglomerates mixed with layers of marls (Figure 10) [44]. At depths shallower than 675 m, the well has penetrated Miocene formations, which consist of conglomerates, sandstones, and marly sandstones (Figure 10). These successions have been interpreted as post-evaporitic and interpreted in the context of the late Messinian Lago–Mare episode in the central Mediterranean Sea (Figure 10).

4.3. *Calibration of the Seismic Section with the Well Data*

The initial group of geological characteristics serves as a reference for calibrating the analyzed seismic data (Figure 9). The association of pyroclastic and alluvial deposits with seismo-stratigraphic units 2a, 3, and 4 elucidates the sedimentary history of the basin in the marine environment adjacent to the Volturno plain. Utilizing these significant indicators allows researchers to acquire important understanding of the geological development of the region.

The volcanic lavas, specifically andesites and basalts, originating from the Villa Literno volcanic complex (refer to Figure 9), are associated with seismic-stratigraphic unit 2a (illustrated in Figures 4 and 5). The stratigraphic connections between the Villa Literno volcanic complex and the neighboring ignimbrites are depicted in Figure 3 (lower geological section).

The lithostratigraphic analysis of the Mondragone 1 well (Figure 10) suggests that the FLS unit, identified as Miocene flysch deposits, is likely comprised of a sequence of alternating shales, sandstones, conglomerates, and marly limestones. These deposits are genetically associated with the Flysch di Frosinone *Auct.* [21–23].

Geological data reveal that the ancient carbonate bedrock beneath the Campania Plain significantly deepens westward, reaching depths over 3 km. This bedrock dates back from the Miocene to the Quaternary, as confirmed by nearby oil wells (Mondragone 1 well; Figure 10). Interestingly, the Castelvolturno 1 well has uncovered Quaternary deposits extending down to 3 km, suggesting the presence of the older Miocene sequence underneath. This intriguing finding opens possibilities for further exploration and understanding of the region's complex geological history.

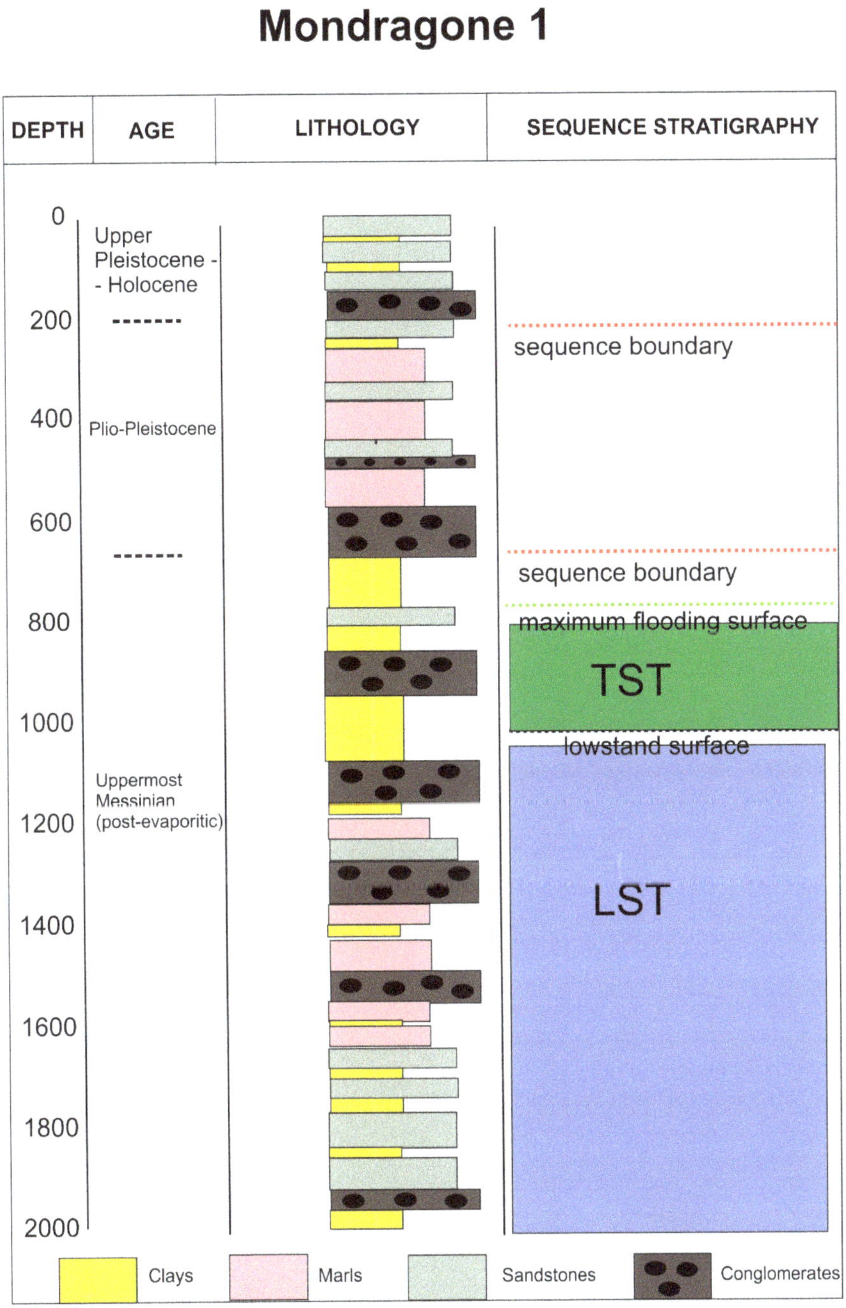

Figure 10. Stratigraphic correlation of the Mondragone 1 well (modified after Cosentino et al. [44]).

5. Discussion

The seismo-stratigraphic data shown in the Volturno basin have confirmed that it represents a half-graben characterized by blocks down thrown by normal faults, mainly involving the top of the Miocene siliciclastic sequences (Figures 4 and 5). The activity period of normal faults probably ranges between the end of the Late Miocene and the

Early Pleistocene. Moreover, the basal seismic sequence (1 in Figure 5) has shown wedging geometries, suggesting that this unit represents a synsedimentary sequence.

The volcanic deposits in the Campania offshore exhibit a notable distinction from other sedimentary basins, as they are significantly developed and interspersed within the sedimentary fill of the basin (see Figures 5 and 9). This phenomenon is exemplified by the volcanic body associated with the Villa Literno volcanic complex, which is identifiable in the interpreted seismic section (denoted as VC in Figure 5) and is correlated with the Quaternary volcanic materials encountered in the Villa Literno 1 and Parete 2 wells, with their stratigraphic relationships illustrated in Figure 9. It appears that the formation of the VC volcanic body occurred subsequent to the extensional activities within the Volturno basin. The presence of normal faults likely provided preferential conduits for the magmatic ascent associated with the VC volcanic complex.

Four distinct seismo-stratigraphic units, which align with the seismic units identified onshore [33,44], have been delineated within the Volturno basin. This classification is based on a comprehensive seismo-stratigraphic analysis of multichannel seismic profiles, with particular emphasis on the stratigraphic interactions with the Massico horst (Figures 6–8). The lithological characteristics of these units have been validated through the examination of lithostratigraphic well data (Figure 9).

The initial seismo-stratigraphic unit (1 in Figure 5) is distinguished by reflectors that range from discontinuous to continuous, exhibiting both parallel and sub-parallel orientations. This unit comprises a mixture of sands, conglomerates, and shales, interspersed with pyroclastic layers, and is dated to the Early Pleistocene. This unit onlaps the side of the VC volcanic seismo-stratigraphic unit, correlating to the volcanites genetically related with the Villa Literno volcanic complex (Figure 5). The seismo-stratigraphic unit 1 is synsedimentary, as suggested by the wedging geometries observed in this unit. Similar geometries have been observed in the correlative seismic units observed onshore [33,44]. The seismo-stratigraphic data indicate that the VC volcanic unit predates the deposition of seismo-stratigraphic unit 1. This conclusion is supported by the observed onlap of unit 1 onto the concealed side of the volcanic formations, as illustrated in Figures 4 and 5.

The second seismo-stratigraphic unit (2 in Figure 5) is subdivided into two primary sub-units, which are designated as occurring on the continental shelf (2a in Figure 5) and within the basin (2b in Figure 5). Sub-unit 2a is distinguished by its prograding clinoforms and is indicative of a relict prograding wedge that overlays the VC volcanic complex. In contrast, sub-unit 2b is characterized by parallel to sub-parallel reflectors and consists of alternating layers of sands and shales, indicative of a deltaic environment and dating back to the Pleistocene.

The third seismo-stratigraphic unit (3 in Figure 5) is distinguished by high-amplitude reflectors that are arranged in parallel to sub-parallel configurations. This unit consists of a succession of sands and shales indicative of a deltaic setting. In contrast, the fourth seismic sequence exhibits reflectors that range from discontinuous to continuous, maintaining a parallel to sub-parallel orientation. This sequence is primarily composed of shales that are characteristic of a coastal environment and date back to the Pleistocene.

This half-graben structure has been confirmed by the seismic profiles recorded onshore in the Garigliano plain (Figure 11) [44]. Four distinct seismic sequences can be identified by the seismic stratigraphic interpretation of these lines. Throughout the examined seismic lines, the topmost seismic horizon (horizon Q) can be easily traced; it separates the topmost seismic sequence (sequence I) from sequence II. Sequence I, the highest seismic sequence, is distinguished by irregular reflections that vary in amplitude and are occasionally parallel. Sand and clayey-sand alternate and pebbles and conglomerates (Upper Pleistocene–Holocene) are frequently intercalated. It correlates to the SP interval D. Four seismic units (U1, U2, U3, and U4 in Figure 11) have been identified within the sequence II. The external form of these seismic units is wedge-shaped at the basin scale, exhibiting an increase in thickness toward the southeast, which must be connected to the extensional syn-sedimentary activity of the fault separating the Massico Mount from the northwest. This trend of faults can be seen also offshore in the interpreted seismic profiles of Zone E (Figures 6–8).

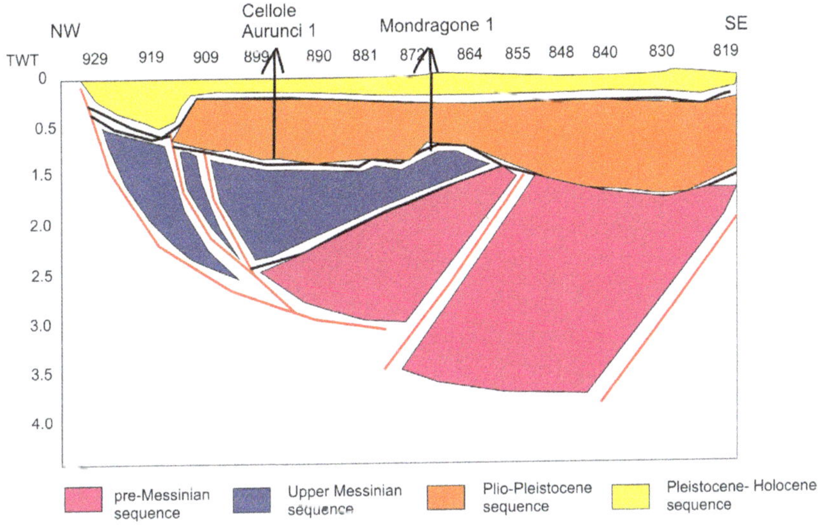

Figure 11. Interpreted seismic section onshore in the Mondragone area, showing the half-graben structures which characterize the area (modified after Cosentino et al. [44]).

The recorded stratigraphic architecture has shown that the tectonic controls are prevalent on the eustatic ones. Normal faults have controlled the individuation of the half-graben structures and downthrown the deep seismo-stratigraphic units (Figure 5). Tectonic control prevailed on the eustatic one also regarding the first seismo-stratigraphic unit, clearly showing wedging geometries and suggesting its synsedimentary nature (Figure 5). This is in overall agreement with the previous papers in the same area, suggesting wedging geometries and synsedimentary tectonics of the seismo-stratigraphic units onshore both in the Volturno and Mondragone areas [33,44]. (Figure 11).

On the other hand, eustatic controls are prevalent in the younger seismo-stratigraphic units, where progradational geometries may have been deposited during eustatic sea-level falls, controlling phases of forced regressions [45]. This is true for the seismo-stratigraphic unit 2b, overlying the Villa Literno volcanic complex (Figure 5). A progressive deepening of the area, coupled with subsidence is suggested by the depositional geometries observed in the other seismo-stratigraphic units, where onlap geometries on both the sides of the basin prevail (Figure 5).

6. Conclusions

1. A geological cross-section of the Volturno basin has been developed through the geological analysis of multichannel seismic profiles. This analysis has been combined with the multichannel seismic data from the ViDEPI project to illustrate the connections between the Volturno basin and the Massico structure.
2. The seismo-stratigraphic analysis has shown four seismo-stratigraphic units, Pleistocene to Holocene in age, overlying deep seismo-stratigraphic units genetically

related to the Campania–Latium carbonate platform and to the Frosinone Flysch. While the deep seismo-stratigraphic units and the first unit of the basin filling are mainly tectonically-controlled, the depositional geometries of the basin filling suggest that the eustatic controls are prevalent from the Late Pleistocene to the Holocene.

3. Important volcanic seismo-stratigraphic units are interlayered in the Volturno basin filling, including the Villa Literno volcanic complex (VC), which can be framed in the volcanological evolution of the ignimbrites of the Northern Campania Volcanic Zone (NCVZ).
4. Seismo-stratigraphic data have shown the offshore prolongation of the Massico structure, as involved by normal faults and flower structures, which suggest the occurrence of basin inversion in the area, according to previous literature data on the Eastern Tyrrhenian margin.
5. Future tasks of this study will include further seismo-stratigraphic interpretation in the adjacent marine areas of the Eastern Tyrrhenian margin to provide detailed seismo-stratigraphic study of the half-graben offshore basins of Southern Italy.

Funding: This research received no external funding.

Institutional Review Board Statement: Not applicable.

Informed Consent Statement: Not applicable.

Data Availability Statement: The data are not available due to privacy.

Conflicts of Interest: The author declares no conflicts of interest.

References

1. Aiello, G.; Marsella, E.; Di Fiore, V.; D'Isanto, C. Stratigraphic and structural styles of half-graben offshore basins in Southern Italy: Multichannel seismic and Multibeam morpho-bathymetric evidences on the Salerno Valley (Southern Campania continental margin, Italy). *Quad. Di Geofis.* **2009**, *77*, 1–33. Available online: https://editoria.ingv.it/archivio_pdf/qdg/77/pdf/qdg_77.pdf (accessed on 24 January 2025).
2. Aiello, G.; Cicchella, A.G.; Di Fiore, V.; Marsella, E. New seismo-stratigraphic data of the Volturno basin (northern Campania, Tyrrhenian margin, southern Italy): Implications for tectono-stratigraphy of the Campania and Latium sedimentary basins. *Ann. Geophys.* **2011**, *54*, 265–283. [CrossRef]
3. Aiello, G.; Marsella, E.; Cicchella, A.G.; Di Fiore, V. New insights on morpho-structures and seismic stratigraphy along the Campania continental margin (Southern Italy) based on deep multichannel seismic profiles. *Rend. Lincei* **2011**, *22*, 349–373. [CrossRef]
4. Milia, A.; Torrente, M.M.; Massa, B.; Iannace, P. Progressive changes in rifting directions in the Campania margin (Italy): New constrains for the Tyrrhenian Sea opening. *Glob. Planet. Change* **2013**, *109*, 3–17. [CrossRef]
5. Conti, A.; Bigi, S.; Cuffaro, M.; Doglioni, C.; Scrocca, D.; Muccini, F.; Cocchi, L.; Ligi, M.; Bortoluzzi, G. Transfer zones in an oblique back-arc basin setting: Insights from the Latium-Campania segmented margin. *Tectonics* **2017**, *36*, 78–107. [CrossRef]
6. Schlische, R.W. Half-graben basin filling models: New constraints on continental extensional basin development. *Basin Res.* **1991**, *3*, 123–141. [CrossRef]
7. Leeder, M.R.; Gawthorpe, R.L. Sedimentary models for extensional tilt-block/half-graben basins. *Geol. Soc. Lond. Spec. Publ.* **1987**, *28*, 139–152. [CrossRef]
8. Lavier, L.L.; Buck, W.R. Half graben versus large-offset low-angle normal fault: Importance of keeping cool during normal faulting. *J. Geophys. Res.* **2002**, *107*, ETG 8-1–ETG 8-13. [CrossRef]
9. Roberts, A.M.; Yielding, G.; Badley, M.E. Tectonic and bathymetric controls on stratigraphic sequences within evolving half-graben. *Geol. Soc. Lond. Spec. Publ.* **1993**, *71*, 87–121. [CrossRef]
10. Milia, A.; Torrente, M.M. Tectonics and stratigraphic architecture of a peri-Tyrrhenian half-graben (Bay of Naples, Italy). *Tectonophysics* **1999**, *315*, 301–318. [CrossRef]
11. Aiello, G.; Budillon, F.; Cristofalo, G.; D'Argenio, B.; de Alteriis, G.; De Lauro, M.; Ferraro, L.; Marsella, E.; Pelosi, N.; Sacchi, M.; et al. Marine Geology and Morphobathymetry in the Bay of Naples (South-Eastern Tyrrhenian Sea, Italy). In *Mediterranean Ecosystems*; Faranda, F.M., Guglielmo, L., Spezie, G., Eds.; Springer: Milano, Italy, 2001. [CrossRef]

12. Aiello, G.; Caccavale, M. The Coastal Areas of the Bay of Naples: The Sedimentary Dynamics and Geological Evolution of the Naples Canyons. *Geosciences* **2023**, *13*, 226. [CrossRef]
13. Trincardi, F.; Zitellini, N. The rifting of the Tyrrhenian Basin. *Geo-Mar. Lett.* **1987**, *7*, 1–6. [CrossRef]
14. Sartori, R.; Torelli, L.; Zitellini, N.; Carrara, G.; Magaldi, M.; Mussoni, P. Crustal features along a W–E Tyrrhenian transect from Sardinia to Campania margins (Central Mediterranean). *Tectonophysics* **2004**, *383*, 171–192. [CrossRef]
15. Bertotti, G.; Marsella, E.; Pelosi, N.; Pepe, F.; Tonielli, R. Sister 99: A seismic campaign to investigate the kinematics of South Tyrrhenian extensional regions. *Giorn. Di Geol.* **1999**, *61*, 25–36.
16. Pepe, F.; Bertotti, G.; Cella, F.; Marsella, E. Rifted margin formation in the South Tyrrhenian sea. A high resolution seismic profile across the north Sicily passive continental margin. *Tectonics* **2000**, *19*, 241–257. [CrossRef]
17. Aiello, G.; Marsella, E.; Sacchi, M. Quaternary structural evolution of Terracina and Gaeta basins (Eastern Tyrrhenian margin, Italy). *Rend. Lincei* **2000**, *11*, 41–58. [CrossRef]
18. Rolandi, G.; Bellucci, F.; Heizler, M.T.; Belkin, H.E.; De Vivo, B. Tectonic controls on the genesis of ignimbrites from the Campanian Volcanic Zone, southern Italy. *Mineral. Petrol.* **2003**, *79*, 3–31. [CrossRef]
19. ViDEPI, 2009. Progetto Visibilità Dati Esplorazione Petrolifera in Italia. © 2009–2010. Ministero dello Sviluppo Economico UNMIG, Società Geologica Italiana, Assomineraria. Available online: https://unmig.sviluppoeconomico.gov.it/videpi/ (accessed on 7 January 2025).
20. Fabbri, A.; Gallignani, P.; Zitellini, N. Geologic evolution of the peri-Tyrrhenian sedimentary basins. In *Sedimentary Basins of Mediterranean Margins*; Wezel, F.C., Ed.; C.N.R. Italian Project of Oceanography; Tecnoprint: Bologna, Italy, 1981; pp. 101–126.
21. D'Argenio, B.; Pescatore, T.; Scandone, P. Schema geologico dell'Appennino meridionale (Campania e Lucania). *Atti Accad. Naz. Dei Lincei* **1973**, *183*, 220–248.
22. Bigi, G.; Bonardi, G.; Catalano, R.; Cosentino, D.; Lentini, F.; Parotto, M.; Sartori, R.; Scandone, P.; Turco, E. *Structural Model of Italy 1:500.000*; CNR Progetto Finalizzato Geodinamica: Roma, Italy, 1992.
23. Vitale, S.; Ciarcia, S. Tectono-stratigraphic setting of the Campania region (southern Italy). *J. Maps* **2018**, *14*, 9–21. [CrossRef]
24. Ippolito, F.; Ortolani, F.; Russo, M. Struttura marginale tirrenica dell'Appennino campano: Reinterpretazione di dati di antiche ricerche di idrocarburi. *Mem. Della Soc. Geol. Ital.* **1973**, *12*, 227–250.
25. Ortolani, F.; Aprile, F. Nuovi dati sulla struttura profonda della Piana Campana a est del fiume Volturno. *Boll. Della Soc. Geol. Ital.* **1978**, *97*, 591–608.
26. Santangelo, N.; Romano, P.; Ascione, A.; Russo Ermolli, E. Quaternary evolution of the Southern Apennines coastal plains: A review. *Geol. Carpathica* **2017**, *68*, 43–56. [CrossRef]
27. Ascione, A.; Aucelli, P.P.C.; Cinque, A.; Di Paola, G.; Mattei, G.; Ruello, M.R.; Russo Ermolli, E.; Santangelo, N.; Ettore Valente, E. Geomorphology of Naples and the Campi Flegrei: Human and natural landscapes in a restless land. *J. Maps* **2021**, *17*, 18–28. [CrossRef]
28. Cuffaro, M.; Martorelli, E.; Bosman, A.; Conti, A.; Bigi, S.; Muccini, F.; Cocchi, L.; Ligi, M.; Bortoluzzi, G.; Scrocca, D.; et al. The Ventotene Volcanic Ridge: A newly explored complex in the central Tyrrhenian Sea (Italy). *Bull. Volcanol.* **2016**, *78*, 78–86. [CrossRef]
29. Milia, A.; Torrente, M.M. Late Quaternary volcanism and transtensional tectonics in the Bay of Naples, Campanian continental margin. *Mineral. Petrol.* **2003**, *79*, 49–65. [CrossRef]
30. Bruno, P.P.G.; Di Fiore, V.; Ventura, G. Seismic study of the 41st parallel fault system offshore the Campanian-Latial continental margin, Italy. *Tectonophysics* **2000**, *324*, 37–55. [CrossRef]
31. Loreto, M.F.; Zitellini, N.; Ranero, C.R.; Palmiotto, C.; Prada, M. Extensional tectonics during the Tyrrhenian back-arc basin formation and a new morpho-tectonic map. *Basin Res.* **2020**, *33*, 138–158. [CrossRef]
32. Bartole, R.; Savelli, D.; Tramontana, M.; Wezel, F.C. Structural and sedimentary features in the Tyrrhenian margin off Campania, Southern Italy. *Mar. Geol.* **1983**, *55*, 163–180. [CrossRef]
33. Mariani, M.; Prato, M. I bacini neogenici costieri del margine tirrenico: Approccio sismico-stratigrafico. *Mem. Della Soc. Geol. Ital.* **1988**, *41*, 519–531.
34. Luiso, P.; Paoletti, V.; Nappi, R.; La Manna, M.; Cella, F.; Gaudiosi, G.; Fedi, M.; Iorio, M. A multidisciplinary approach to characterize the geometry of active faults: The example of Mt. Massico, Southern Italy. *Geophys. J. Int.* **2018**, *213*, 1673–1681. [CrossRef]
35. Florio, G.; Fedi, M.; Cella, F.; Rapolla, A. The Campanian Plain and Phlegrean Fields: Structural setting from potential field data. *J. Volcanol. Geotherm. Res.* **1999**, *91*, 361–379. [CrossRef]
36. Romano, P.; Santo, A.; Voltaggio, M. L'evoluzione geomorfologica della Pianura del fiume Volturno (Campania) durante il tardo Quaternario (Pleistocene medio-superiore-Olocene). *Il Quat.* **1994**, *7*, 41–56.
37. Mitchum, R.M., Jr.; Vail, P.R.; Sangree, J.B. Stratigraphic interpretation of seismic reflection patterns in depositional sequences. In *Seismic Stratigraphy–Applications to Hydrocarbon Exploration*; Payton, C.E., Ed.; AAPG Memoir: Tulsa, Oklahoma, 1977; Volume 26, pp. 117–133.

38. Vail, P.R.; Mitchum, R.M.; Todd, R.G.; Widmier, J.M.; Thompson, S.; Sangree, J.B.; Bubb, J.N.; Hailelid, W.G. Seismic stratigraphy and global changes in sea level. In *Seismic Stratigraphy—Applications to Hydrocarbon Exploration*; Payton, C.E., Ed.; AAPG Memoir: Tulsa, Oklahoma, 1977; Volume 26, pp. 49–212.
39. Van Wagoner, J.C.; Posamentier, H.W.; Mitchum, R.M.; Vail, P.R.; Sarg, J.F.; Loutit, T.S.; Hardenbol, J. An overview of sequence stratigraphy and key definitions. In *Sea Level Changes—An Integrated Approach*; Wilgus, C.K., Hastings, B.S., Kendall, C.G.S.C., Posamentier, H.W., Ross, C.A., Van Wagoner, J.C., Eds.; SEPM Special Publication: Tulsa, Oklahoma, 1988; Volume 42, pp. 39–45.
40. Catuneanu, O.; Abreu, V.; Bhattacharya, J.P.; Blum, M.D.; Darlymple, R.W.; Eriksson, P.G.; Fielding, C.R.; Fisher, W.L.; Galloway, W.E.; Gibling, M.R.; et al. Towards the standardization of sequence stratigraphy. *Earth-Sci. Rev.* **2009**, *92*, 1–33. [CrossRef]
41. Zecchin, M.; Catuneanu, O. High-resolution sequence stratigraphy of clastic shelves I: Units and bounding surfaces. *Mar. Pet. Geol.* **2013**, *39*, 1–25. [CrossRef]
42. Catuneanu, O. Model-independent sequence stratigraphy. *Earth Sci. Rev.* **2019**, *188*, 312–388. [CrossRef]
43. Zecchin, M.; Catuneanu, O.; Caffau, M. Wave-ravinement surfaces: Classification and key characteristics. *Earth Sci. Rev.* **2019**, *188*, 210–239. [CrossRef]
44. Cosentino, D.; Federici, I.; Cipollari, P.; Gliozzi, E. Environments and tectonic instability in central Italy (Garigliano Basin) during the late Messinian Lago–Mare episode: New data from the onshore Mondragone 1 well. *Sediment. Geol.* **2006**, *188–189*, 297–317. [CrossRef]
45. Posamentier, H.W.; Allen, G.P.; James, D.P.; Tesson, M. Forced regressions in a sequence stratigraphic framework: Concepts, examples, and exploration significance. *AAPG Bull.* **1992**, *76*, 1687–1709.

MDPI AG
Grosspeteranlage 5
4052 Basel
Switzerland
Tel.: +41 61 683 77 34

Journal of Marine Science and Engineering Editorial Office
E-mail: jmse@mdpi.com
www.mdpi.com/journal/jmse

www.ingramcontent.com/pod-product-compliance
Lightning Source LLC
Chambersburg PA
CBHW040748250926
51850CB00007BB/1441
* 9 7 8 3 7 2 5 8 7 4 8 0 4 *